Stars

Their Birth, Life, and Death

Stars

Their Birth, Life, and Death

Iosif S. Shklovskii

Sternberg Astronomical Institute, Moscow University
Institute for Space Research, USSR Academy of Sciences

Translated by

Richard B. Rodman

Harvard College Observatory

Foreword by Carl Sagan
Prologue by J. P. Ostriker

W. H. FREEMAN AND COMPANY
SAN FRANCISCO

Original edition published by Nauka, the Central Press for
Literature in Physics and Mathematics, Moscow, 1975.
This edition, specially revised by the author, is published
under agreement with the USSR Copyright Agency.

Library of Congress Cataloging in Publication Data
Shklovskiĭ, I   S
    Stars, their birth, life, and death.

    Rev. translation of the author's Zvezdy, ikh rozh-
denie, zhizn' i smert'.
    Bibliography: p.
    1. Stars—Evolution.   I.   Title.
QB806.S5413   1978   523.7   77-13889
ISBN 0-7167-0024-7

Printed in the United States of America

9   8   7   6   5   4   3   2   1

In memory of
Ivan Georgievich Petrovskii

Contents

Foreword

We live in a universe of stars. Stars are born, live—often for billions of years—and die, sometimes quietly and sometimes in a more spectacular manner. The end products of the life cycle of the stars may lead to exotic inhabitants of the cosmic bestiary—white and black dwarfs, planetary nebulae, supernovae, neutron stars, and, perhaps, black holes. Most of the mass which makes up the Milky Way Galaxy may be in stars. Most of the radiant energy which permeates our galaxy comes from stars. The nearest star provides the light and heat which is absolutely essential for life on Earth. And stars grace our night sky and provoke our sense of wonder.

This book treats stellar evolution in the broadest possible perspective. It effortlessly crosses disciplinary and national boundaries, and it is written clearly and engagingly, with an occasional dash of humor. The book has been expertly translated from the Russian by Richard B. Rodman, who has added an important and unique detailed bibliography for those who wish to dig further. *Stars* will appeal to many people. There is much in this book which will attract the intelligent layman with no professional interest in astronomy; but also, in the breadth of presentation and occasional insight into the underlying mathematics, there is much that will make this book essential for students of astronomy as well as for professional astronomers.

The book has many virtues, but its chief attraction is its author, I. S. Shklovskii. Shklovskii has been for more than a quarter century one of the leading astrophysicists in the world. His contributions have run through virtually all of modern astronomy, and much of the excitement and enthusiasm he brings to this book derive from the simple fact that he played a major role in many of the discoveries described. Shklovskii's talents as a popularizer of science are well known in the Soviet Union and are reflected in this book.

My own contact with Shklovskii began in the early 1960s when, after an exchange of letters, he suggested that we collaborate in writing a book subsequently published as *Intelligent Life in the Universe* (Holden-Day, 1966; Dell Publishing Co., 1967, and many later printings). Because of restrictions placed on Shklovskii's freedom to travel, the book was completed before the authors ever met. Since then I have had many opportunities, both in the United States and in the Soviet Union, to get to know this extraordinary, brilliant and humane man. His students have played a major role in the scientific space program of the USSR. His accomplishments have been recognized throughout the world—as a corresponding member of the USSR Academy of Sciences, one of only six honorary members of the American Astronomical Society, a foreign member of the National Academy of Sciences of the United States, a recipient of the Bruce Gold Medal of the Astronomical Society of the Pacific, and a laureate of the Lenin Prize. He is one of the leading minds of our age and deserves to be better known to the general public of all nations.

Shklovskii is also well known for his sense of humor. We were once attending an international scientific meeting which, as commonly happens, included a presentation which was particularly poorly reasoned. The author of the paper argued that scientific creativity depended on the phase of the solar cycle. Near solar maximum, when there are many sun spots, the author maintained, were made the principal creative insights of Newton, Darwin, of Einstein . . . "Yes," commented Shklovskii to me in a loud stage whisper, "but this paper was conceived in a deep solar minimum."

Stars: Their Birth, Life, and Death is the result of a lifetime of work in the stellar realm, embracing the full solar cycle but evincing none of the alleged deficiencies of composition at solar minimum. It is a pleasure to read.

March, 1978

CARL SAGAN

*Professor of Astronomy and
Space Sciences*

*Director, Laboratory for Planetary
Studies, Cornell University,
Ithaca, New York*

Preface to the American Edition

Astronomy is the most ancient, yet the youngest of the sciences. From time out of mind people have been baffled as to what may lie beyond their immediate experience. A prominent Soviet astronomer puts it well: "On occasion people will lift up their eyes to the sky—that's one thing distinguishing them from the animals."

Now I would agree that the utilitarian, sporting interest a wide public has lately been taking in "space" (a term that often mistakenly replaces the word *universe*) can mask a dispassionate awareness of the harmony, the beauty, the eternal nobility of the world around us. Even so, the decades just past have witnessed an exceptionally vigorous and fruitful development of astronomy. These have been years of drastic reshaping of our ideas about the universe. Only yesterday, before the war, the universe was thought static, unvarying; today we see it as everchanging and unfolding. Modern astronomy is thoroughly evolutionary, and evolution is observed at all levels, from the expanding universe to a fragmenting swarm of meteoroids.

This book deals with stars, the most interesting and important objects, I contend, in nature. Recent discoveries through the astronomy of the invisible include a new class of objects ranked as stars: the pulsars and x-ray sources. Their properties are startling, their role in stellar evolution tremendous—and so not only astronomers but people just interested in science are greatly attracted to them. But so commonplace an object as a classical star warrants the most intent examination, for without comprehending its traits you cannot expect to gain a real grasp of the marvels that the newest astronomy holds in store.

For this version of the book in English I have supplied a variety of additions to reflect the progress of astronomy over the three years since the original Russian edition appeared. Some necessary corrections have also been made. I am especially delighted that *Stars: Their Birth, Life, and Death* is being published in America, a country that has made so distinguished a contribution to the growth of modern astronomy, and to the study of the stars in particular. This monograph belongs to the category of serious popular books, and I trust it will find its share of readers in your great land.

January, 1978   I. S. SHKLOVSKII

Moscow

Prologue

I. S. Shklovskii has few peers among the world's astrophysicists. His interests have spanned an extraordinary diversity of fields, with major contributions to science ranging from the nearest astrophysics, the earth's ionosphere, to supernovae in our galaxy, to the most distant observed objects in our universe, the quasars. Equally at home in detailed technical investigations of physical processes and in philosophical discussions of the origin of life, he brings to all topics an ability to penetrate to the essentials of the problem at hand, imagination, and the gift of lucid exposition. In this English edition of *Stars: Their Birth, Life, and Death,* he introduces the general reader to the most fully developed part of modern astrophysics.

Outside of the familiar objects in our solar system, the stars have historically been the central subjects of astronomy, and that emphasis has been maintained into the modern era. We know that most of the light emitted by normal galaxies is starlight, and we think we understand the processes by which the energy is generated in the stellar interiors and carried to the radiating stellar surfaces. We also know that most of the mass (at least in the inner parts of galaxies) is locked up in ordinary stars, and we believe, with some moderate conviction, that most of the elements heavier than hydrogen and helium (including most of the atoms in our bodies!)

were made in thermonuclear "pressure cookers" within stars and then expelled into the interstellar medium during supernova explosions, to be incorporated in subsequent generations of stars and planets. Although the weird and wonderful galactic stars that are being discovered by new techniques present some completely puzzling phenomena (such as x-ray and γ-ray "bursters") and the origin of even many old and well-studied objects from close binaries to type I supernovae remains mysterious, we do feel that we have attained a hard-won and hopefully not too precarious understanding of the life cycles of most ordinary stars.

This understanding has developed quite recently. When Shklovskii was a student at Moscow University, no one could answer so basic a question as "Why do the stars shine?" Kelvin had written that none of the then existing "natural theories can account for the solar radiation continuing like the present rate for many hundred millions of years," and it was in 1938, the year of Shklovskii's university graduation, that several fundamental papers of C. F. von Weizsäcker and H. Bethe first pointed to the thermonuclear processes as providing most stellar energy. Since that time our knowledge has grown at a truly astonishing rate.

Shklovskii's survey begins with several chapters on the interstellar medium—the environment within which stars are born—taking us into the dark, relatively dense molecular clouds which are probably the real sites of star formation and then presenting some theoretical calculations concerning the hidden processes by which gas clouds condense into stars. Chapters 6 to 14 outline the theories of stellar structure and evolution in a clear and not too technical manner, describing both the successes achieved and the remaining paradoxes, such as our troubling inability to detect neutrinos from the sun. Several chapters follow on the exciting and important subject of the death throes of stars—supernova explosions. It was here that Shklovskii made one of his most significant contributions to astrophysics in forwarding the idea that the light emitted by the Crab Nebula (the remnant of an explosion seen by Chinese observers in A.D. 1054) is produced not by an ordinary thermal mechanism but by synchrotron radiation emitted by charged particles spiralling at nearly the speed of light through a relatively large and well-ordered magnetic field. In the remaining six chapters, Shklovskii focuses on the enormously condensed remnants of explosive stellar deaths known as pulsars (neutron stars) or black holes. These objects, under intensive scrutiny at present, typically manifest themselves as celestial radio and x-ray sources, but given the current pace of development γ-ray and even gravitational wave detection may become a reality in the not too distant future.

January, 1978

JEREMIAH P. OSTRIKER

Princeton University Observatory

Introduction

Light, as we all know, occupies just a tiny part of the vast spectrum of electromagnetic waves that are emitted—and absorbed—by manifold objects in the sky. Clearly, then, astronomers could gain only a biased idea of the universe when they were limited to the narrow wavelength band of visible rays. After World War II the state of affairs changed radically: astronomical research began to spread over the whole range of electromagnetic waves. To begin with, radio astronomy was created and intensively developed; over the past three decades it has enriched astronomy with many discoveries of extraordinary importance. Radio astronomy today is armed with the largest antennas in the world (even hundreds of meters across) and the most sensitive radio receivers yet designed. It has expanded the horizons of deep-space research to an enormous degree. What prospects does radio astronomy now face?

Generally speaking, the capabilities of observational astronomy are effectively limited by the weakest radiant intensity that can be measured, by the sharpest spectral resolution (as described by the ratio $\Delta\nu/\nu = -\Delta\lambda/\lambda$, where $\Delta\nu$ and $\Delta\lambda$ denote respectively the smallest distinguishable frequency and wavelength differences), and by the narrowest angular resolution—the possibility of measuring a minimal angular size for cosmic emission sources

1

or the features comprising them. In none of these respects does optical astronomy have any advantage over radio astronomy.

As a matter of fact, the largest optical telescopes today (the 5-m reflector on Palomar Mountain in California and the 6-m reflector in the Caucasus) can, near the limit of their power, record a star or galaxy of stellar magnitude $23^{m}.5$. A celestial object this faint sends to the earth optical radiation whose total flux density is about 3×10^{-14} erg cm^{-2} s^{-1}, the lowest perceptible value. On the other hand, giant radio telescopes are now so sensitive that at their limit they can record a flux density per unit frequency interval of roughly 10^{-29} W m^{-2} Hz^{-1}, or 10^{-3} jansky. Taking $\Delta\nu \approx 300$ MHz as the reception bandwidth, we find that the weakest radiation currently detectable by radio astronomy (at wavelengths in the 100-cm range) has a flux density close to 3×10^{-21} W/m$^{2} = 3 \times 10^{-18}$ erg cm^{-2} s^{-1}—a threshold 10,000 times lower than in optical astronomy! One should, of course, keep in mind that in optical astronomy the brightest stars outshine the faintest—and vastly more numerous—ones by a factor of around 10^{10}, whereas in radio astronomy the emission reaching us from the strongest sources in the sky has a flux density about 10^{6} times that of the weakest. Understandably enough, far fewer cosmic radio sources are accessible to observation than optical objects, but even so the extremely high sensitivity of large, modern radio telescopes inevitably staggers the imagination.[1]

Just as impressive is the capacity of radio astronomy to achieve a frequency resolution of individual features in the spectrum. For instance, sources of cosmic maser emission at 18-cm wavelength (the OH molecule) exhibit radio-line profiles in which details narrower than 1 kHz can easily be resolved. Hence the spectral resolution $\Delta\nu/\nu \approx 3 \times 10^{-7}$, while in optical stellar spectroscopy $\Delta\nu/\nu \approx 10^{-4}$ is considered good.

During the early years the "Achilles' heel" of radio astronomy was the low resolving power of the radio telescopes. As a natural limit for angular resolution, one may take the angular size λ/D of the central diffraction disk, where λ is the wavelength of the radiation and D is the diameter of the telescope mirror. In optical astronomy, with λ typically 4×10^{-5} cm, diffraction limits the resolving power of large telescopes to a few hundredths of a second of arc. But the scintillation of the earth's atmosphere and imperfections in the mirror surface prevent this limit from being attained. In practice, large optical telescopes rarely have a resolving power better than

---

[1]The total energy that all operating radio telescopes have ever "caught" from all cosmic sources (except the sun) throughout the history of radio astronomy is perhaps 10^{3} erg. That is scarcely enough energy to heat a glass of water by 10^{-7} degree. The reader might like to estimate what this much energy would cost.

0".5. And what is the angular resolving power of radio telescopes? Even for a very large one with a reflector perhaps 100 m in diameter, the angular size of the diffraction disk at 10-cm wavelength is about three *minutes* of arc. Indeed, in the early 1950s, when the development of radio astronomy was just beginning, the angular resolution was measured in many degrees.

Matters changed fundamentally once the interference technique was applied to radio astronomy research. With this method cosmic radio waves are received simultaneously by *two* radio telescopes. The expression for the angular size of the diffraction disk should now contain not the antenna diameter but the distance between the pair of telescopes, which can be made very long. Record angular resolutions have been achieved through interferometric observations with antennas separated by intercontinental distances.[2] For example, interferometers have been operated with baselines extending from the United States to the Crimea and to Australia. At the shortest wavelength used for such observations, an angular resolution of 10^{-4} second of arc has been reached! Paradoxically, the angular resolving power is now far higher in radio than in optical astronomy!

One should, however, recognize that this exceedingly high radio resolving power has thus far been attained only in a few particular experiments, and for a special class of sources of very small angular size (the nuclei of quasars and radio galaxies, and cosmic maser sources). Ordinarily, radio sources are more or less extended objects for which it is vital to know the brightness distribution in as much detail as possible. In other words, we need a *radio image* of extended objects with maximum angular resolution in both coordinates. The best installation currently available for such purposes is the Westerbork Synthesis Radio Telescope in the Netherlands; the system consists of 12 reflectors each 25 m in diameter, suitably spaced and connected in a line up to 1.6 km long. At 21-cm wavelength the resolving power of the system is about 20". Astronomers are putting great hope in the similarly designed but much bigger American VLA (Very Large Array) system, scheduled for completion by 1981 (See Figure 3.3). It will comprise 27 paraboloids, again 25 m in diameter, arranged in a Y shape on a New Mexico site 21 km in radius. The system's resolving power at 6-cm wavelength will be 0".6, twice that of the famed optical *Sky Survey*, the atlas prepared at the Palomar Mountain Observatory after years of diligent effort.

A distinctly new step in the progress of astronomy has been its entry into space, an outgrowth of the vigorous development of rocket technology during the postwar years. This new extra-atmospheric astronomy differs just as strikingly from the classical science as does radio astronomy. By

---

[2]The concept of such an interferometer with independent, widely spaced antennas originated with Soviet radio astronomers.

mounting scientific instruments such as photon counters and telescopes on space platforms, astronomers have breached the powerful armor of the earth's atmosphere, which completely absorbs short cosmic electromagnetic waves (ultraviolet radiation, x rays, and gamma rays). As a result study of the short-wavelength ("hard") emission of the sun, stars, nebulae, and galaxies has become possible, and the scope of our information on the nature of these objects has enlarged immensely. For example, almost all of the "resonance" spectral lines of various elements and their ions occur in the ultraviolet part of the spectrum. Analysis of these lines is absolutely necessary if we are to acquire a detailed understanding of the chemical composition of stars and the interstellar medium.

To illustrate the opportunities afforded by modern space astronomy, let us consider briefly the features of a telescope carried by a specialized American earth satellite placed in a rather high ($\approx$ 750 km), nearly circular orbit. Officially named the *Third Orbiting Astronomical Observatory* (OAO-3), this satellite is more often called *Copernicus* because it operated during 1972–73, when astronomers marked the 500th birthday anniversary of their great Polish precursor. *Copernicus* represents the last word in refinements. Its design includes as the principal instrument a telescope serving as an ultraviolet spectrometer. The primary mirror in the Cassegrain optical system is 80 cm in diameter, a respectable size even for an observatory on the ground. The spectrograph uses a concave diffraction grating that provides a wavelength dispersion of 4.2 Å/mm in the first order (1 angstrom unit $= 10^{-4} \mu = 10^{-8}$ cm). In the 950–1450 Å range the spectral resolution $\Delta\lambda$ reaches about 0.05 Å. The star whose spectrum is being scanned is tracked by the telescope with astonishing accuracy: after 10 min of observation its line of sight drifts by an angle of no more than $0''.02$! Upon command from the earth the telescope can be set on any star (brighter than 5^m) of interest to astronomers; a spectrum is then obtained and telemetered to the ground. *Copernicus* has secured results of exceptional scientific value; some of these will be discussed in Part I of this book. Even larger space telescopes are planned, with mirrors up to 3 m in diameter; they will, in particular, enable important research to be conducted on the ultraviolet spectra of considerably fainter stars and nebulae, among which are some most interesting objects.

Fresh in the memory of all are the outstanding investigations carried out in 1973–74 in the manned American space laboratory *Skylab*. The work of its intrepid astronauts has proved singularly fruitful for specialists in solar physics.

Studies of the radiation of cosmic objects in the x- and γ-ray spectral regions are noteworthy for astronomy. X-ray astronomy has made impressive progress. A wealth of information has been acquired on the radiation of certain cosmic objects both in the soft x-ray range (photon energies on the order of a few hundred electron volts) and for harder x rays (thousands to hundreds of thousands of electron volts per photon). The chief importance of x-ray astronomy is that it permits inspection of objects under extreme physical conditions—gas at temperatures of tens and hundreds of millions of degrees, or powerful eruptive processes, examined later in the book. Just as at radio wavelengths, in the x-ray range many sources emit not thermal radiation but distinctive "nonequilibrium" radiation that accompanies the motion of electrons at enormous, ultrarelativistic energies. The possibilities of optical astronomy are very limited in this respect.

We see, then, that optical, radio, and x-ray astronomy do not duplicate but vitally complement one another. While some objects, including most of the ordinary stars, radiate primarily in the optical part of the spectrum, there are others whose principal emission falls in the radio or the x-ray range. Especially interesting are celestial bodies that radiate in all three ranges simultaneously, although with differing power; complex research is needed to study them.

The renowned American satellite *Uhuru*, launched into an equatorial orbit in December 1970 and continuing to operate in the years following, exemplifies the capabilities of x-ray astronomy. This special-purpose satellite was designed to carry two x-ray detectors, each with an area of 840 cm^2. The detectors are proportional x-ray counters with beryllium windows about 0.1 mm thick; their fields of view are $5° \times 5°$ and $0°.5 \times 5°$, and they point in opposite directions. *Uhuru* is provided with a magnetic guidance system that incorporates coils for aligning the longitudinal axis of the satellite relative to the geomagnetic field, and a flywheel that stabilizes the satellite as it slowly spins around its axis at a rate of $0°.1$–$0°.5$ each second. The guidance system also includes star sensors for determining the angle of rotation about the spin axis. Discrete sources of x rays with a flux density as low as 0.005 photon cm^{-2} s^{-1} in the 2–20-keV energy range can be investigated with this satellite. Each source is measured with an eight-channel pulse-height analyzer so that its spectrum can be established.

Gamma-ray astronomy is still in its infancy. It demands highly specialized techniques. Very hard photons are ordinarily detected by spark chambers; these devices are quite heavy, a circumstance that hinders experiments with spacecraft. Actually, the chambers need not be lifted into space: balloons rising to a height of 25–40 km may also be used. Even so, a special satellite is better. In the United States a *Small Astronomy Satellite* of this type, the SAS-2, was launched in November 1972 for the purpose of studying cosmic γ rays of energy above 30 MeV with a flux density down to 10^{-6} photon cm^{-2} s^{-1}. A diffuse general γ-ray background with an enhanced intensity along the Milky Way has been observed with confidence, and also a discrete source recognized as a special kind of nebula, the remnant of an exploding star (see Chapter 16). Astronomers are impatiently awaiting further confirmation of "point" sources of cosmic γ rays reported from balloon experiments.

One other important branch of space astronomy consists of infrared and "submillimeter" surveys. To a limited extent, measurements in this difficult range can be made from observatories on the ground by taking advantage of certain "transparency windows" in the earth's atmosphere (for example, at wavelengths of 8–13 μ and 20–25 μ). Observing programs using high-altitude aircraft laboratories and balloons have furnished valuable information. But it would be very profitable to have a satellite designed especially for infrared astronomy; none has yet been orbited. The main achievement in this field of astronomy has been the development of new types of high-sensitivity receivers, made possible only by rapid progress in electronics, semiconductor physics, and cryogenic technology. This wavelength range is of paramount significance because a large part of the radiation of the universe is concentrated there. Active galaxy and quasar nuclei, giant stars and protostars, clouds of cosmic dust—all emit primarily infrared and submillimeter radiation. In addition, the primordial background radiation of the universe has its maximum spectral flux density in the submillimeter range. These wavelengths are also highly important for the crucial problem of the origin of stars and planetary systems (see Chapter 3). Astronomers are eagerly looking forward to further progress in this difficult part of the spectrum.

Modern astronomy is not, however, restricted exclusively to the study of electromagnetic radiation emitted at all frequencies by various cosmic objects. The first steps have already been taken in neutrino astronomy; we shall discuss this topic in some detail in Chapter 9. Next comes the possible discovery of gravitational waves from exploding stars and their remnants, and perhaps also from other celestial bodies, including extragalactic objects (see Chapter 24). Primary cosmic rays have been under investigation for many years, and there have been major recent developments in this area. Finally, with advances in rocket and space technology, direct measurements can now be made of the magnetic fields and gas density in the interplanetary medium. As for studies of the planets in the solar system and their satellites, a very singular state of affairs has arisen: planetary astronomy is no longer just one branch of astronomy, but has spilled over to other fields. In the past a comparable situation came about with geophysics.

Thus our experience during the past two decades of scientific and technical revolution has propelled astronomy into a new era. Most importantly, astronomy has encompassed the full range of waves, so that its possibilities have vastly expanded. Progress has also been aided by the application of electronic computers to the acquisition and analysis of observations. In particular, the high sensitivity of modern radio telescopes could never have been achieved without computers. Electronics and automation have come into wide use not only in the novel fields of radio and space astronomy but also in the classical optical range, where they have made possible a marked improvement in the parameters of optical telescopes.[3]

We would nevertheless emphasize that although such developments have radically changed the face of astronomy, its essential character has remained the same. Astronomy still depends mainly on observing electromagnetic radiation from various cosmic objects by means of suitable receivers. In the course of its history of many centuries, astronomy has survived more than one revolution that has completely changed its appearance without altering it in a fundamental way. One will recall that for a long period astronomy was conducted without any telescopes at all. Could we now imagine an observatory without a telescope? The new generation of astronomers consider radio and space astronomy an organic part of their science. And today, just as in the seventeenth and eighteenth centuries but on an incomparably higher level, we are witnessing a genuine convergence—even a merger—of astronomy and physics.

While the inherent character of astronomy has remained unchanged for centuries, its basic concepts and its very approach to the solution of both old and newly formulated problems have undergone profound transformations. Of greatest significance, modern astronomy has become thoroughly *evolutionary.* In this respect it differs from physics, whose laws expressing the fundamental properties of elementary particles and fields are eternal—independent of time. Attempts by some physicists, such as Paul Dirac, to introduce the idea of a gradual time variation in the universal constants (the electron charge, the velocity of light, the Planck constant) have not met with success. Just in the past few years astronomy has afforded striking evidence that the universal constants had practically the same values billions of years ago that they have now.

In the eighteenth and early nineteenth centuries, when astronomy dealt largely with the celestial mechanics of the solar system, it was not an evolutionary subject. Matters are different today, for we now recognize evolution at all levels. Active formations on the solar surface, comets, planetary atmospheres, gaseous nebulae, stars, pulsars, galaxy nuclei, galaxies them-

---

[3]As one example, the replacement in astronomical research practice of photographic emulsions by the latest electronic imaging devices and video radiation detectors has been equivalent to a fivefold enlargement of telescope apertures. Even if such an increase in the diameter of telescope mirrors were technically feasible (a most doubtful proposition), billions of dollars would have to be spent for major instruments, as the cost of a telescope is roughly proportional to the cube of its mirror diameter.

selves, and finally the universe as a whole—all are evolving on time scales characteristic of each particular case. For example, as we study the universe at earlier stages in its evolution we will inevitably come to an epoch when there were neither stars, nor galaxies, nor clusters of galaxies; there would only have been a more or less homogeneous plasma of hydrogen and helium. This conclusion is by no means merely theoretical. Ten years ago the primordial background microwave radiation of the universe was discovered; this represents the radiation of hot plasma that has "degenerated" because of the red-shift effect but has survived from the time when no stars or galaxies were yet in existence.

The underlying evolutionary character of astronomy makes it an ally of biology and geology. In biology, natural selection and mutation act as a moving force for evolution from the simple to the complex. How does evolution in the universe compare? Here too we can discern a general tendency for evolution from the simple to the complex. If we look back to the era in the history of the universe when the radiation observed today as a primordial relic was created, we find that the universe was then a moderately hot plasma (temperature about 4000°K) confined to a small region with a radius of about 15 million light years.[4] This primordial plasma also was only moderately dense, with perhaps 3000 particles per cubic centimeter, but all space was filled with equilibrium radiation corresponding to the temperature of $\approx$ 4000°K.

In its chemical composition the plasma was very simple: a mixture of hydrogen and helium. There were hardly any heavy elements in it.[5] And as this remarkably simple plasma proceeded to evolve, the universe we observe at the present time began to develop in all its diversity. In some fashion that was not simple in the least, heavy elements were formed, particles of very high energy—the cosmic rays—made their appearance (modern astronomy has shown that cosmic rays could not have originated during the plasma-cloud era), and galaxies emerged, as well as stars with their wonderfully different properties, planets, and eventually the great variety of living things. To be sure, it is only for the earth that we so far have clues as to how life has evolved. Presumably at least some of the other planets revolving about other stars ought to have living organisms of their own, whose development may have differed radically from that of our terrestrial organisms.

---

[4]Today the center of the Virgo cluster of galaxies, to which our own Galaxy belongs, is two or three times farther away than that, while the distance of the most remote galaxies observed is measured in tens of billions of light years.

[5]The primordial plasma also contained deuterium, which was presumably created from protons and neutrons when the universe was just a few minutes old (see end of Chapter 2).

But what is the mechanism behind this grand process of evolution, whereby the universe has continually grown more complex? We can point to at least one such factor that must certainly have provoked the evolution of matter in the universe. This mechanism is gravitational instability, a concept recognized long ago by Newton. A theory was worked out for it early in our century by the famed British astronomer Sir James Jeans. Because of universal gravitation, matter cannot be distributed at uniform density over an indefinitely large volume. Matter will ineluctably become unstable, breaking up into separate condensations or clumps. In our case, the plasma that originally was almost homogeneous should first have broken into huge pieces, from which clusters of galaxies later formed. The same mechanism would have caused the clusters to fragment into "proto-galaxies," and from these aggregates "protostars" would have arisen in a natural way—a process that we shall describe in detail at the end of Part I of this book. Indeed, the formation of stars from the diffuse interstellar medium is still in progress at the present day.

In addition to gravitational instability, other factors of fundamental importance might very well be contributing toward the evolution of matter in the universe. This problem, laden with deep philosophical meaning, calls for thorough investigation.

Just as the question still remains open of how the first primitive DNA molecules arose on the pristine earth, a shroud of mystery covers the primal cause of the expansion of the universe. As we regress back toward epochs when the universe was younger, we come to the stage when the primordial radiation originated. The age of the universe then amounted to a few million years, and as we have said it consisted of a rather simple, moderately hot plasma. Well, what happened before that? This exceedingly important problem, one of enormous cognitive significance for mankind, is now occupying the minds of many theoreticians.

Regrettably, astronomy has not yet supplied any means whereby this remote epoch might be made "observable"; there is nothing comparable to our study of the primordial radiation enabling us to observe the universe in the ancient past, when it was a thousand times younger than now. By taking advantage of indirect evidence and recognizing that the primordial radiation has nearly the same intensity throughout the sky, we can none-theless trace the history of the universe back to the time when its age was about a day; its temperature, $\approx 10^8$ °K; and its density, $\approx 10^{-5}$ g/cm^3. Yet we can envision a still earlier epoch when the universe was very much smaller, denser, and hotter. For instance, when the average density of the universe was of the same order as the density of an atomic nucleus ($\approx 10^{15}$ g/cm^3), its radius would have been rather similar to the distance of the earth from the sun, or $\approx 10^{13}$ cm.

In scientific papers devoted to this intriguing problem, one hears of far higher densities for the universe during the first instants of its existence: as great as 10^{91} g/cm^3! And at this density the radius of the universe would have been 10^{-12} cm, about the same as the classical electron radius. It is hard to avoid the impression that such a universe somehow resembles an elementary particle. Or might it be better to compare the embryonic universe to a "supergene" having a vast choice of potential alternatives that may come into play as its evolution proceeds? It would seem, though, that ordinary concepts and laws of physics would be completely inapplicable even for systems whose density is not quite so extreme. In particular, the concept of *time* might very well lose its familiar meaning altogether. We therefore feel that there is probably no scientific content in such natural questions as these: "Then what happened still earlier?" "Did the universe have a *beginning?*" One can't help but recall that memorable imagery of the hapless bureaucrat Poprishchin in Gogol's immortal story.[6]

The possibility remains that further developments in astronomy may permit direct observation of the dawning stages in the evolution of the universe. Only neutrinos, which can pass freely through a great thickness of matter, could reach us from such remote times. Today neutrino astronomy is taking just its first modest steps; but who knows what the trends of the future will be? Yet the very first instants in the existence of the universe would not even furnish us with neutrinos. We are here confronting the Unknown. And still, the thought keeps coming to mind: Is our universe really the only one anywhere? Might there not be other universes about which the tools of modern science can tell us nothing? And maybe, when our universe was very tiny, exceedingly dense, and had not set about expanding at all, could it have collided with another such universe and begun to expand—whereupon, so to speak, what happened, just happened? Of course such a primitive, decidedly "anthropocentric" model for the start of the expansion in the universe is very naïve; yet who knows but that this picture might contain that notorious raw grain of truth?

But enough of that. In this book we shall not discuss the evolution of the universe at all. We shall hardly even touch on the captivating problems of modern extragalactic astronomy. Our topic in this book is stars. If you were to pose the child's innocent question, asking what the chief objects in the universe are, I would unhesitatingly respond: *Stars.* Why?

Well, they are paramount if only because 97 percent of the matter in our Galaxy is concentrated in stars. In many if not most other galaxies, star material comprises over 99.9 percent of the mass. The highly rarefied

---

[6]*Diary of a Madman* (1834), a grotesque persiflage of human vanity. Driven to derangement by the pettiness of his career, Poprishchin expresses comic anxiety about the moon in an incoherent world where time loses all meaning.—R.B.R.

intergalactic gas, which has not yet even been detected with any confidence, seems to have such a low density that the bulk of the matter in the universe resides in galaxies, and accordingly in stars. Some hold the view, to be sure, that in the nuclei of many galaxies and in quasars most of the material is a dense and quite hot gas. Even then our claim would stay the same, for the nucleus of a galaxy has only a small mass compared to the galaxy itself. Thus at the present stage of evolution in the universe, the matter in it is mainly in the stellar state.

It follows that most of the matter in the universe is hidden in the interiors of stars, where its temperature reaches tens of millions of degrees, its density is very high, and the physical conditions differ little from thermodynamic equilibrium. The principal evolution of matter in the universe has occurred and is occurring in stellar interiors. That is where the "crucible" was (and still is) located—the place that has directed the chemical evolution of matter in the universe by enriching it with heavy elements. It is there that matter, according to the regular laws of nature, is transformed from a perfect gas into a very dense degenerate gas, and even into "neutronized" material. Indeed, at pivotal stages in their evolution some stars may enter a state that is still far from being well understood: such a star may become a black hole. Despite their central importance, the stars surrounding the nuclei of the 10 billion galaxies occupy a combined volume (apart from singular regions not yet adequately studied) amounting to only about 10^{-31} of the volume of the universe.

One topic of deep significance concerns the interaction between stars and the interstellar medium, including the problem of continuous star formation from clouds condensing inside that medium. The very existence of stars emphasizes the *irreversibility* of the processes whereby matter evolves in the universe. Stars essentially radiate by the irreversible process of converting hydrogen into heavier elements, mainly helium. The continual accumulation in the universe of the final "inert," or "dead," products of stellar evolution—white dwarfs, neutron stars, and probably black holes—further underscores the irreversible character of evolution in the universe.

In the world of stars we come across an enormous variety of phenomena in all wavelength ranges. X-ray stars, cosmic masers, pulsars, dwarf flare stars, planetary nebulae with their peculiar nuclei, Cepheids, and indeed just "ordinary" stars that one might think are not remarkable at all—how wonderful nature is! To gain some insight into the universe, we must begin by seeking to understand stars and how they evolve. In this book the author has relied on the achievements of modern astronomy in an attempt to deal with pending questions. We should keep in mind, though, that many questions cannot as yet be answered comprehensively. The frontiers of science in this field are ever advancing. But perhaps that is what lends to debate on our captivating subject so much charm.

Part I

Stars Are Born

. . . so simple a thing as a star.

SIR ARTHUR STANLEY EDDINGTON, 1926

1

Stars: Their Basic Observational Properties

Professor Martin Schwarzschild, a founder of modern stellar evolution theory, opens his distinguished monograph *Structure and Evolution of the Stars* with a very deep thought:

> If simple perfect laws uniquely rule the universe, should not pure thought be capable of uncovering this perfect set of laws without having to lean on the crutches of tediously assembled observations? True, the laws to be discovered may be perfect, but the human brain is not. Left on its own, it is prone to stray, as many past examples sadly prove. In fact, we have missed few chances to err until new data freshly gleaned from nature set us right again for the next steps. Thus pillars rather than crutches are the observations on which we base our theories; and for the theory of stellar evolution these pillars must be there before we can get far on the right track.[1]

You could scarcely put it better. Generations of astronomers have painstakingly gathered an enormous store of facts touching on the most diverse properties of stars. Just which properties can be established by analyzing observational material?

---

[1]Martin Schwarzschild, *Structure and Evolution of the Stars* (Princeton, N.J.: Princeton University Press, 1958), p. 1.

First of all one has to understand that, with very rare exceptions, stars are observed as point sources of radiation. In other words, their angular sizes are immeasurably small. Even the largest telescopes cannot show the "real" disks of stars. We emphasize the word *real,* because purely instrumental effects, and particularly the restless state of the atmosphere, combine to produce a spurious star image shaped like a disk. This disk is seldom less than a second of arc across, whereas the true disks of even the closest stars would measure only a few thousandths of a second.

Thus the largest telescope in the world would not be capable of "resolving" a star, as astronomers say. Accordingly, we can do no better than to measure the flux of radiation emitted by stars in various parts of the spectrum. The radiant flux reaching the earth is expressed by the apparent magnitude, a term whose definition will be considered familiar.[2] One need only remember that the faintest observable stars have an apparent magnitude $m \approx 23.5$, whereas Sirius, the brightest star, has $m = -1.5$. If we know the difference between the magnitudes of two stars we can find their flux ratio F_1/F_2 from the simple expression

$$F_1/F_2 = 2.512^{m_2 - m_1}. \tag{1.1}$$

It is also helpful to know that the sun has an apparent visual magnitude $m_\odot = -26.73$. But direct measurements show that the average flux of solar radiation per unit area at the earth, expressed in absolute units, is

$$F_\odot = 1.37 \times 10^6 \text{ erg cm}^{-2} \text{ s}^{-1}. \tag{1.2}$$

This quantity is called the solar constant. From the observed apparent magnitude of any star having the same color as the sun, we can easily estimate its flux density F_m at the earth in absolute energy units. Suppose that a star has an apparent magnitude $m = 20$. By Equation 1.1, the logarithmic ratio l_m of the flux of a star to the solar flux is given by

$$l_m = \log (F_m/F_\odot) = 0.4(m_\odot - m); \tag{1.3}$$

thus for our star $l_m = -18.7$, and $F_m = 3 \times 10^{-13} \text{ erg cm}^{-2} \text{ s}^{-1}$.

In the event that we can estimate the distance r of the star by some method or other, we can at once find its luminosity—its total radiant power—from the simple relation

$$L = 4\pi r^2 F. \tag{1.4}$$

If, in our example, the star is 100 pc away (1 parsec = 3.26 light years = 3.085×10^{18} cm), its luminosity will be $L = 3 \times 10^{29}$ erg/s. Another

---

[2]Concepts such as this are fully explained in any elementary introduction to astronomy. Some representative suggestions are listed in "Further Reading" at the end of this book.

quantity worth remembering is the sun's luminosity $L_{\odot} = 4 \times 10^{33}$ erg/s. Our star is therefore radiating ten thousand times more weakly than the sun; it is a faint dwarf star.

Equation 1.4 states that for a star of given luminosity L, the flux density F of its radiation is inversely proportional to the square of its distance r. A comparatively close star of low luminosity (a dwarf) and a distant star of high luminosity (a giant) may both have the same apparent magnitude m. We consequently describe the luminosity of a star by its absolute magnitude, denoted by the capital letter M. This is the apparent magnitude that a star of interest to us would have if it were located at a standard distance of 10 pc. The absolute and apparent magnitudes are related by the simple equation

$$M = m + 5 - 5 \log r, \tag{1.5}$$

where r is expressed in parsecs.

Thus one of the fundamental properties of a star, its luminosity, is established if its apparent magnitude and distance are known. Astronomers have thoroughly reliable methods at their disposal for determining the apparent magnitude, but it is not so easy to measure the distance of a star. For stars quite close to us, no more than a few dozen parsecs away, distances can be determined by the trigonometric method, which has been used ever since the early nineteenth century. This method requires measurement of the extremely small angular shift in the position of a star when it is observed from different points along the earth's orbit, that is, at different times of the year. Trigonometric distances are the most accurate of all and are very trustworthy.

But for the overwhelming multitude of more remote stars the trigonometric method is of no value: one would have to measure too tiny a shift— less than a hundredth of a second of arc. Other methods, considerably less exact but still fairly reliable, come to our rescue. In many cases the absolute magnitude of a star can be determined indirectly by taking advantage of certain characteristics of the radiation we observe, rather than by measuring the distance itself. As we cannot pause to review these methods here, we suggest that the interested reader consult one of the standard texts mentioned at the end of this book. Generally speaking, the determination of the distances of remote objects in the universe (stars, nebulae, galaxies) has always been and remains today one of the central problems in astronomy.

An extraordinary treasure of information comes to us from stellar spectra. Astronomical spectroscopy is now making use of very delicate and refined techniques. In particular, the latest advances in electronics and other branches of modern applied physics are routinely employed. We lack the space here to describe these developments in any detail.

The spectra of the great majority of stars have long been classified into types. Spectral types are designated in order by the sequence of letters O, B, A, F, G, K, M. The system devised to classify stellar spectra is so accurate that a spectrum can be specified to within one-tenth of a spectral type. For example, the portion of the sequence of spectra from type B to type A is written as B0, B1, . . . , B9, A0. To a first approximation, the spectrum of a star resembles the spectrum of a black body radiating at a certain temperature T. These temperatures range smoothly from 40–50 thousand degrees for stars of spectral type O to 3000°K for type M stars. Most of the radiation of type O and B stars falls in the ultraviolet part of the spectrum, which is inaccessible to observation from the earth's surface. Over the past several years, however, special purpose satellites have carried telescopes above the atmosphere, enabling us to study ultraviolet stellar radiation as well.

One distinctive feature of stellar spectra is their vast number of absorption lines, representing various chemical elements (Figure 1.1). Careful analysis of these lines has yielded exceptionally valuable information about the nature of the outer layers of stars. Through much effort astronomers have succeeded in performing a quantitative chemical analysis of these layers. Even though stellar spectra differ tremendously from one another, the chemical composition of all stars is remarkably similar in the first approximation. The distinctions among the spectra result mainly from differences in the temperatures of the outer stellar layers. Temperature differences strongly affect the state of atomic ionization and excitation of the various elements in those layers, and this circumstance is strikingly reflected by the spectra.

The outer layers of stars, which are responsible for radiation sent to us "directly," have a chemical composition noteworthy for a great preponderance of hydrogen. Helium takes second place, with the other elements comparatively low in abundance. For every 10,000 hydrogen atoms there are roughly 1000 helium atoms, about 10 oxygen atoms, almost as many carbon and nitrogen atoms, and only one iron atom. The abundance of the other elements is quite insignificant. Without exaggeration we may regard the outer layers of a star as a gigantic hydrogen–helium plasma containing a minor admixture of heavier elements. This result, as we shall see presently, is of vital importance for the whole problem of stellar structure and evolution.

Although the chemical composition of stars generally is much the same, individual stars nevertheless show decided peculiarities in this respect. Some stars, for instance, have an abnormally high abundance of carbon; and unusual objects occur that are overabundant in rare earth elements. While the abundance of lithium is quite negligible ($\approx 10^{-11}$ that of hy-

18

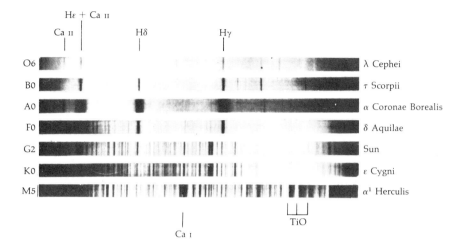

Figure 1.1 A sequence of stellar spectral types.

drogen) for the great majority of stars, one now and then comes across "unique" objects in which this rare element is fairly abundant. Here are two other singular phenomena. In the spectra of certain stars, lines of technetium have been detected—an element that does not exist in a natural state on the earth. This element does not have a single stable isotope. Its most durable isotope survives only about 200,000 years—a minute time span on a stellar scale. So astonishing an anomaly in chemical composition presumably means that nuclear reactions producing technetium are taking place in the outer layers of these stars, which are still puzzling in many ways. Finally, a star has been found in which the helium in the outer layers occurs primarily in the form of He^3, a very scarce isotope on the earth.

All these anomalies in chemical composition are interesting and certainly important, but to examine them in this book would take us too far afield. Fortunately, though, these particularly rare exceptions, caused by very special processes in both the outer and inner layers, have little significance for the problem of chief concern to us—the evolution of stars.

A good indicator of the temperature in the outer layers of a star is its color. Hot stars of spectral types O and B are bluish in color; stars resembling our sun, whose spectral type is G2, are yellow; and type K and M stars are red. A perfectly objective system of color measurement has been developed in astrophysics. It entails a comparison between the magnitudes of stars observed through several accurately standardized filters. Quantitatively, the color of a star is commonly specified by the magnitude difference recorded with two filters, of which one (the *B* filter) transmits mainly blue light rays, whereas the other (the *V* filter, for *visual*) has a spectral sensitivity curve similar to that of the human eye. Color measurement techniques are precise enough that from the color index $B - V$ one can determine the spectrum of a star to one subtype. For faint stars the analysis of colors provides the sole opportunity for classifying the spectra. In Chapter 12 we shall find that comprehensive color determinations for faint stars belonging to clusters have served as an observational underpinning of modern stellar evolution theory.

Knowledge of the spectral type or color of a star immediately yields the temperature of its surface. We have mentioned that a star radiates approximately like a black body at some temperature T. Hence the power radiated by a unit surface area is given by the Stefan–Boltzmann law as

$$\pi B = \sigma T^4, \tag{1.6}$$

where $\sigma = 5.67 \times 10^{-5}$ erg cm^{-2} s^{-1} deg^{-4} is the Stefan–Boltzmann constant. The radiant power of the whole stellar surface, or the luminosity, will evidently be

$$L = 4\pi R^2 \cdot \sigma T^4, \tag{1.7}$$

where R is the radius of the star. Thus to determine the radius of a star we must know its luminosity and surface temperature. We are speaking here of the bolometric luminosity, the power of the radiation emitted throughout the entire range of electromagnetic waves, including ultraviolet and infrared light. The bolometric luminosity is derived from the absolute bolometric magnitude of the star, which in turn is obtained from the "ordinary" absolute magnitude by applying a bolometric correction that depends only on the star's surface temperature.

We have yet another property of a star to determine, one that is quite the most important of all: its mass. Needless to say, measuring the mass is not the simplest of tasks. Indeed, not many stars have a reliable mass determination. The mass can be estimated most readily if a pair of stars form a binary system in which the major semiaxis a and period P of the relative orbital motion are known. In this event the masses of the component stars can be found from Kepler's third law, which may be written in the form

$$\frac{a^3}{P^2(\mathfrak{M}_1 + \mathfrak{M}_2)} = \frac{G}{4\pi^2}. \tag{1.8}$$

Here $\mathfrak{M}_1$ and $\mathfrak{M}_2$ are the masses of the components, and $G = 6.67 \times 10^{-8}$ cm^3 g^{-1} s^{-2} is the constant in Newton's law of universal gravitation. Equation 1.8 gives the combined mass of both components in the system. If we also know the ratio of the orbital velocities of the two components, we can determine their masses separately. Unfortunately, however, there are comparatively few binary systems in which the mass of each star can be established in this manner. Indeed, one cannot do so at all for "close binaries," wherein the two stars are such tight companions as to be indistinguishable. For example, if in the case of a spectroscopic binary star (see the beginning of Chapter 2) the spectrum of just one component is observed, we can determine only the mass function, a combination of the masses of both components and the sine of the inclination angle i of the orbit plane relative to the line of sight:

$$\varphi(\mathfrak{M}_1, \mathfrak{M}_2, i) = \frac{\mathfrak{M}_2{}^3 \sin^3 i}{(\mathfrak{M}_1 + \mathfrak{M}_2)^2} \odot \tag{1.9}$$

If the spectrum of each component is available, a fairly rare occurrence, we can determine the two quantities $\mathfrak{M}_1 \sin^3 i$ and $\mathfrak{M}_2 \sin^3 i$. When it comes to estimating the mass of a single star, the state of affairs deteriorates further.

As a matter of fact, astronomers have not had and still do not have a method providing a direct and independent determination of the mass of an *isolated* star (one not belonging to a double or multiple system). This is a most serious defect in our scientific study of the universe. If such a method existed our knowledge would progress much more rapidly. In this situation astronomers tacitly assume that stars of the same luminosity and color all have the same mass. But masses can actually be determined only for binary systems. Any claim that an isolated star with a certain luminosity and color has the same mass as a sister star belonging to a binary system should be viewed with some reserve. Indeed, as we shall see at the end of Part II, stellar evolution does not proceed in the same fashion in close binary systems as it does for isolated stars. Mass determinations can therefore only be "representative" for stars that are widely separated and presumably evolving independently. But caution should be exercised here as well (see Chapter 14). Matters are in a particularly unsatisfactory state with regard to mass estimates for unusual (or, as astronomers say, "peculiar") isolated stars. But let us not talk of such eccentric objects for the time being. We would like to think that ultimately astronomers will learn how to determine the masses of individual stars by some method that we cannot even imagine today.

Notwithstanding all the reservations just expressed, the masses of normal stars can still be determined with a reasonable degree of confidence.

Modern astronomy, then, has procedures at hand for ascertaining the basic properties of stars: luminosity, surface temperature (or color), radius, chemical composition, mass. An important question arises: Are these properties independent? It turns out they are not. To begin with, there is a functional relationship among the radius of a star, its bolometric luminosity, and its surface temperature. This relationship, expressed by the simple equation 1.7, is a trivial one. In addition, however, the luminosity of stars has long been recognized to depend on their spectral type (or, equivalently, on their color). Early in our century the eminent Danish astronomer Ejnar Hertzsprung and American astronomer Henry Norris Russell separately established an empirical dependence of this kind on the basis of extensive statistical data. If we plot the positions of many different stars in a diagram with the spectrum (or the corresponding $B - V$ color index) along the horizontal axis and the luminosity (or absolute magnitude) along the vertical axis, we find that the points are by no means distributed at random, but instead form definite *sequences.*

Figure 1.2 displays this diagram, called the Hertzsprung–Russell diagram, for stars located within 5 pc of the sun. The figure shows that nearly all the stars are grouped along a fairly narrow band extending from the upper left corner of the diagram to the lower right. This band is termed the main sequence. The spectral types of the stars on the main sequence

Figure 1.2   The Hertzsprung–Russell diagram for the stars closest to the sun.

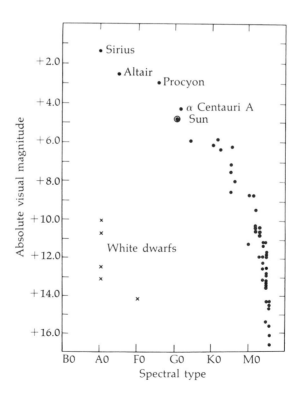

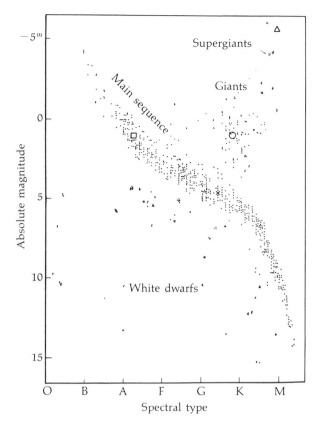

Figure 1.3 The Hertzsprung–Russell diagram for stars with known luminosities and spectra. The × marks the sun.

range progressively from B to M. Besides this sequence we find a small grouping of five stars in the lower left part of the diagram. These stars belong to a rather "early" spectral type (to use a traditional phrase), and their absolute magnitude may be 10 to 12 or fainter; thus they are generally hundreds of times less luminous than the sun, and seem white in color. Stars such as these have therefore come to be called white dwarfs.

However, the diagram of Figure 1.2 is not really a representative one. Figure 1.2 includes in succession all the stars close to the sun; hence less common types of stars that are more than 5 pc distant from the sun cannot appear in such a diagram, because there simply are none of them in the solar neighborhood. In Figure 1.3 we present a Hertzsprung–Russell diagram for many more stars with known luminosities and spectra. Along with nearby objects the diagram now includes quite remote stars of high luminosity. We notice that this diagram has a different appearance from the one shown in Figure 1.2. Common to the two diagrams is the familiar main sequence. But in Figure 1.3 the sequence reaches even further toward the upper left because it contains distant, brilliant, and very rare stars of spectral type O. Although both diagrams clearly show the white dwarf group, in Figure 1.3 it extends toward cooler stars. Figure 1.3 also reveals a sparse sequence of stars situated below the main sequence; these are called the subdwarfs. A very curious property has emerged from spectroscopic research. Subdwarfs differ sharply in chemical composition from main sequence stars in that they have a low abundance of heavy elements, especially metals. This circumstance, as we shall see later, serves as a key to understanding the nature of these interesting stars.

But the most significant difference between the two diagrams is the presence in Figure 1.3 of a sequence (or, properly speaking, a group) of giants in the upper right corner. These are very luminous stars with comparatively low surface temperatures (spectral types K and M). The radii of these stars are accordingly very large, tens of times larger than the sun's radius. They are known as red giants, while the most luminous objects of all belonging to this group of stars are called supergiants.

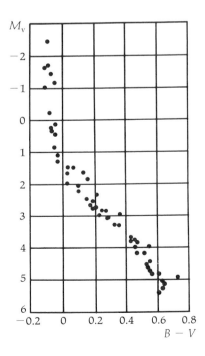

Figure 1.4 The Hertzsprung–Russell diagram for the Pleiades star cluster.

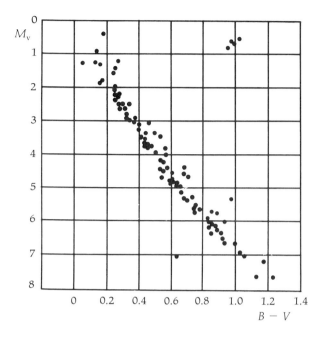

Figure 1.5 The Hertzsprung–Russell diagram for the Hyades star cluster.

Of special interest for the problem of stellar evolution, as will become clear in Chapter 12, are Hertzsprung–Russell diagrams of the more or less compact aggregations of stars known as clusters. Two kinds of clusters are distinguished: open (or galactic) and globular. Apart from their very regular, spheroidal shape, globular clusters are noteworthy for the enormous number of stars belonging to them (perhaps hundreds of thousands) and for their very characteristic distribution in space. They are not concentrated at all toward the galactic plane, but they do show a striking concentration toward the center of our stellar system. Studies of their spectra demonstrate that the member stars of globular clusters are deficient in metals and in heavy elements generally. In this respect as well as many others, the stars belonging to such clusters are identical to subdwarfs, which incidentally exhibit the same space distribution in the Galaxy.

It is important to construct Hertzsprung–Russell diagrams for star clusters, since all the members of a single cluster will have been formed from the same original cloud of gas and dust in the interstellar medium, and will therefore have approximately the same age. We immediately notice that the Hertzsprung–Russell diagrams differ greatly in appearance for different clusters. For instance, the main sequence starts at various spectral

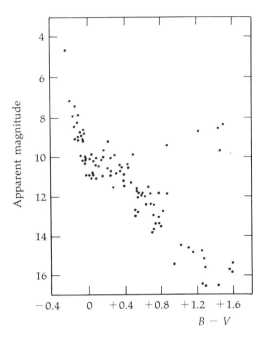

Figure 1.6 The Hertzsprung–Russell diagram for the young open star cluster NGC 2264 in the constellation Monoceros (M. F. Walker).

types, depending on the cluster. The general form of the diagrams for open and globular clusters (Figures 1.4–1.8) also shows prominent differences.[3] These remarkable differences will be explained in Chapter 12.

We again wish to emphasize that the compilation of such diagrams, which requires great effort in measuring precisely the apparent magnitudes and colors of a large number of stars, is of lasting value for our science. It is also worth noting that the diagrams can be prepared without knowing the distances of the clusters. The only factor of importance here is that all the stars in a cluster are located at practically the same distance from us.

We have pointed out the very specific distribution of globular clusters and subdwarfs in space. These objects form a structure in our Galaxy resembling a nearly spherical corona or halo with a strong concentration toward the galactic center. Other objects, however, are distributed in a

---

[3]In these diagrams the horizontal axis is labeled with the color index $B - V$ rather than the spectral type. Indices $B - V = -0^{m}4$ to $-0^{m}2$ are representative of hot O and B stars, while $B - V = +1^{m}6$ corresponds to a type M star.

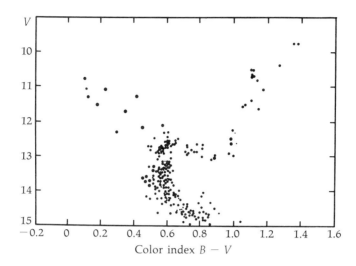

Figure 1.7 The Hertzsprung–Russell diagram for the old open star cluster M67 in the constellation Cancer (H. L. Johnson and A. R. Sandage).

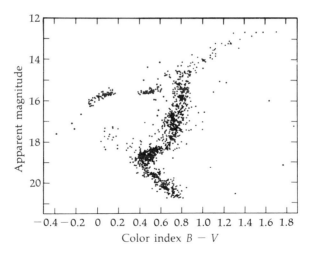

Figure 1.8 The Hertzsprung–Russell diagram for the old globular star cluster M3 in the constellation Canes Venatici (H. L. Johnson and A. R. Sandage).

manner that is not spherical at all. For example, hot and massive main sequence stars, as well as interstellar gas clouds (Chapter 2), comprise within our Galaxy a markedly flattened system concentrated toward the plane of the galactic equator. Very few such objects occur at distances appreciably more than 100 pc away from the galactic plane. The bulk of the main sequence stars with a moderate or low mass have a space distribution somewhere in between the two extreme cases we have described. These stars are at the same time concentrated toward both the galactic center and the galactic plane, forming great disk systems a few hundred parsecs thick.[4]

The distinction in space distribution between different types of stars has a very deep physical meaning. It is extraordinary to realize that stars differing in their space distribution show a significant disparity in their chemical composition. We have already mentioned that subdwarf atmospheres are unusually deficient in heavy elements. The same is true of stars belonging to globular clusters. Thus we may conclude that the objects comprising the galactic halo have a lower heavy element abundance than the objects forming the flat component and disk in our stellar system. This circumstance arises from a substantial difference in the age of the stars populating the spherical and flat components of the Galaxy. Since interstellar gas clouds have practically the same space distribution as hot, massive stars, we may infer that these objects are genetically related. In fact, astronomers have further evidence to support the basic premise that stars are continually being formed in the Galaxy through the contraction of clouds in the interstellar medium (see Chapter 3). The relation between the age of stars and their chemical composition will be discussed in Chapter 12, where we shall consider some general aspects of stellar evolution.

Stars belonging to the galactic halo are often called population II stars, whereas objects concentrated toward the galactic plane are referred to as population I. In the neighborhood of the sun (which is located in the outer part of the Galaxy, very close to its plane of symmetry), population I objects predominate. This is why the subdwarfs, which are population II stars, are represented by such a sparse branch in the Hertzsprung–Russell diagram. On the other hand, in the nuclear region of our stellar system, where the space density of stars is tens of times greater than in the solar neighborhood, the dominant objects are of population II, and they are primarily subdwarfs.

---

[4]Stars with a spherical space distribution possess high random velocities (up to $\approx$ 100 km/s), but objects with a flat distribution are moving at low velocities ($\approx$ 10 km/s). These differences in random velocity are closely associated with the dissimilar space distribution; the situation is analogous to the familiar barometric formula whereby a hotter gas will occupy a larger volume and form a more extended atmosphere.

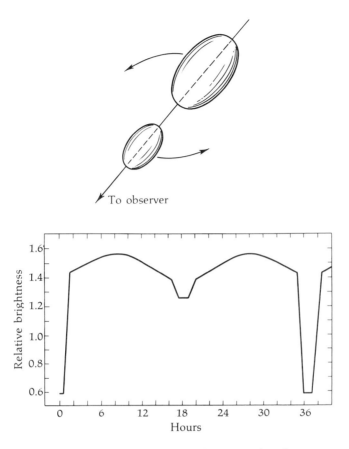

Figure 1.9 Schematic diagram of an eclipse in a close binary system with ellipsoidal component stars, together with the corresponding light curve.

There are roughly 100 billion subdwarfs in the Galaxy; they form a majority among all stars.

We have now recounted some very general facts about the basic characteristics of stars. Of course these objects have many properties to which we have not alluded at all. Certain stars deviate sharply from normalcy. Stars of unusual chemical composition have already been mentioned. The Galaxy also contains variable stars, whose brightness changes from time to time. These are remarkably diverse in their behavior. Sometimes the variability is caused by purely geometric factors: in a close binary system, if our line of sight forms only a small angle with the orbit plane we will periodically observe eclipses, as one star passes behind the other (Figure 1.9). But stellar variability most often results from perfectly genuine luminosity variations, ordinarily accompanied by fluctuations in the surface temperature and radius.

Among all the variable stars, objects that vary periodically in luminosity, radius, and temperature because of physical pulsations are especially interesting. Such stars are called Cepheids. They have played a prominent role in the history of astronomy, as they can help us to estimate the distances of very remote objects (galaxies), which might not be measurable in other ways. How is this done? Astronomers have discovered empirically that the longer the pulsation period of a Cepheid, the more luminous it is.[5] Very faint Cepheids have been observed in distant galaxies; by recording their period one can estimate their luminosity and hence their absolute magnitude. Their distance is then given by Equation 1.3. As the luminosity of Cepheids (particularly those of long period) is very high, they are visible from distances of intergalactic scale. No wonder Cepheids have been called "the beacons of the universe."

Stellar variability of nonperiodic character is encountered far more often. On occasion stars of this kind are observed to increase their level of radiation by some amount or other; frequently this brightening resembles a flare. Flare activity is very widespread among red dwarf stars. A sizable proportion if not the majority of red dwarfs belonging to spectral type M are flare stars. During a flare, which typically lasts some tens of minutes, the brightness of such a star will increase by tens of times; concurrently a burst of radio emission may be detected. Evidently we are here observing a phenomenon analogous to solar flares but on an incomparably greater scale. This type of stellar variability generally involves unstable processes taking place in the surface layers.

"Exploding" stars—novae and supernovae—comprise a group all their own. While nova flares do not entail any fundamental change in the structure of the star itself (see Chapter 14), supernova outbursts—extremely rare occurrences—are accompanied by catastrophic transformations of the stellar structure. This exceedingly uncommon event is so important for astronomy that we are devoting a separate part of this book to it.

Most of the stars in the Galaxy, though, do not show any appreciable signs of instability. Their luminosity is notable simply for remaining so constant. Naturally they do change their properties as they evolve. But those changes proceed ever so slowly.

---

[5]Studies of this sort have been carried out for Cepheids located in the Magellanic Clouds, the galaxies closest to our own. Since all the Cepheids there are at practically the same distance, which is known, their apparent magnitude immediately determines their luminosity.

2

A Survey of the Interstellar Medium

Long before humanity as we know it first appeared on the earth, there were those who were aware of the sun, moon, and planets, and of the stars as well. I suspect that even animals, and not just the higher ones, have some primitive astronomical information. Science had to advance for thousands of years, though, before people recognized the simple yet majestic fact that stars are objects more or less resembling the sun but located incomparably farther away from us. Not even such outstanding thinkers as Johannes Kepler understood this to be so. Sir Isaac Newton was the first to estimate the distances of the stars correctly.

Two centuries after the time of the great British scientist, nearly everyone tacitly assumed that the enormously vast reaches of space containing the stars are a *perfect vacuum.* For eighteenth- and nineteenth-century astronomers this question was never a pressing one; their range of interests was altogether different from ours today. Only an occasional astronomer raised the possibility that light could perhaps be absorbed by an interstellar medium. It was not until the very beginning of the twentieth century that the German astronomer Johannes F. Hartmann showed persuasively that the space between the stars is not some mythical void at all. It is filled with gas—at a very low density, to be sure, but gas is definitely present. This noteworthy discovery, like so many others, was made by spectroscopic analysis.

Hartmann investigated the spectra of binary stars. Because of the orbital motion of these stars the wavelengths of their spectral lines vary with a strict period by a small amount, first in one direction, then in the other. The period of these variations is exactly equal to the relative orbital period of the two stars. Periodic variations such as these in the wavelengths of spectral lines result from the Doppler effect, which is familiar from laboratory physics. When a source of radiation is moving toward the observer at a velocity v, the wavelength λ of a line will decrease by the amount $\lambda v/c$, where c is the velocity of light; if the source is receding from the observer at the same velocity, the wavelength will increase by the same amount. Clearly a star executing a periodic motion as it travels along its orbit will alternately approach us and recede from us, causing the wavelengths of its spectral lines to shift back and forth periodically.

The discovery by the German astronomer was that the spectra of certain binary stars exhibit two absorption lines whose wavelengths remain unchanged despite the periodic variation in the wavelengths of all the other spectral lines. These fixed lines, which belong to ionized calcium, have been called stationary lines. They are formed not in the outer layers of the stars but somewhere along the way from the stars to the observer. This is how interstellar gas was first detected; starlight passing through it experiences absorption in narrow intervals of the spectrum.

For almost half a century interstellar gas was investigated mainly by analyzing the absorption lines formed in it. One finding was that the interstellar lines quite often have a complex structure: they are made up of several components located close together. Each of these components represents the absorption of starlight by a particular cloud in the interstellar medium, and these clouds are moving relative to one another at velocities of roughly 10 km/s. The Doppler effect thereby produces slight shifts in the wavelengths of the absorption lines.

Even though interstellar gas was first detected from its calcium-line absorption, we cannot, of course, infer that calcium is the most abundant element in interstellar space. The gas is manifest in other absorption lines as well, such as the prominent yellow sodium line. Certainly the intensity of the absorption lines is not always governed by the abundance of the corresponding chemical element. Far more often the intensity is determined by a felicitous arrangement of the atomic energy levels between which electrons make transitions and thus form a line. It is a very important fact that in interstellar space practically all atoms, ions, and molecules should be on their "lower," unexcited energy level. Processes whereby atoms might become excited, generally involving either the absorption of radiation or collisions between particles, take place incredibly rarely in the interstellar medium. If after an electron recombines with an ion the resulting neutral atom happens to be excited, it will always succeed in making a spontaneous transition to its "deepest" state, emitting one or more photons as it does so. Collisions with other particles will not interfere with this emission.

After having waited an indefinitely long time on its "ground" level, an atom may absorb radiation at certain frequencies. The lowest such frequency is called the resonance frequency, and the spectral line corresponding to it is the resonance line. Resonance lines are usually the strongest. Calcium, like sodium, has the spectroscopic property that its resonance lines are located in the visible part of the spectrum. But for the great majority of other elements the resonance lines are in the far ultraviolet region. The classic examples are the most abundant elements in the universe, hydrogen and helium. The wavelength of the hydrogen resonance line (the renowned Lyman-α line) is 1216 Å, and for helium it is still shorter, 586 Å. But all extraterrestrial radiation at wavelengths shorter than 2900 Å is completely absorbed by the earth's atmosphere.

Until balloon, rocket, and satellite technologies were developed, the ultraviolet spectral region of all cosmic objects was entirely inaccessible to astronomers. It is only rather recently that scientists have been able to obtain stellar spectra in the far ultraviolet and to record the interstellar Lyman-α line and the resonance lines of oxygen (at 1300 Å) and of other interstellar atoms. To avoid any misunderstanding one should recognize that spectral lines of hydrogen, helium, oxygen, and other elements have long been observed in the spectra of the sun and stars. These lines, however, were not resonance lines but lines formed through transitions between excited atomic levels. In hot, dense stellar atmospheres filled with radiation, the excited levels may have high enough populations to permit the formation of absorption lines, but in the interstellar medium the physical conditions are altogether different.

The chemical composition of the interstellar gas turns out to be fairly similar, in the first approximation, to the chemical composition of the solar and stellar atmospheres. Hydrogen and helium are the predominant elements, while other elements may be regarded as "impurities." Curiously enough, in the interstellar gas calcium is about a million times less abundant than hydrogen.

Over the past few years a genuine revolution has come about in studies of the interstellar medium by optical techniques as a result of the impressive achievements of space astronomy. The chemical composition of interstellar gas clouds relatively close to us has been most fully analyzed thus far by the specialized American astronomical satellite *Copernicus,* mentioned in the Introduction. As we have pointed out, the most abundant elements generally have their resonance lines in the ultraviolet part of the spectrum. By observing bright, comparatively nearby stars, astronomers using satellites can detect in the ultraviolet stellar spectra the interstellar resonance absorption lines of such elements as hydrogen (the Lyman-α line at 1216 Å), carbon, nitrogen, oxygen, magnesium, silicon, sulfur, argon, and manganese. Lines of both neutral interstellar atoms and their ions have been observed. And as a result perfectly definite distinctions in chemical composition have been established between individual clouds and the sun.

Research on the interstellar medium has thereby been lifted to a higher plane: although in the first approximation, relying simply on the very limited observations possible from the ground, one could regard the chemical composition of the interstellar gas as more or less like that of the solar atmosphere, authentic differences in composition can now be distinguished with confidence, even from one cloud to another. For example, the abundance of magnesium, chlorine, and manganese relative to hydrogen is four to ten times lower in interstellar clouds than in the solar atmosphere. Figure 2.1 illustrates the departures of chemical composition from solar values for five different clouds projected against bright stars. On a logarithmic scale, this diagram gives a graphic idea of the distinctions in chemical composition between individual clouds and the sun. We notice, in particular, that most of the bars in the figure tend to lie below the horizontal lines—evidence that the clouds have a deficiency in the corresponding elements relative to the sun.

Along with the atoms and ions, the interstellar gas normally contains a very small proportion of molecules (roughly 10^{-7} of the number of hydrogen atoms). Through optical astronomy the simple diatomic molecules CH, CH$^+$ (the plus sign indicates an ionized molecule), and CN have been detected in the interstellar medium. In laboratory physics molecular spectra ordinarily consist of a great many lines merging into *bands;* but interstellar molecules as a rule exhibit just a single line, since they are all in their deepest vibrational, rotational, and electronic state. The interstellar CN molecules are an exception: nearly 40 years ago *two* CN lines were discovered. Accordingly the second rotational level must also have an appreciable population, for in the CN molecule this level is considerably closer to the first level than in the CH and CH$^+$ molecules.

Is it really worth mentioning such a trifle? Well, this "trifle" has lately been found to have a very deep origin: the second rotational level of the CN molecule is excited by the primordial background radiation filling the whole universe. This radiation, as observers have learned, has a Planck spectrum corresponding to a temperature of 2°7 on the absolute Kelvin scale and represents a relic or vestige of an ancient state when the universe was tens of thousands of times younger than now and 1400 times smaller in size! The discovery of the vestigial radiation, an event of inestimable importance for astronomy, is at least as significant as the discovery of the

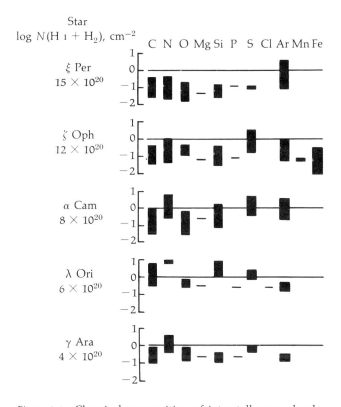

Figure 2.1   Chemical composition of interstellar gas clouds.

red shift in the spectra of galaxies. How astonishing that this radiation had been detected indirectly 25 years before its discovery, but alas not recognized for what it was! There have been other such cases, though, in the history of science, and we shall meet some of them later in this book.

A further finding of exceptional value has been the discovery of hydrogen molecules in the interstellar gas. Since the electronic resonance band of the H_2 molecule is located in the ultraviolet region near 1092 Å, only astronomical observations from outside the atmosphere could solve the problem. Here again the *Copernicus* satellite has supplied the best information. Special studies have been made of the ultraviolet spectra from highly reddened stars, which must be situated behind dense clouds of gas and dust because such clouds absorb blue wavelengths especially strongly, as we shall see in a moment. It is in these clouds that we would expect to find a measurable amount of molecular hydrogen. Spectrograms reveal very strong interstellar H_2 lines for the reddened stars. As the Lyman-α resonance line of atomic hydrogen has been measured concurrently in the spectra of the same stars, it has been possible to evaluate directly the abundance ratio of molecular and atomic hydrogen in the clouds. The ratio fluctuates over a very wide range, from a few tenths to less than 10^{-7}, the value that represents the limiting sensitivity of the spectrograph to the faintest lines.

Our discussion of the interstellar medium has been thus far confined to the interstellar gas. But the medium contains another component: interstellar dust. We have pointed out that even in the nineteenth century the transparency of interstellar space was a matter of debate. Not until around 1930 was it definitely proved that interstellar space is not quite transparent. The substance absorbing light is concentrated in a rather thin layer near the galactic plane. Blue and violet rays are absorbed most strongly, while absorption in the red is comparatively slight. Interstellar absorption is therefore accompanied by a *reddening* of the light of distant objects located in the Milky Way belt. The particular amount of absorption varies rather irregularly in different directions. There are whole parts of the sky where the absorption is small, but there are also regions in the Milky Way where light is absorbed to an enormous extent. Regions such as these are sometimes called coalsacks; a famous one is shown in Figure 2.2. Clearly, then, the absorbing substance is distributed very nonuniformly in interstellar space, forming separate condensations or clouds.

Just what is this substance? It now appears certain that light is being absorbed by interstellar dust—microscopic particles of solid material less than a micron across. These grains have a complex chemical structure: graphite, silicates, and impure frozen substances called "dirty ices" are representative samples. The particles are known to be rather elongated in shape, and to some extent they are aligned; that is, their long axes tend to be oriented more or less parallel to one another in a given cloud. As a result, starlight traversing a medium with thinly scattered dust particles will become partially polarized, by 1 to 2 percent; the amount of polarization is correlated with the reddening of the color due to absorption. The alignment of the grains results from the presence of very weak magnetic fields in interstellar space. To account for the polarization observed in the light of distant stars, the strength of the field ought to be about 10^{-6} to 10^{-5} gauss. We shall return time and again to the interstellar magnetic field; at present we would merely note that other, more refined methods of measuring it confirm the estimate just given.

The ionization of the interstellar gas and the related question of its temperature are matters of outstanding importance. It must be emphasized, however, that the concept of temperature is by no means elementary when applied to the interstellar gas. Strictly speaking, the concept of temperature applies only to bodies in a state of thermodynamic equilibrium. Such a state

Figure 2.2   The Horsehead Nebula in the constellation Orion.

presupposes that a variety of conditions are satisfied simultaneously. For instance, the spectral density of the radiation should be as described by the Planck formula; the total energy density should be as given by the Stefan–Boltzmann law, which calls for a proportionality to the fourth power of the temperature; the velocity distribution of the various atoms and ions and also the electrons should accord with the Maxwell law; and the distribution of the atoms, molecules, and ions over their quantum states should reflect the Boltzmann formula. Each of these laws contains an important parameter having the meaning of a temperature. The *kinetic* temperature appears in the Maxwell velocity distribution, the *excitation* temperature in the Boltzmann formula, and so on. If a body (or system) is in a state of thermodynamic equilibrium, then all these "temperature" parameters should be equal to one another and would simply be called the temperature of the body.

One can easily understand that even under the natural conditions to which we are accustomed on the earth, thermodynamic equilibrium does not as a rule prevail. When we speak, for example, of the air temperature, we should always add the qualification "in the shade." The complete absence of thermodynamic equilibrium may be demonstrated strikingly by the following simple illustration. Let us pose the question: What is the temperature of our room on a sunny day? The answer might appear to be simple—say 68°F, or 293°K. But I am equally justified in claiming that the room temperature is no less than 5700°K. Why? Because the whole room is filled with direct and scattered sunlight whose spectral composition is much the same as that of solar radiation. And the spectrum of the sun is very similar to the spectrum of an ideal black body heated to a temperature of 5700°K. On the other hand, the energy *density* of the sunlight in the room may be a hundred thousand times smaller than that on the solar surface, for as one recedes from the sun the density of its radiation flux falls off in proportion to the square of the distance.

What, then, is the significance of our intuitive impression that the room temperature is 68°F? We are implicitly referring here to the kinetic temperature, the parameter in the Maxwellian velocity distribution of the air molecules confined in our room. The 5700°K value, on the other hand, represents the *color* temperature of the radiation filling the room. This very simple example shows how great the departures from thermodynamic equilibrium may be, even under the most ordinary conditions. As a matter of fact, life itself is possible only if thermodynamic equilibrium is violated. Strict thermodynamic equilibrium means death.

Can we speak of a temperature in interstellar space, where the deviations from thermodynamic equilibrium are exceedingly great? Indeed we can, if each time we stipulate what temperature we have in mind. Usually we will be concerned with the kinetic temperature of the interstellar medium, which, as we shall see, may vary over a fairly wide range. Interstellar space is, however, filled with radiation from a vast number of stars. This radiation therefore has the same color temperature as the stars; it is measured in thousands or tens of thousands of degrees. If we consider, for instance, the region of interstellar space a few dozen light years away from a hot star such as a giant of spectral type O or B (see Chapter 1), then the color temperature there will be 20,000 to 40,000°K. But at the same distance from a red giant the color temperature may be about 3000°K.

In contrast, the radiation density in interstellar space is extraordinarily low. It is smaller than the radiation density at the surface of the closest star by the same factor that the solid angle subtended by the stellar disk at a given point in space is smaller than 2π steradians.[1] Upon calculating this ratio we find that it is of order 10^{-15}. In interstellar space the mean radiant energy density is about 1 electron volt per cubic centimeter, or 10^{-12} erg/cm^3. Since the energy of each photon of light is about 3 electron volts, there

---

[1]Not a full 4π, as one can show, because at the stellar surface the radiation flux is directed only outward; thus even there the condition of thermodynamic equilibrium is violated.

should be less than one photon per cubic centimeter of interstellar space. Yet the energy of these photons is approximately the same as in stellar atmospheres, where the photon density is incomparably higher. In this sense we may figuratively say that the radiation field in interstellar space is extremely dilute. In fact, the radiation in our sunny room, and on the earth generally, is also dilute. The temperature of the interstellar medium, as determined from the density of the radiation occupying it, is exceptionally low, just a few degrees on the Kelvin scale. That should be the surface temperature of the solid grains dispersed over interstellar space in thermal equilibrium with the dilute radiation field surrounding them, for such grains ought to absorb just as much radiation as they emit.

The great disparity between the high color temperature of the radiation permeating the interstellar medium and its very low density is assuredly the chief factor determining the characteristic physical conditions in this medium. Let us take a specific example, one that will be important for the discussion to follow. Consider the photoionization of interstellar atoms when they absorb photons of rarefied ultraviolet radiation. In such an ionization process the electrons detached from the atoms will acquire a kinetic energy given by a fundamental equation of Einstein:

$$h\nu = \chi + \frac{1}{2}m_e V^2, \tag{2.1}$$

where ν is the frequency of a photon of the radiation absorbed, and the ionization potential χ represents the binding energy of an electron in an atom. This equation, resting on the basic premises of quantum theory, implies that the kinetic energy of the photoelectron is determined solely by the frequency of the absorbed photon; it does not depend at all on the density of such photons in the surrounding space.

Accordingly, the kinetic energies of photoelectrons in interstellar space will be the same as in the atmospheres of stars—quite high, of the order of several electron volts. By colliding with each other, these electrons will fairly rapidly establish a Maxwellian velocity distribution, so we may refer to their *kinetic* temperature. On the other hand, the electrons will continually lose energy because of inelastic collisions with atoms. As a result of the balance between the energy lost in this way and the energy acquired by photoionization, the temperature of the interstellar medium in the vicinity of hot stars will stabilize at the fairly high level of about 10,000°K.

The low radiation density in interstellar space, together with the extremely low gas density, has another important consequence to which we have already alluded. Processes whereby atoms absorb radiation will take place very rarely, so that if atoms and molecules should be excited in some manner they will return to their ground state without any hindrance, emitting photons as they do so. This emission will occur even when the excited levels are "metastable"—persistent for an abnormally long time. In the terrestrial laboratory, collisions and absorption processes would meanwhile raise atoms to higher levels, so to speak, and a transition from a metastable level down to the ground level would not be accompanied by the emission of photons of the corresponding frequency. But under the conditions of the interstellar medium an atom can wait patiently on its metastable level, for there will be no collisions or absorptions to disturb it; eventually it will descend to its ground level, emitting a photon of what spectroscopists call a forbidden line.[2]

Since virtually no interactions of excited atoms with matter or radiation will have a chance to occur, almost all atoms, ions, and molecules can only execute downward transitions to the ground state, emitting the appropriate photons. Upward transitions to a state of higher energy will be possible only for atoms already in their deep ground state. Such processes will generally require the absorption of ultraviolet photons, because the resonance-line frequencies and ionization potentials of atoms and ions are quite high. Thus the very significant process of photon redistribution should take place in the interstellar medium: atoms will absorb ultraviolet photons and then, after they recombine on excited levels and experience several downward "cascade" transitions to the ground level, they will emit less energetic photons whose wavelengths fall within the optical range. In laboratory physics, a process of this kind is called fluorescence.

The following situation would be typical in interstellar space. An interstellar gas cloud located fairly close to a hot star (and therefore emitting strongly in the ultraviolet spectral region) absorbs photons capable of ionizing hydrogen. Such photons should have wavelengths shorter than 912 Å. Through their absorption of these photons the great majority of the hydrogen atoms in the cloud become ionized, releasing electrons. Some electrons recombine with the protons (the hydrogen nuclei), emitting light in the visible and infrared regions, particularly in the lines of the Balmer series. Other electrons will collide with atoms and ions of nitrogen, oxygen, sulfur, and so on, exciting their metastable levels. These excited particles will then be able to radiate away their excess energy freely in the form of forbidden lines.

Masses of interstellar gas situated close enough to hot type O and B giant stars should without question be fully ionized. But will all interstellar gas be ionized? Calculations supported by observational evidence (see below) demonstrate that throughout much of the interstellar medium hydrogen will not be ionized. Hot stars are capable of ionizing the hydrogen around them only out to a certain distance that depends both on the power of the ultra-

---

[2]Some important exceptions to this rule will be discussed below.

violet radiation from the star and on the density of the interstellar medium. As a result the ionization of the interstellar medium will seem to have a most peculiar topology: hot stars will be surrounded by closed cavities (spheres, in the ideal case of an interstellar medium with constant density) where the hydrogen is ionized, while between the cavities the hydrogen will be neutral. Regions of the interstellar medium in which hydrogen is ionized are called H II zones, and neutral hydrogen regions are H I zones.

The radius of an H II zone is determined by the ionization balance within it: the number of ultraviolet photons (emitted by the hot star) absorbed by the zone per unit time should be equal to the number of recombinations between protons and electrons. Since every photon absorbed will give rise to a new pair of ions, while every recombination event will destroy a pair of ions, our condition merely states that the degree of ionization should remain constant with time. Expressed mathematically, this condition reads

$$\frac{4}{3}\pi R^3 \alpha N_e N_i = \frac{L(T)}{h\bar{\nu}},$$
(2.2)

where R is the radius of the ionization zone (we shall regard it as spherical), $\alpha N_e N_i$ is the number of recombinations per unit volume each second, $N_e = N_i$ is the numerical density of the electrons and of the ions, α is the recombination coefficient, $L(T)$ is the power of the star's ultraviolet radiation (depending on its surface temperature), and $h\bar{\nu}$ is the average energy of the ultraviolet photons. Equation 2.2 implies that the radius has the functional form

$$R = \varphi(T) \cdot N_e^{-2/3}.$$
(2.3)

One finds by calculation that if $N_e \approx 1$ cm^{-3} (a fairly realistic value—see below) for stars of spectral types O and B, then R may extend to many tens of parsecs. This huge region may enclose tens of thousands of stars. It is interesting to note that the transition between H II and H I zones is very sharp: within just a few hundredths of a parsec interstellar hydrogen changes from a state of nearly 100 percent ionization to a neutral state.

An H II zone reprocesses all the ultraviolet radiation it absorbs from the hot central star into visible and infrared photons belonging to the Balmer and Paschen series of hydrogen lines and to forbidden lines, as well as into ultraviolet photons of the Lyman-α line. But regions of this kind are nothing other than gaseous nebulae, the brightest of which (such as the one in the constellation Orion) have long been familiar to astronomers. In such a nebula the emission of an elementary volume is produced by various types of collisions between electrons and ions, yielding atoms and ions in excited states. The emission should therefore be proportional to the squared density N_e^2. The principal quantity determining the observability of a nebula is its *surface brightness,* which is proportional to the product of the emission per unit volume by the distance R that the emitting region extends along the line of sight. Hence the surface brightness I of the nebula will be proportional to the quantity N_e^2R, known as the emission measure.

Photographs of several H ɪɪ regions, or gaseous nebulae, are displayed in Figures 2.3 to 2.5. These photographs were taken through filters transmitting the red Hα line of hydrogen. Notice the complexity of the brightness distribution for these nebulae. One should recognize, however, that the patchy structure of absorbing dust clouds projected against the nebulae or located inside them will seriously alter the actual brightness pattern.

Knowing the surface brightness of a nebula from astronomical observations, we can always obtain the corresponding emission measure. If we also know the extent R of the nebula along the line of sight we can at once establish the density N_e of the interstellar gas. However, the distribution of inter-

Figure 2.3   The Great Orion Nebula.

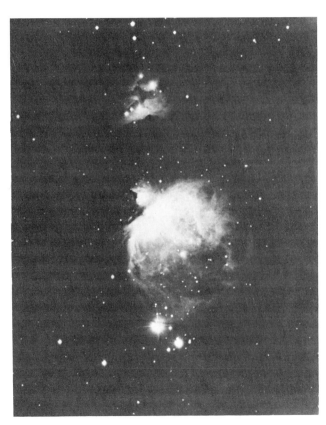

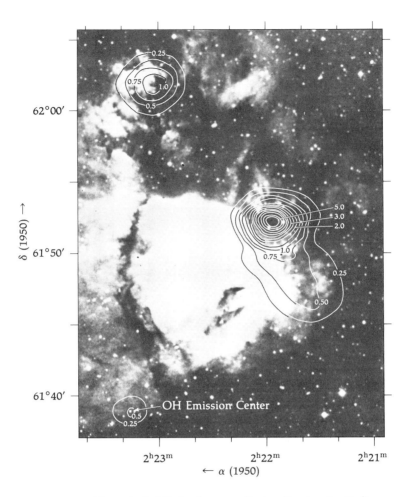

Figure 2.4 The nebula W3 in the constellation Cassiopeia. Radio observations disclose several compact H ɪɪ regions in the vicinity; the contours are labeled with the antenna temperature at 2-cm wavelength (Lick Observatory Photograph).

stellar gas is so irregular that the density determined in this fashion merely represents some average value. The mean density in interstellar gas clouds turns out to be about 10 ionized hydrogen atoms per cubic centimeter. Isolated, very dense clouds may have several thousand atoms, or even more, per cubic centimeter. These dense clouds are observed as particularly bright nebulae. In the interstellar space between the clouds the density of atoms is at least a hundred times lower than inside the clouds. Estimates for

43

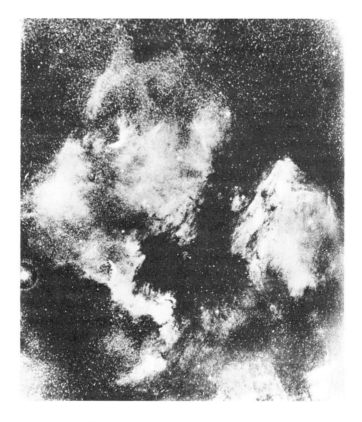

Figure 2.5 The North America and Pelican Nebulae in the
constellation Cygnus.

the density of atoms in interstellar gas clouds where hydrogen is not
ionized (H I zones) can be made with great confidence by analyzing the
ultraviolet absorption lines of this gas in stellar spectra recorded with
orbiting astronomical observatories. Indeed, the spectrograms obtained
with the *Copernicus* satellite permit a quantitative chemical analysis of the
interstellar medium. For clouds that have been investigated in this manner, viewed in projection against stars comparatively close to us, the hydrogen density is found to be of the order of several hundred atoms per
cubic centimeter.

   A careful analysis of the *Copernicus* spectra of nearby stars (20–150 pc
away from us) devoid of any reddening caused by cosmic dust has enabled
the physical properties of the very rarefied interstellar medium located
between the clouds to be established. In this situation the interstellar absorption lines are of very low intensity. Resonance lines observed have mainly

been those of singly ionized atoms. One gains the impression that heavy elements are less abundant, relatively speaking, in the intercloud medium than in the clouds. The hydrogen density in the intercloud medium varies over a fairly wide range, from 0.2 to 0.02 cm^{-3}.

Interstellar gas is concentrated in a very thin layer near the plane of symmetry of the Galaxy. The layer is no more than 200 pc thick, and the average number density of particles in it is about 1 cm^{-3}, corresponding to an average mass density of about 10^{-24} g/cm^3. Interstellar dust has an average mass density about a hundred times smaller. It is interesting to realize that the density of heavy elements (all those except hydrogen and helium) in the interstellar gas is about 10^{-26} g/cm^3. Since interstellar grains consist primarily of heavy elements, it follows that approximately half of all heavy elements in the interstellar medium are bound up in solid grains, with the other half being in a gaseous state. This surprising circumstance, which has yet to be explained, should have a significant bearing on the origin of the interstellar dust.

Atoms in the interstellar gas, then, have a number density at least a billion billion times lower than in the earth's atmosphere. Our assertion that the interstellar gas is definitely no vacuum would seem all the more paradoxical! As a matter of fact, what is a vacuum? Surely, not even a highly rarefied gas will necessarily be a vacuum. We may not speak of a vacuum unless the mean free path of the gas particles exceeds the size of the volume filled by the gas. For example, in a gas discharge tube the number density of gas atoms might be 10^{12} cm^{-3}. Then the mean free path $l \approx 1/n\sigma \approx 10$ m, where $\sigma \approx 10^{-15}$ cm^2 is the collisional cross-section of the atoms. In interstellar space, where $n \approx 1$ cm^{-3}, we have $l \approx 10^{15}$ cm $= 3 \times 10^{-4}$ pc, whereas the gaseous disk in the Galaxy is about 200 pc thick. Under these conditions we are not entitled to call interstellar gas a vacuum.

The interstellar gas is a continuous, compressible medium—a continuum. The laws of gas dynamics are not fully applicable to it. Waves such as shock waves are propagated through this continuous medium. One important type of shock wave in the interstellar medium, the shock generated by the explosion of a star, will be discussed in Chapter 16. The medium is pervaded by a complex, turbulent motion, with "ripples" generally passing through it; we shall return to this phenomenon in Chapter 21.

It is also important to remember that the continuous medium has quite a high electrical conductivity, for it is ionized either fully (in H II zones) or partially (in H I zones). Because of the high conductivity of the interstellar medium, interesting effects arise from the magnetic fields it contains. The magnetic lines of force are in effect glued to, or "frozen" into, the interstellar gas, and they follow the clouds in their capricious motions. Often the interstellar magnetic field, if it is strong enough, tends to control the motion of the clouds, preventing them from crossing the lines of force. Magneto-hydrodynamics, a vital branch of modern physics of great practical significance, grew out of astronomy, particularly out of research into the nature of the interstellar gas.

Although before World War II astronomers were limited to the study of certain interaction processes between the interstellar gas and the rarefied radiation field, in the postwar period the magnetohydrodynamic aspect of the problem began to take on increasing importance. This aspect has a special meaning for the central problem in which we are interested: the formation of stars through condensation of the interstellar medium. The next chapter will be devoted to this problem.

Our discussion of the interstellar gas has thus far focused primarily on H II zones, which emit spectral lines in the optical wavelength range and which have therefore been examined with particular care by the methods of optical astronomy. Before the war information (and very little of it) about H I zones could be gleaned only by studying interstellar absorption lines. This approach was developed in a major way during the postwar years with the advent of space astronomy. Radio astronomy was also pursued after the war, opening a new era in research on the interstellar gas.

Back in 1944 a Dutch astronomy student, Hendrik van de Hulst (later director of the Leiden Observatory), put forward a brilliant idea whose principle is the following. If two atomic levels happen to differ very little in energy, then the transition of an atom from its upper to its lower level will be accompanied by the emission of a photon whose wavelength falls in the radio range. And as the most important example of such a transition, the young Dutch astronomer pointed to the hydrogen atom in its deepest quantum state. That two very close levels correspond to this state had long been recognized. The energy difference between the two levels results from an interaction between the intrinsic magnetic moments of the proton and the electron that make up the hydrogen atom. The magnetic moments in turn are associated with the spins of the corresponding elementary particles. This effect, familiar in spectroscopy for a long time, manifests itself as a splitting of spectral lines into several components grouped very tightly (called the hyperfine structure). By van de Hulst's estimate the transition between the upper and lower hyperfine-structure levels of the hydrogen atom ought to be accompanied by the emission of a line at a wavelength of 21 cm.

Four years later, having come across van de Hulst's idea by chance and taken great interest in it, the author of this book carried out a detailed theoretical analysis of the situation. First of all an estimate had to be made of how long a hydrogen atom would stay in its upper hyperfine-structure level before it would descend spontaneously to its lower level, emitting a

46

photon of 21-cm wavelength. Such an estimate is needed because it determines the intensity of this 21-cm radio line, and thereby the likelihood of observing it, the point in which astronomers are mainly interested. The waiting time τ turned out to be remarkably long—fully 11 million years! Compare that with the typical lifetime in the excited state for atoms emitting "optical" lines—about a hundred-millionth of a second!

There is a far higher probability that a hydrogen atom in its upper hyperfine-structure level will descend to the lower level without emitting a 21-cm photon. It will do so because of the ordinary collisions that take place between hydrogen atoms. For a hydrogen atom in an interstellar gas cloud the time interval between two such collisions will be a few hundred years—a relatively insignificant span. On the other hand, collisions will similarly serve to excite the upper hyperfine-structure level. Accordingly, an equilibrium distribution of the atoms with respect to their hyperfine-structure levels will tend to be established, and there will be three times as many atoms in the upper level as in the lower.

With this circumstance in mind we can write the expression

$$\varepsilon = \frac{3}{4} n_H A_{21} h\nu \tag{2.4}$$

for the emission per unit volume of photons in the 21-cm line. Here $A_{21} = 1/\tau$ is the probability of a transition accompanied by the radiation of a 21-cm photon, $h\nu$ is the energy of the photon, and n_H is the numerical density of hydrogen atoms. The intensity of this radiation is given by the customary equation

$$I = \frac{1}{4\pi} \varepsilon R, \tag{2.5}$$

where as before R designates the extent of the emission region along the line of sight. Equation 2.5 holds only if the radiation is not absorbed by the emitting atoms themselves. Indeed, in our case this condition is not satisfied. But even if we allow for self-absorption, the 21-cm line was expected to be intense enough that the sensitivity of postwar radio astronomy equipment would be perfectly adequate to detect it.

The 21-cm line, it was felt, should have a well-defined profile and not be infinitely narrow. The atoms of neutral interstellar hydrogen emitting this line would take part in various motions which would act to broaden the line by the Doppler effect. In the first place, interstellar hydrogen atoms have thermal velocities corresponding to their kinetic temperature; moreover, separate interstellar gas clouds move bodily at speeds of about 10 km/s. Finally, the interstellar gas, just like the stars, participates in galactic rotation. The velocity of galactic rotation is very high, about 200 km/s in the solar neighborhood, and the rotation itself is not at all like that of a rigid body but has a rather complicated behavior. Profiles of the 21-cm radio line ought to be affected by this differential galactic rotation, or more explicitly, by the difference between the radial velocities of a given interstellar cloud and of the sun resulting from this mode of rotation. Differential galactic rotation depends on the galactic longitude.

After it had been predicted theoretically and calculated, the 21-cm line was discovered in 1951 in the United States, Australia, and the Netherlands. Figure 2.6 shows some profiles of the 21-cm hydrogen radio line. Typically the line is a few tens of kilohertz wide on a frequency scale. Profiles such as these can furnish an extraordinary wealth of information on H I zones. It was soon found that the kinetic temperature in these zones is about 100°K, although in places it drops to a few tens of degrees.[3] The low temperature of H I zones may be attributed to the absence of processes for photoionizing hydrogen. Photoionization in a gas would generate a substantial number of quite energetic photoelectrons; these would collide with the atoms and ions, transferring energy to them and "heating" them (see above). H I zones contain no such strong heat sources.

One should not think, however, that H I zones have no free electrons. There are some, but they are thousands of times less common than in H II zones. The electrons in H I zones are formed mainly through ionization of the atoms by cosmic rays of comparatively low energy (of the order of a few million electron volts), which are quite abundant there,[4] and by the soft x rays that permeate the whole Galaxy (see Chapter 23). Furthermore, electrons are formed in H I zones by ordinary photoionization of elements whose ionization potential is lower than that of hydrogen. Carbon is the principal element of this class.

A major role in the heat balance of H I zones is played by carbon, because it acts as a very efficient refrigerator there. If the energy of the electrons formed through ionization did not ultimately leave an interstellar cloud as radiation, then even a tiny amount of ionization operating for a prolonged time would heat the cool gas to a high temperature, as given by the condition $kT = \varepsilon$ (where ε is the mean energy of the photoelectrons). The electrons resulting from ionization would collide with atoms, contin-

---

[3]Approximately the same temperature has recently been obtained with the *Copernicus* satellite from an analysis of ultraviolet absorption lines.

[4]Primary cosmic rays observed near the earth have energies exceeding 1 billion electron volts. But these observations do not imply that cosmic rays of energy less than 1 billion volts are absent from interstellar space. The comparatively soft cosmic rays simply fail to reach the earth. They cannot enter the solar system because they are repelled from it by magnetized clouds of highly rarefied plasma ejected from the sun (the "solar wind").

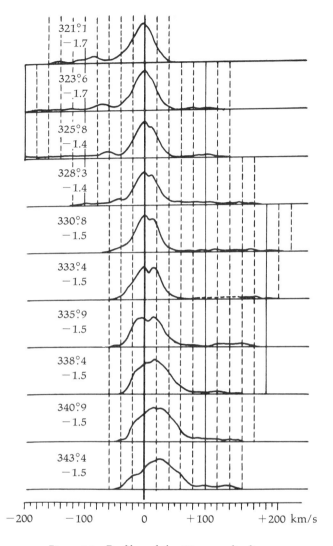

321°.1
−1.7

323°.6
−1.7

325°.8
−1.4

328°.3
−1.4

330°.8
−1.5

333°.4
−1.5

335°.9
−1.5

338°.4
−1.5

340°.9
−1.5

343°.4
−1.5

−200   −100   0   +100   +200 km/s

Figure 2.6   Profiles of the 21-cm radio line.

ually transferring kinetic energy to them and thereby heating them. But this sequence of events does not happen. Instead, not only will elastic collisions take place between electrons and atoms, accompanied by a transfer of kinetic energy from the electrons to the atoms, but inelastic collisions will also occur, leading to excitation of the atoms with subsequent emission of photons. The inelastic collisions will transform the kinetic energy of the electrons into radiation.

Not all atoms are of equal worth for inelastic collisions. Clearly, if the excitation energy for a given kind of atom is too high, only a negligible proportion of the electrons will have enough kinetic energy to excite the atom. Hence the surrender of energy through excitation of such atoms will be an inefficient mechanism. The atoms (or molecules) most effective for cooling the gas will be those with an excitation energy close to the thermal energy of the electrons, even though there may be comparatively few atoms of this kind. Atoms of carbon, both ionized and neutral, have just the right properties. In H I zones, as we have mentioned, the carbon atoms are ionized. Their level of excitation corresponds to the thermal energy of particles at a temperature of 92°K. Thermal equilibrium should prevail in the interstellar medium in H I zones: the gas should acquire just as much energy by ionization heating as it loses due to the radiation of carbon atoms excited by collisions. As a result of this equilibrium, a constant kinetic temperature of some tens of degrees will become established. And this is the very temperature obtained by analyzing profiles of the 21-cm radio line in clouds. Carbon atoms, then, act as a thermostat for the clouds.

Similarly, thermal equilibrium will be present in the "hot" H II zones. But in this case the role of the thermostat will be played by ionized nitrogen and oxygen atoms, whose excited levels lie considerably higher than for carbon. When these levels are excited, the forbidden lines referred to earlier will be radiated. The thermal equilibrium in H II zones will maintain the kinetic temperature at a level of about 10,000°K, which corresponds to the average kinetic energy of the particles (ions and electrons) there, roughly 1 eV. The electrons formed through ionization of hydrogen by ultraviolet photons have a mean kinetic energy several times as high.

Let us return, however, to H I zones, where the heating of the gas is attributable primarily to its ionization by "soft" cosmic rays and x rays. If we knew the numerical density of cosmic-ray particles and x-ray photons, we could accurately calculate the dependence of the kinetic temperature and degree of ionization of the gas on the gas density. But we can proceed in the opposite sense: the temperature and density of the clouds are known from radio observations, so it is not very difficult to estimate the cosmic-ray and x-ray densities. From the temperature and density of the gas we can find its pressure. The relation established in this manner between the pressure of the interstellar gas and its density (actually the numerical density of gas particles, which is proportional to the mass density) appears in Figure 2.7. This curve has a rather distinctive shape resembling the van der Waals curve familiar from molecular physics. We shall now show that the resemblance is far from being accidental.

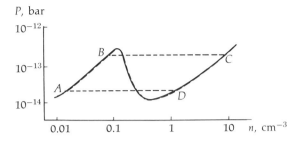

Figure 2.7 Pressure in interstellar gas clouds as a function of density.

The curve of Figure 2.7 indicates that at low interstellar gas densities (up to 0.1 cm^{-3}) the pressure increases with the density; the kinetic temperature stays at the 7000 to 10,000°K level typical of H II zones. As the density rises above 0.1 cm^{-3} the gas temperature drops abruptly to a value typical of an H I zone, so that the pressure decreases toward higher densities. With a further rise in density the gas temperature diminishes only gradually, almost reaching its minimum value; hence the rise in density outweighs the decline in temperature and the pressure again begins to increase.

It is evident from the curve that a range in pressure exists (10^{-14} to 3 × 10^{-13} bar) wherein three values of the gas density correspond to a given value of the pressure (as along line *BC*). The state of a gas is, of course, regarded as specified if its pressure and density (or temperature) are known. We may therefore conclude that three different states correspond to a definite pressure of the interstellar gas. On the portion of the curve of Figure 2.7 where the pressure decreases toward higher densities, the gas is in an unstable state: any small accidental condensation in some part of the gas will grow vigorously, because in such a condensation the internal pressure will diminish in that part of the gas, while the uncompensated external pressure of the surrounding gas (which has remained unchanged) will begin to force it to contract. The contraction will proceed until the point describing the state of the contracting gas has moved along the curve of Figure 2.7 into the region where the pressure begins to increase with rising density.

Thus the interstellar gas is in a state of thermal instability. Initially homogeneous, it will inevitably separate into two phases: comparatively dense clouds and the highly rarefied medium surrounding them. The thermal instability of the interstellar gas is thus among the most important factors responsible for its patchy cloud structure. This sort of structure is clearly apparent at 21-cm wavelength. The size, density, and velocity of the neutral hydrogen clouds are comparable to the same parameters for the ionized hydrogen clouds in H II zones. Hence the cloud structure should be the same in regions of the interstellar medium where the hydrogen is neutral as it is in ionized hydrogen regions. The picture we have outlined for thermal instability in the interstellar gas was developed through the efforts of the late Solomon B. Pikel'ner, the eminent Soviet astrophysicist, and it provides a fully satisfactory explanation.

A most important result of investigations in the 21-cm line is the finding that comparatively dense clouds of neutral interstellar hydrogen, in particular the "gas–dust complexes" (see Chapter 3), are grouped along arms of the galactic spiral structure. A similar distribution may hold for the H ɪɪ zones observed optically, but in this case the absorption of light by cosmic dust prevents the spiral structure of the Galaxy from being traced to large distances from the sun. The fact that the comparatively dense H ɪɪ zones are concentrated in the spiral arms implies that the hot, massive type O and B stars are also grouped in the spiral arms. This behavior certainly is no accident, and as we shall see in the next chapter it is directly relevant to the problem of the origin of the stars.

Now what are these spiral arms? How did they originate? We cannot gloss over the question of the origin of spiral structure in our own and other stellar systems, for it is empirically clear that the spiral arms are the site of the star formation process. For a long while disparate and quite incorrect explanations were given for the origin of the spiral structure in galaxies. The presence of spiral structure was generally ascribed to a stretching of interstellar gas clouds by the differential rotation of the galaxy. Our stellar system is known to be rotating about an axis perpendicular to its plane, not like a solid body but in a considerably more complex fashion. The central regions of the Galaxy are rotating much faster than the periphery. Hence interstellar gas clouds flowing out of the galactic-center region should presumably become twisted and spread along some spiral.

Apart from the problem of how interstellar gas clouds are ejected from the central regions of the Galaxy, a matter that is very far from being understood, we shall mention just one insuperable difficulty associated with the concept described above. Over the life of the Galaxy (about 10 billion years) the spiral arms would be expected to have become twisted around the galactic center dozens of times, since the period of galactic rotation in the solar neighborhood is about 200 million years. Yet the spiral arms are wound only a few times around the center, as in the galaxy illustrated below. The arms therefore exhibit a remarkable stability against differential galactic rotation.

This old problem was solved scarcely more than ten years ago by the Chinese-American astrophysicist Chia-chiao Lin, who developed ideas of the Swedish astronomer Bertil Lindblad. The fundamental Lindblad–Lin concept is that every spiral arm represents not some material formation but a *wave*. The distinction between the old and new approaches is of vital importance. According to the old concept, identical clouds would in a sense remain attached to a particular arm, while in the new interpretation interstellar clouds are only temporary residents of an arm. Interstellar gas flows into an arm, stays there for a fairly long time, but then emerges from the confines of the arm and other interstellar gas clouds arrive to take its place. The same process applies to stars. It is for this reason that the spiral shape of an arm is so stable despite the differential galactic rotation. In the inner parts of an arm the galactic rotation will be faster, so that the clouds and stars comprising the arm will be renewed more rapidly. The arm itself should here be regarded as rotating about the galactic center as a whole, at a constant angular velocity.

Figure 2.8 schematically illustrates the motion of stars through a spiral arm in the inner region of the Galaxy. Because the stars there are moving at a higher angular velocity than the arm, they will overtake it on the inside. When they enter the arm, the stars already in the arm will attract them, compelling them to leave their circular orbits about the galactic center and to move considerably more slowly through the arm. Properly speaking, the velocity component of the stars perpendicular to the axis of the arm will decrease, so that the stars will travel at a comparatively small angle from the axis and will stay in the arm a relatively long time. As a result the star density in the arm will increase, strengthening the gravitational attraction upon new stars coming into the arm. After the stars have emerged from the clouds in the arm they will resume their more rapid motion around the galactic center until they overtake another arm.

Gas clouds flowing into an arm will behave similarly. The gas will also condense. Spiral arms contain both comparatively dense clouds and quite tenuous intercloud gas; the pressure in these two phases is the same on the curve of Figure 2.7, where the state of gas in the clouds and in the intercloud medium is marked by the points B and C. After the interstellar gas has left

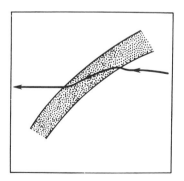

Figure 2.8 Stars moving through a spiral arm.

the arm its density will drop significantly, but both phases—the clouds and the intercloud medium—will be preserved. The corresponding states are marked by the points A and D in Figure 2.7. Thus clouds as well as an intercloud medium will exist between arms. But although the mean gas density for clouds inside an arm is about 3 to 5 cm^{-3}, between arms the clouds have a density of about 0.2 to 0.3 cm^{-3}. And the corresponding values for the density of the medium between the clouds are 10 times lower still, so that this medium can be detected only by the techniques of ultraviolet space astronomy discussed above.

New gas entering an arm will be slowed down rather abruptly by the gas already present there. In a situation of this kind shock waves can develop. The gas density will jump sharply. At the inner edge of a shock wave the gas will be heated, but somewhat farther on its temperature will return to the normal value as indicated in Figure 2.7. Compression of the gas in the shock wave is of course an additional factor serving to raise the gas density. And this factor, as we shall find in Chapter 3, facilitates rapid operation of the star formation process.

A striking illustration to justify this new concept of the nature of spiral arms in galaxies is provided by Figure 2.9, a photograph of the galaxy M51. This picture clearly shows dark, narrow lanes extending along the inner edges of the arms. These lanes represent cosmic dust which a shock wave has compressed along with the gas belonging to that part of the corresponding arm.

Radio observations at 21-cm wavelength have been used to study in detail the rotation of the Galaxy, and dynamical models have been worked out. The radio data have the inestimable advantage of being free from the influence of absorption by cosmic dust. As a result one can observe interstellar gas clouds in the most remote regions of the Galaxy. Of special interest is the examination of the nucleus of our stellar system and the areas surrounding it, which are completely inaccessible to optical astronomy because of the nearly complete absorption of light in that direction. We have merely touched on the fundamental importance of the results gained after more than 20 years of research in the 21-cm line. It is no exaggeration to say that modern astronomy would simply be inconceivable without all the diversified applications of this outstandingly effective method.

Another fortunate circumstance is that the 21-cm radio line is emitted by the most widespread element in the universe. Hyperfine structure in the deepest atomic level is not so common an occurrence among atoms. For example, no such structure is present for helium, carbon, or oxygen. The author of this book nevertheless pointed out as long ago as 1948 that the radio spectrum of the Galaxy ought to contain another line of the same kind belonging to deuterium at a wavelength of about 92 cm. This faint line was

Figure 2.9 The galaxy M51, the Whirlpool, in the constellation Canes Venatici.

not detected until 24 years later. Deuterium is tens of thousands of times less abundant than ordinary hydrogen in the interstellar medium. There are some grounds for believing that interstellar deuterium may be vestigial: it might have been formed during the first 15 minutes of existence, when the universe was an extremely hot and dense mixture of protons, electrons, neutrons, neutrinos, and photons of light.[5] If this is the case, then the mean density of the universe today should be about 10^{-31} g/cm^3 and the universe could not be a closed system. Look what far-reaching conclusions can be drawn from the discovery of the very weak radio line of interstellar deuterium!

---

[5]We are presuming here that during the subsequent evolution of matter in the universe no deuterium has been formed in stellar interiors, an assumption which is by no means obvious (see Chapter 8).

As with any plasma, H II zones are sources of thermal radio emission with a continuous spectrum. At low frequencies H II zones are opaque to their thermal radiation, and their radio spectrum is described by the Rayleigh–Jeans law according to which the intensity is proportional to the temperature and the square of the frequency. These zones are transparent at high frequencies and, as in the optical range, their intensity is proportional to the emission measure. But while the observed optical intensity is seriously distorted by interstellar absorption, at frequencies in the radio range absorption has a negligible effect. With good radio images of H II zones their true structure can be established.

In addition to the continuous spectrum the H II zones also emit radio lines. These lines are distinctive indeed. They represent transitions between adjacent, very highly excited levels in atoms of hydrogen and of other elements as well. These levels are so high that their principal quantum number $n \approx 100$ to 200 or even more. Such levels become "populated" upon recombination of electrons with protons.[6] So high an excitation of atoms is never achieved in laboratory plasmas or even in stellar atmospheres; it is prevented by interactions of excited atoms with surrounding charged particles. The recombination radio lines can best be observed in the centimeter and millimeter wavelength ranges.

Lines convey much more information than the continuous spectrum, for the analysis of line profiles makes it possible to study the motion of radiating clouds. At present the scrutiny of H II zones by means of recombination radio lines—not only of hydrogen but helium, carbon, and other elements as well—is surely the most effective method.

---

[6]It is fascinating to note that highly excited atoms are about 10^{-2} mm across; they are tens of thousands of times larger than normal atoms, as the diameter of the Bohr orbit is proportional to n^2.

3

Interstellar Gas–Dust Complexes: The Cradle of Stars

One very typical property of the interstellar medium is the great diversity of the physical conditions prevailing there. It contains both H I and H II zones, whose kinetic temperatures differ by two orders of magnitude. There are comparatively dense clouds with thousands of gas particles per cubic centimeter, and a very tenuous intercloud medium where the density is no more than 0.1 particle/cm^3. Finally, vast regions exist through which strong shock waves from exploding stars are propagated (see Chapter 16). In this chapter we shall concentrate our attention on the relatively dense, cool complexes of gas and dust, with their very distinctive physical processes.

Along with separate clouds of both ionized and nonionized gas, we observe in the Galaxy much larger, denser, and more massive aggregates of cold interstellar matter that are sometimes called gas–dust complexes.[1] Quite a few such complexes in various parts of the sky have long been known to astronomers. One of the closest to us, and perhaps the best studied, is located in the constellation Orion (see Figure 2.3). It comprises the famed Orion Nebula, some dense, absorbing clouds of gas and dust,

---

[1]The terms "dark" and "black" clouds have lately come into increasingly wide use. Black clouds are denser and absorb light more strongly.

and several other remarkable objects. The most essential fact for our purposes is that a very important process takes place in these gas–dust complexes: stars condense from the diffuse interstellar medium.

We shall return to this process later; for now, let us consider the interesting question of how such complexes originate. Of course one might sidestep the question and simply accept the gas–dust complexes as something actually observed. But a purely empirical approach of this kind, however useful it may be, cannot help us to gain a deep insight into the phenomenon or to grasp its inherent inevitability. The Introduction emphasized the thoroughly evolutionary nature of modern astrophysics. It is not possible to comprehend fully how stars are formed from the diffuse interstellar medium unless we understand the origin of the dense, massive gas–dust complexes. They could not have resulted from the thermal instability of the interstellar medium that we discussed in Chapter 2. Such instability would merely lead to the formation of separate clouds dispersed through a considerably more rarefied medium. The key to the origin of the massive gas–dust complexes is provided by certain properties of the interstellar magnetic field.

Of these properties the most important is the "elasticity" of the magnetic lines of force. The field lines run mainly in a direction parallel to the plane of the galactic equator. Since the clouds formed through thermal instability of the interstellar medium are more or less strongly ionized and therefore constitute a conducting medium, they cannot move across the lines of force, as they would then bend the lines and induce a force opposing the motion. The clouds would thus be halted rather rapidly. They can, therefore, move only by "sliding" along the magnetic field lines.

Now suppose that for some reason, perhaps even by chance, a depression or hollow develops in a system of field lines stretching "horizontally." Clouds will then slide down into the hollow by gravity. The mass of gas in the depression will increase, and because of its weight the hollow will sag even more. The slopes of the hollow will become steeper, and clouds of interstellar gas will flow in still faster. As a result of this special type of instability (known as Rayleigh–Taylor instability) in the magnetized interstellar plasma, deep wells filled with quite dense gas form within the system of interstellar lines of force (Figure 3.1). In this way a gas–dust complex is born.

The lines of force in the well do not sag all the way to the galactic plane itself. Some distance away from it they become so compressed that their elasticity balances the mass of interstellar gas in the well. At the sides of the well the magnetic field lines are quite high and rise steeply above the galactic plane, forming gigantic arches.

A fact worth stressing is that the kinetic temperature of gas–dust complexes is substantially lower than the average value for H ɪ regions. This circumstance results from the comparatively high density of the gas and the

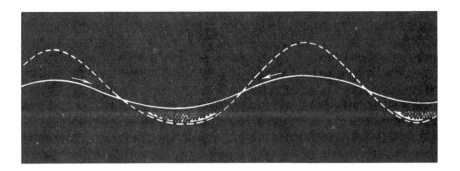

Figure 3.1   Diagram of Rayleigh–Taylor instability.

cosmic dust associated with it. Dense gas serves to reduce the ionization, because it absorbs the ionizing soft x rays. Dust absorbs the stellar ultraviolet radiation that ionizes carbon. The absorption will, in the first place, diminish the ionization and thereby the heating of the gas; second—and this is perhaps the more important effect—it will render the carbon neutral, sharply changing the thermal balance of the interstellar gas. In fact, the excited energy levels of neutral carbon atoms lie even closer to the ground level than those of ionized carbon. Hence the equilibrium temperature in the new thermal balance that sets in when carbon ionization has stopped will be considerably lower: only 5–10°K. As we shall see, recent observations fully confirm this theoretical result.

Gas–dust complexes are of great importance for modern astrophysics. Indeed, largely through intuition, astronomers long ago associated the formation of condensations in the interstellar medium with the vital process whereby stars arise out of the diffuse, comparatively rarefied medium of gas and dust. What are the grounds for believing that a relation exists between gas–dust complexes and the star formation process?

First of all, it was clear to astronomers nearly 40 years ago that stars should continually (literally "before our eyes") be formed in the Galaxy from some qualitatively different substance. By 1939 it had been established that the source of stellar energy is thermonuclear synthesis taking place in the interiors of stars (Chapter 8 deals with this process). Speaking roughly, we may say that the great majority of visible stars radiate because groups of four protons in their interior combine, after several intermediate steps, into a single α-particle. Since the four constituent nucleons each have an average mass of 1.0080 atomic units while the mass of a whole helium nucleus (an α-particle) is only 4.0027, the excess mass of 0.0073 atomic unit per proton should be released as energy. This quantity determines the energy reserve in the star that is continuously expended as radiation.

In the most favorable case of a pure hydrogen star, the initial energy supply should be

$$E = 0.007 \ \mathfrak{M}c^2 = \frac{\mathfrak{M}}{\mathfrak{M}_\odot} \times 10^{52} \ \text{erg,} \qquad (3.1)$$

where $\mathfrak{M}$ is the mass of the star and $\mathfrak{M}_\odot = 2 \times 10^{33}$ g is the mass of the sun. But the bolometric luminosity of a star with a mass of 20 $\mathfrak{M}_\odot$ is 10^{38} erg/s (see Chapter 1), so that the nuclear energy reserve of such a star would last no longer than 100 million years. This is actually an overestimate; the lifetime of a real star evolving with a mass of 20 $\mathfrak{M}_\odot$ turns out to be an order of magnitude shorter. Ten million years, however, is a negligible period compared to the evolution of our stellar system: the Galaxy surely is at least 10 billion years old. Massive stars are so young that their age is commensurate with the time humans have spent on the earth! Accordingly, stars (at any rate, massive ones of high luminosity) could not have been in the Galaxy from the epoch of its very origin. The star formation process must therefore operate *permanently*.

In later parts of this book we shall discuss the very important topic of the "death" of stars—the end of their evolutionary road. We shall find that every year at least one star in the Galaxy dies. In order that the stellar tribe should not become extinct, just as many new stars, on the average, must be formed annually in our Galaxy. If over some billions of years the Galaxy is to preserve without change its basic properties (such as the distribution of stars with respect to mass or, almost equivalently, spectral type), a dynamic equilibrium should automatically be maintained in it between newborn and perishing stars. In this regard the Galaxy resembles a primeval forest with all kinds of trees of various ages, but with the trees much younger than the forest. Admittedly, there is an important distinction between the Galaxy and a forest. Stars less massive than the sun will have a lifetime exceeding the present age of the Galaxy. We would expect, then, a continual increase in the number of stars of comparatively low mass, for such stars have not yet managed to die but are still being born. However, for more massive stars the state of dynamic equilibrium will inescapably prevail.

Just where in our Galaxy are the young and fledgling stars formed? From long ago, by an established tradition that can be traced to the hypothesis of Immanuel Kant and the Marquis de Laplace, astronomers have assumed that stars develop out of a scattered, diffuse medium of gas and dust. There has been only one strict theoretical basis for this conviction: gravitational instability in an initially homogeneous diffuse medium. The fact is that small density perturbations, or departures from perfect homogeneity, are unavoidable in such a medium. Under the action of universal gravitational forces the small perturbations will grow, and the initially homogeneous medium will break up into several condensations. If the mass of such a condensation exceeds a definite limit it will continue to contract by gravity, and presumably it will ultimately turn into a star.

Let us examine this matter more fully, taking a special but important example so as to obtain a numerical estimate. Suppose that we have a cloud with a constant radius R and a constant density ρ. If the total energy of the cloud is negative, the cloud will begin to contract under its own gravitation. The energy is composed of the negative gravitational energy W_g of interaction among all the particles in the cloud, and the positive thermal energy W_t of the particles. The negative sign of the total energy means that the gravity forces impelling the cloud to contract exceed the gas-pressure forces tending to disperse the cloud into the space surrounding it.

For the two components of the energy we have

$$W_t = \frac{A}{\mu}\rho T \cdot \frac{4}{3}\pi R^3, \tag{3.2}$$

where $A = 8.3 \times 10^7$ erg mole^{-1} deg^{-1}, μ is the molecular weight, and ρ is the density of the cloud; and

$$W_g = -\frac{G\mathfrak{M}^2}{R} \approx -16G\rho^2 R^5. \tag{3.3}$$

We see that, for a constant density ρ and temperature T in the cloud, W_t increases with the radius R like R^3, while $W_g \propto R^5$, a much stronger dependence on R. Thus if ρ and T are specified, a radius R_1 will exist such that for any $R > R_1$ the cloud will inescapably contract because of its own gravitation. If, on the other hand, the mass $\mathfrak{M}$ of the cloud is specified, then R_1 will be given by the relation

$$R_1 = \frac{\mu G\mathfrak{M}}{AT} \approx \frac{0.2}{T}\frac{\mathfrak{M}}{\mathfrak{M}_\odot} \text{ pc.} \tag{3.4}$$

In this case, where the mass and temperature of the cloud are fixed, the cloud will contract if its radius $R < R_1$.

One readily finds that an ordinary interstellar gas cloud with $\mathfrak{M} \approx \mathfrak{M}_\odot$ and $R \approx 1$ pc will not contract by its own gravitation, whereas a gas–dust complex with $\mathfrak{M} \approx 10^3$–$10^4$ $\mathfrak{M}_\odot$, $T \approx 50°$K, and a radius of some tens of parsecs will contract. Under the conditions that hold in the great majority of stars, such a contraction will automatically raise the temperature and thereby the pressure. The rising pressure will balance the force of gravity and the cloud will stop contracting. We shall return to this process in Chapter 6. But in contracting interstellar gas clouds the conditions are different: the temperature will not rise during the contraction process, at least not during the initial and most important phase. The reason is that such clouds contain a very efficient "refrigerator."

We shall see presently that in these dense clouds hydrogen, like most other elements, is in a molecular state. Collisional excitation of the rotational levels of hydrogen molecules followed by the emission of an infrared line at a wavelength of 28 μ will maintain the gas temperature at a nearly constant level. A contracting cloud will temporarily be transparent to this infrared radiation, which will thus leave the cloud. As a result the gravitational energy released by the contraction of the cloud will not be expended in heating its material but will be transformed into infrared radiation that escapes into the space beyond. Indeed, the temperature of the cloud will actually drop somewhat, for as it becomes denser the x-ray photons pervading the Galaxy that would otherwise heat the cloud will be absorbed in its outer layers. Moreover, the number of molecules cooling the gas will increase.

Let us return now to the condition 3.4 limiting the gravitational contraction of a cloud. Suppose the mass of the cloud is equal to the solar mass and its temperature is $10°K$. Then by Equation 3.4 the cloud will contract if it is smaller than 0.02 pc in radius. The density of such a cloud will be 2×10^{-18} g/cm^3, and it will contain about 10^6 gas particles per cubic centimeter, quite a substantial number. But if the mass of the cloud is 10 $\mathfrak{M}_\odot$, then we find that the cloud will begin to contract at a considerably lower mean number density of gas particles—about 10^4 cm^{-3}. As we shall see, clouds with this gas density are actually observed.

Thus the criterion for gravitational contraction set by Equation 3.4 is significantly weaker for more massive clouds. It is therefore natural to regard the condensation of interstellar gas clouds into stars as taking place in several stages. First a massive, extended gas–dust complex contracts; its mass may be thousands of times the solar mass. When this complex has compressed sufficiently and its mean density has become high enough, separate parts of it will begin to contract independently, and the complex will break up into several less massive condensations. This process explains qualitatively why stars tend to be born in clusters (or "associations") rather than individually, although under certain conditions isolated stars can also emerge.

A serious difficulty arises at once with the mechanism we have described for the formation of stars from dense clouds of the interstellar medium. As is directly apparent from profiles of the 21-cm radio line, separate pieces of interstellar gas clouds move relative to one another at a velocity of about 1 km/s. For this reason the clouds should possess a certain amount of angular momentum. In view of the great size of the clouds the angular momentum would in fact be very large. According to the laws of mechanics, if a cloud were isolated, then as it contracts by its own gravitation its angular momentum should be conserved. But this circumstance implies that the cloud ought to spin faster and faster about its axis as it continues to contract. Its axial rotation velocity would reach the speed of light even before the cloud has turned into a star!

This result, however, depends on the assumption that the contracting cloud is isolated. Clouds are not isolated, of course; they are surrounded by other clouds and the magnetic field lines associated with them. And at least 99 percent of the angular momentum of a cloud will leak out along these lines of force. So long as the cloud material has a high enough electrical conductivity (it need only be slightly ionized), the magnetic field lines will be "frozen" into it. The angular momentum will therefore be "pumped over," as though along flexible strings, from the contracting cloud to the surrounding interstellar medium. The process of angular-momentum pumping will continue until, because of the rising density, the ionization of the cloud material diminishes sharply and its electrical conductivity also drops much lower. Then the magnetic connection between the cloud and the surrounding medium will be broken.

Stars formed in this way will preserve a fairly high angular momentum, just as we observe in comparatively massive stars, beginning with spectral type O. Less massive stars such as our sun, however, can in principle shake off their excess angular momentum in a rather distinctive fashion, by forming planetary systems around themselves.[2] But a more probable mechanism for liberating angular momentum from such stars is an outflow of material from their atmospheres (a "stellar wind") in the presence of magnetic fields.

We can estimate the characteristic time required for a cloud to contract to the size of a protostar by applying the simple equation of mechanics that describes the free fall of a body under the action of an accelerating force. As the cloud contracts its particles will be subject to an acceleration of increasing amplitude, but to simplify that analysis we shall regard the acceleration as constant; our estimate will not be affected. With this simplifying assumption, then, the path length R traversed by the surface layers of the contracting cloud after a time span t will be given by

$$R = \frac{1}{2}gt^2, \tag{3.5}$$

where the acceleration $g = G\mathfrak{M}/R^2$. Accordingly,

$$t = \left(\frac{2R^3}{G\mathfrak{M}}\right)^{1/2} = \left(\frac{2}{3}\pi G\bar{\rho}\right)^{-1/2}, \tag{3.6}$$

---

[2]The author has described the process more fully in Chapter 13 of his popular book with Carl Sagan, *Intelligent Life in the Universe* (San Francisco: Holden-Day, 1966).

where we have introduced the mean density $\bar{\rho}$ of the cloud such that $\mathfrak{M}$ $= (4/3)\pi R^3 \bar{\rho}$.

Equation 3.6 shows that the time required for a substantial contraction of the cloud depends only on its initial mean density. We can write Equation 3.6 in a different way by substituting into it the value of $\mathfrak{M}$ from the gravitational instability condition 3.4:

$$t = \left(\frac{5\mu}{4AT}\right)^{3/2} G\mathfrak{M} = 8 \times 10^6 \frac{\mu^{3/2}}{T^{3/2}} \frac{\mathfrak{M}}{\mathfrak{M}_\odot} \text{ yr.} \tag{3.7}$$

If the molecular weight $\mu = 2$ and $T \approx 10°K$, then a cloud of solar mass will contract in about a million years.

During this first phase—called the free-fall stage—of the condensation of a gas–dust cloud into a star, gravitational energy will be released amounting to about $G\mathfrak{M}^2/R_1$, where R_1 is the radius at the end of this phase, when the cloud has become opaque to its own infrared radiation. Half of this liberated energy should leave the cloud in the form of infrared radiation, and the other half will heat the material in the cloud (see Chapter 7). To estimate the energy we shall merely consider an approximate value of R_1. The estimate may be made as follows. By the time the free-fall stage has ended a considerable part of the liberated gravitational energy will have been used to heat the gas in the cloud and thereby to dissociate the hydrogen molecules of which the cloud mainly consists. To dissociate one hydrogen molecule requires 4.5 eV of energy, or 7×10^{-12} erg. Hence in dissociating 1 g of hydrogen, which comprises 3×10^{23} molecules, an energy $E = 2.1 \times 10^{12}$ erg will be consumed, and $\mathfrak{M}$ times as much energy will be needed to dissociate all the hydrogen molecules in the cloud, where $\mathfrak{M}$ is the mass of the cloud in grams.

If we set the energy expended in dissociating molecular hydrogen equal to half the gravitational energy released by the contraction of the cloud, we find that

$$R_1 \approx \frac{G\mathfrak{M}^2}{2E\mathfrak{M}} \approx 500 \frac{\mathfrak{M}}{\mathfrak{M}_\odot} R_\odot, \tag{3.8}$$

where $R_\odot$ is the radius of the sun. To estimate the infrared luminosity of the contracting cloud we divide half the liberated gravitational energy by the contraction time. With Equation 3.7 we have

$$L = \frac{E}{2G}\left(\frac{4AT}{5\mu}\right)^{3/2} \approx 0.002\left(\frac{T}{\mu}\right)^{3/2} L_\odot, \tag{3.9}$$

in which T is the temperature of the cloud material by the time the process of hydrogen dissociation has ended, and $L_\odot = 4 \times 10^{33}$ erg/s is the luminosity of the sun. The temperature T should be of the order of a few thousand degrees, so that $L \approx 100$ to $200\ L_\odot$. This is a very high value, but it represents only the average luminosity over the whole period of contraction.

Actually the bulk of the radiation from the release of gravitational energy will be emitted during the closing phases of the free-fall stage, when the radius of the cloud has decreased nearly to R_1. Even though most of the contraction time is spent in the initial phase of the process, the cloud will emit very little radiation during that period. Theory thus predicts that the cloud will experience a flare of infrared radiation. Estimates show that the flare ought to last perhaps a few years, and the cloud should then have an infrared luminosity thousands of times the sun's bolometric luminosity.

As soon as the contracting cloud becomes opaque to its own infrared radiation its luminosity will drop abruptly. It will continue to contract, but far more slowly now than by the free-fall law. After all the molecular hydrogen has been dissociated, the temperature in the interior of the cloud will rise steadily because half the gravitational energy released in the contraction will serve to heat the cloud (see Chapter 7). We are no longer entitled, as a matter of fact, to call such an object a cloud; it has become a genuine protostar.

Simple laws of physics suggest, then, that gas–dust complexes in the interstellar medium may evolve, by a natural and regular process, first into protostars and subsequently into stars. Yet possibility is not the same thing as reality. The paramount task here facing observational astronomy is to study real interstellar clouds and to ascertain whether they are capable of contracting under their own gravitation. Their size, density, and temperature must be established for this purpose. A second major need is to develop additional arguments for the "genetic coupling" of clouds and stars (such as fine details in their chemical and even isotopic composition, or an organic relationship of stars to clouds). Third, it is important to derive from the observations irrefutable evidence that the earliest phases in the evolution of protostars do occur (for example, the infrared flares toward the end of the free-fall stage). Moreover, entirely unexpected phenomena enter the picture (see Chapter 4). We should finally investigate protostars in detail. But in order to do so we must first of all be able to distinguish them from "normal" stars. How we can observe protostars evolve into stars will be discussed in Chapter 5.

Empirical support for the process of star formation from clouds in the interstellar medium is afforded by a fact long known to astronomers. The hot, massive, high-luminosity stars of spectral types O and B are not distributed uniformly over the Galaxy but are grouped into extensive, separate aggregates that have come to be called associations. Yet, as we have emphasized, such stars ought to be young objects. Thus the experience of astronomical observation has itself suggested that stars are born not in isolation but in "nests," an inference in qualitative accord with the theory of gravitational instability. Young stellar associations (which contain not only hot, massive giant stars but also other remarkable and unquestionably young objects which we shall meet in Chapter 4) are intimately related to the large gas–dust complexes of the interstellar medium. It is natural to suppose that the connection should be a genetic one, with the stars in the associations being formed from condensing clouds of gas and dust.

Nevertheless, as we have said, it is one thing to adhere to the cosmogonic ideas described above, and quite another to give concrete astronomical proofs, based on observations, that young stars condense out of the diffuse medium. And even though new, important facts elicited in recent years lend firm support to the classical cosmogonic interpretation of stars developing from the interstellar medium, the problem has not yet been definitively solved. We have no hesitation in making this statement. Indeed, the situation has turned out to be all too complicated.

It seems that much more progress has been made with questions concerning the problem of stellar death than with the wealth of questions touching on the birth of stars. The reason is apparently that the death of stars is accompanied by such impressive events as supernova outbursts (see Part III) and the formation of planetary nebulae (Chapter 13). These phenomena are very striking, cannot be confused with anything else, and are theoretically tractable. The birth of stars is a different affair. This process as a rule takes place unnoticed, because it is concealed from us by a shroud of light-absorbing cosmic dust.

Only radio astronomy, we can now say with full assurance, has radically altered the experimental study of the birth of stars. Interstellar dust does not absorb radio waves. Moreover, radio astronomy has revealed completely unexpected phenomena in the gas–dust complexes of the interstellar medium, phenomena which presumably are directly related to the star formation process. We shall explain these findings in the next chapter. Infrared astronomy has proved to be invaluable for our problem; to a large extent this wavelength range also is free from the effects of absorption by cosmic dust. New technological developments such as these, permitting astronomical observations in previously inaccessible spectral regions, may ultimately lift the star formation problem from the realm of pure speculation and make it an exact science.

What new information have we gleaned about the comparatively dense gas–dust complexes of the interstellar medium over the past dozen years? First of all there are the remarkable achievements of molecular radio spectroscopy in these clouds. Early in Chapter 2 we alluded to the fact that the interstellar gas contains not only atoms but a very small representation

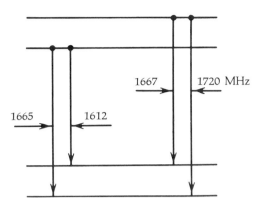

1667   1720 MHz

1665   1612

Figure 3.2 Diagram of the transitions in which four radio lines of the OH molecule are formed.

of the diatomic molecules CH, CH⁺, and CN. These molecules have been detected by the methods of optical astronomy. Recently H_2 molecules have been discovered by means of space astronomy. But the author of this book pointed out the opportunity of making spectroscopic observations of interstellar molecules in the radio range as long ago as 1949, and published more explicit calculations in 1953.

In some molecules the rotational levels happen to be split by the lambda doubling effect, wherein the motion of the electrons in a molecule interacts with the mutual rotation of its nuclei. This doubling of molecular rotational levels is a very delicate effect, so that a transition from the upper to the lower sublevel of the fine structure yields a spectral line with a wavelength long enough to fall in the radio range. Actually the picture is a more complicated one, for each of the Λ-doublet sublevels is itself split into even tighter levels because of interaction with the magnetic moment of the nuclei. Again we encounter a spectrum with hyperfine structure.

My most detailed calculations were performed in 1953 for the hydroxyl molecule OH, as the appropriate molecular constants were quite well established at that time. Apart from its hyperfine structure, this molecule has a Λ-doubling wavelength of 18 cm. If the hyperfine structure is taken into account (as was done soon afterward by the noted American physicist Charles H. Townes, one of the inventors of masers and lasers), four lines should appear; Figure 3.2 shows how they are arranged. The frequencies of these lines are 1612, 1665, 1667, and 1720 MHz. Also in 1953, I made similar calculations for some other molecules such as CH, but the computed wavelengths for those were considerably less accurate than for the OH molecule.[3]

---

[3]Not until late 1973 was a very weak radio line of the CH molecule detected. Its wavelength was 9.4 cm, quite close to the value I had computed 20 years earlier.

It is important to understand that although these new molecular lines would not be expected to be so intense as the celebrated 21-cm line, none-theless they should be strong enough to be observable. At first glance this situation may seem paradoxical: OH molecules, which have not been de-tected optically in the interstellar medium, ought to be many millions of times less abundant than hydrogen atoms. The fact is, however, that unlike the 21-cm hydrogen line, the molecular lines arising from transitions be-tween Λ-doublet components are *permitted* lines; thus their transition probabilities are nearly a million times higher, largely compensating for the low abundance.

Not until 1963, or 10 years after my calculations, were the four lines of the interstellar hydroxyl molecule in the 18-cm range discovered by American radio astronomers; the frequencies agreed accurately with the computed values. This discovery marked the opening of a new chapter both in radio astronomy and in the study of the interstellar medium. Since that time a rather large number of further molecular radio lines have been found in the decimeter, centimeter, and millimeter wavelength ranges. Almost all of them arise from transitions between rotational levels in various molecules. After another 10 years, by 1973, a total of 19 new molecules had been detected in the interstellar medium by radio methods, not counting the three already known from optical observations (CH, CH$^+$, CN) and the hydrogen molecule H$_2$, whose lines in the ultraviolet part of the spectrum can be observed with instruments in space. Altogether 40 molecules have now been identified.

A noteworthy feature of radio surveys of the interstellar medium is the opportunity of observing separately lines that belong to different isotopes of a given molecule, since these sets of lines are quite widely spaced in the radio spectrum. Thus we are now able to make an isotopic analysis of the interstellar medium. The first 19 new molecules discovered by radio astronomy are found in 34 isotopic combinations. In addition to the lines of the ordinary $O^{16}H^1$ molecule, considerably weaker lines of $O^{18}H^1$ are ob-served. In the case of interstellar carbon monoxide, molecules have been found in the isotopic combinations $C^{12}O^{16}$, $C^{13}O^{16}$, and $C^{12}O^{18}$ (see below).

While some molecules (such as OH) are encountered in many clouds of interstellar gas, most of the molecules—especially the multiatomic ones—occur in the huge gas–dust complex located in the galactic-center direction and called Sagittarius B. To a lesser extent, they also occur in the Orion Nebula. Certain molecules (such as CO, which has a radio line at a wave-length of 2.64 mm) are observed in both H I and H II zones; others occur only in the dense, cool gas-dust clouds. Of special note are the numerous multiatomic molecules with rather complicated chemical structures. For example, radio lines of H_2HCO, CH_3HCO, CH_3CN, and other molecules have been found in the Sagittarius B complex.

In an important discovery, clouds of the gas–dust interstellar medium have been located where the absorption lines of the OH molecule are fairly strong and the 21-cm neutral hydrogen line is very weak. This circumstance can have only one meaning: in such clouds hydrogen is in a *molecular* state, whereas in "ordinary" H I clouds it is primarily in an atomic state. Theoretical calculations show that in order for hydrogen to become molecular the gas density in the cloud should be high (over 100 cm^{-3}) and the kinetic temperature relatively low. The process whereby hydrogen atoms combine into molecules takes place on the surface of solid grains inside the cloud. At the same time the grains shield any hydrogen molecules that may be formed from dissociation by ultraviolet radiation coming from hot stars. The H_2 molecule unfortunately produces no radio lines, so that the details of the process remain hidden from us—all the more so because in such clouds even the ultraviolet H_2 lines that can be viewed from outside the atmosphere are completely absorbed by cosmic dust.

Research on molecular radio lines is valuable because it allows us to make a quantitative analysis of the physical conditions in interstellar clouds with a completeness that would have seemed incomprehensible a few years ago. This is primarily true of the dense, cold H I clouds that are especially interesting to us because of the star formation problem. The molecules in these clouds serve as probes, so to speak, with which astronomers can feel out the physical state of the medium surrounding the molecules.

Analysis of this kind demonstrates above all that the cold clouds in gas–dust complexes have total masses tens of thousands of times the solar mass. Indeed, the mass of the gigantic gas–dust complex Sagittarius B reaches $3 \times 10^6\ \mathfrak{M}_\odot$, and it is as much as 50 pc across. There are several thousand hydrogen molecules per cubic centimeter in such clouds. In the densest clouds, such as the Orion Nebula, the molecular hydrogen density attains 10^7 cm^{-3}. At this high a density a cloud is, in a sense, midway between an ordinary cloud of the interstellar medium and the extended atmosphere of a red giant star. Astronomers have not yet been able to estimate the total number of such dense molecular clouds in the Galaxy. But an important conclusion can already be drawn: a substantial part of the interstellar gas in the Galaxy may be in the form of comparatively dense molecular clouds.

The kinetic temperature of the gas in such clouds is low, and it varies over a fairly wide range. In the coldest molecular clouds the temperature is about 5°K. At most the clouds scarcely reach a kinetic temperature of 50°K. The temperature of the Sagittarius B complex is about 20°K, and it remains practically constant throughout this vast volume. In conjunction with a rather high density and a large mass, the low temperature renders such aggregates of matter very unstable against the force of gravity (see above). They will necessarily contract because of their gravity, and everything indicates that these condensations will evolve quite rapidly into stars.

In the near future it may become possible to observe directly such clouds fragmenting into dense little protostars. Detailed radio observations of the molecular clouds at very high resolving power, better than 1", will be needed. The radio telescope must not only have high resolution but also be very sensitive, since the radio flux density from such condensations will be small. The best approach toward solving this fundamental problem is offered by the giant VLA radio telescope project (see the Introduction), a portion of which is shown in Figure 3.3.

It is not too early to speak of the quantitative chemical analysis of interstellar molecular clouds, the "dark" and "black" clouds. If the mean number density of H_2 molecules is about 10^4 cm^{-3}, the OH density will be close to 10^{-2} cm^{-3}. The ammonia (NH_3) density will be about the same. Carbon monoxide (CO) will have a very high density, up to 1 cm^{-3}. In view of the fact that the cosmic abundance of carbon is about 10^{-4} that of hydrogen, we at once obtain the important result that practically all the carbon is bound to the more abundant oxygen. Oxygen probably occurs in the form of O_2 molecules. This interesting problem awaits a solution.

The comparatively high abundance of compound molecules is noteworthy. For example, the density of CH_3OH molecules is about 10^{-3} cm^{-3}, which is only one order of magnitude lower than the OH number density. Compound interstellar molecules are probably formed by successive ion–molecule reactions of the type $C^+ + H_2 \rightarrow CH^+ + H$, $CH^+ + H_2 \rightarrow CH_2^+ + H$, $CH_2^+ + H_2 \rightarrow CH_3^+ + H$, and so on. The corresponding neutral molecules would be formed by recombination, such as $CH^+ + e^- \rightarrow CH$.

Another interesting subject is the isotopic composition of the interstellar gas in molecular clouds. The isotopic composition of carbon has been the most reliably determined, because of the high abundance of the CO molecule. Analysis of the radio lines representing various isotopes of this molecule reveals that the abundance ratio $C^{12}O^{16}/C^{13}O^{16} \approx 90$, which is nearly the same as the C^{12}/C^{13} isotope ratio on the earth. Similarly, the O^{16}/O^{18} abundance ratio in interstellar molecular clouds is about the same as on the earth. The isotopic composition of nitrogen can be determined by analyzing the HCN^{14} and HCN^{15} radio lines; again the ratio is practically the same in molecular clouds as on the earth. Since the isotopic composition of matter is established through thermonuclear reactions taking place in stellar interiors (see Chapter 8) as well as in supernova explosions, we may conclude that interstellar matter has had the same thermonuclear history

Figure 3.3   The VLA radio telescope under construction in New
Mexico (courtesy National Radio Astronomy Observatory).

as the matter from which the earth and planets were formed. In particular,
it follows that neither terrestrial matter nor the material in molecular clouds
took part during the past in the carbon–nitrogen cycle whereby energy is
processed in the interiors of sufficiently massive stars (Chapter 8). The
similarity of the isotopic composition of matter on the earth and in interstellar molecular clouds represents strong evidence that our solar system as
well as other stars originated out of the interstellar medium.

Curiously enough, though, the ratio of the deuterium and hydrogen
abundances, derived from analysis of the radio lines of the DCN and HCN
molecules, turns out to be 40 times higher in interstellar space than on the
earth. It is also significant that the interstellar D/H ratio is 80 times as great
as that obtained from direct analysis of the intensity of the 92.5-cm interstellar deuterium radio line (see the end of Chapter 2). The disparity is
evidently attributable to purely chemical processes for the formation of
these molecules and has nothing to do with the nuclear history of the
interstellar medium.

A remarkable distinction is observed between the dark molecular
clouds and the denser black clouds. Whereas the OH and CO radio lines
in the dark clouds are very narrow and correspond to random velocities of
less than 1 km/s for the absorbing molecules, in the black clouds the ab-

sorption lines are considerably broader, corresponding to velocities of 5–10 km/s as a rule. The most natural explanation for this difference would be that the more compact black clouds are contracting through the influence of gravity forces, and are in their free-fall stage (see above). There is, however, one difficulty with this interpretation: in a formal sense the broad lines might equally well be evidence for an expanding cloud!

How, then, can we show that such clouds are indeed contracting? A proof does appear to be available in one important case. We refer to the black cloud located inside the nebula NGC 6334. At the center of this cloud is a bright maser source emitting OH lines (see Chapter 4). The wavelengths of the OH absorption lines observed in the cloud are shifted toward the red relative to the maser emission lines. Thus we may conclude that the cloud material is moving toward its center; the cloud must be contracting.

In a word, application of astronomical radio spectroscopy to the study of clouds in the interstellar medium has yielded exceptionally rich results. These investigations have disclosed the existence of a new class of interstellar clouds: molecular clouds in which a considerable proportion of all interstellar matter has accumulated. A recent analysis of the strong CO absorption lines in the region of the galactic center has in fact demonstrated that at least 90 percent of the interstellar gas there is in a molecular state. Furthermore, a very large number of molecular clouds are concentrated in the most prominent inner arm of the Galaxy, about 4 kpc away from the galactic center.

Detailed study of the radio lines of many molecules and their isotopes has opened the way to an understanding of the physical and chemical processes occurring in these clouds. Without exaggeration we would claim that the problem of the condensation of interstellar matter into stars is only now being placed on a firm scientific footing. Were it not for radio astronomy we would still be simply marking time with this important subject. But the powers of radio astronomy methods have not been limited to this area alone. Investigators here were in for a surprise.

4

Cosmic Masers

Shortly after the first radio lines of interstellar hydroxyl had been discovered, in the midst of a routine observing program where various interstellar gas clouds were being examined in the OH line at 18-cm wavelength,
astronomers detected a new phenomenon, wholly unexpected and pro-
foundly impressive. Interstellar hydroxyl lines had customarily been observed in absorption in the spectrum of bright radio sources. As a rule these
lines were very weak; their absorption depth rarely exceeded a few percent.
How amazed radio astronomers were when, upon inspecting some nebulae
not thought peculiar in any way, they encountered OH lines in *emission,* and
with an exceptional brightness! The investigators literally did not believe
their eyes; perplexed, they decided that the lines were not emitted by the
commonplace OH molecule but by an unknown substance for which they
even selected a suitable name: "mysterium." But it took only a matter of
weeks for mysterium to share the fate of its optical brothers, nebulium and
coronium. Decades had passed before these optical lines were dethroned,
yet mysterium lost its spell in just a couple of weeks. A good illustration of
how the development of science has quickened over the past century!

First of all, any doubt that OH molecules were responsible for the
remarkable phenomenon observed was set to rest for the simple reason that

all four hydroxyl lines appeared at the very frequencies where they were predicted to be. Their relative intensities, however, were altogether unlike the ratios suggested by the simple theory that had been confirmed by observations of weak absorption lines. This theory predicts the intensity ratios $9:5:1:1$ for the OH lines at frequencies of 1667, 1665, 1612, and 1720 MHz. But the very first observations of the OH *emission* lines in the strange new sources showed that the 1665-MHz line was generally the strongest, while the satellite lines at 1612 and 1720 MHz were either extremely weak or wholly absent. Yet other sources of the same type were soon discovered in which the satellite lines were the strongest: in some cases 1612 MHz, in others 1720 MHz. The chief peculiarity of the "mysterium" lines, then, is their enormous intensity, and a second feature is the distortion of the relative intensities of the various lines.

Another interesting property of these lines was recognized at once. Their spectral profile comprises a fairly large number of extraordinarily narrow maxima spread over an interval a few tens of kilohertz wide (Figure 4.1). The profile of any spectral line, including radio lines, is determined by the Doppler effect due to the motion of the radiating particles (atoms or molecules) along the line of sight. An inspection of the spectral profiles of the unusual OH emission lines shows that the emission region consists of several sources moving relative to one another at velocities of several kilometers or even tens of kilometers per second. Thus it was all the more astonishing that the maxima are so very narrow—less than 1 kHz wide on a frequency scale! Astronomers had never dealt with such narrow lines before. If the spectral width of each maximum could be attributed to thermal motions of the radiating OH molecules, then the great sharpness of these features in the spectrum would imply an exceedingly low gas temperature in the emission region, only a few degrees Kelvin. But such an interpretation conflicts with the brightness of the lines, which inescapably signifies a very high temperature (provided, of course, that the radiation is considered to be thermal). It became apparent that no "mysterium" exists in nature; the radiation comes from ordinary OH molecules which are subject to unusual conditions.

Further observations disclosed more interesting properties of this singular radiation. For example, it was found to be strongly polarized, and as a rule circular polarization was recorded. Within the same source, individual narrow maxima in the profile are nearly 100 percent polarized, but some maxima exhibit left- and others right-circular polarization.

Even the earliest observations showed that the OH emission sources have an unusually small angular size. This finding was reinforced when study of the sources with radio interferometers began. The measurements demonstrated that the emission sources were about a second of arc across,

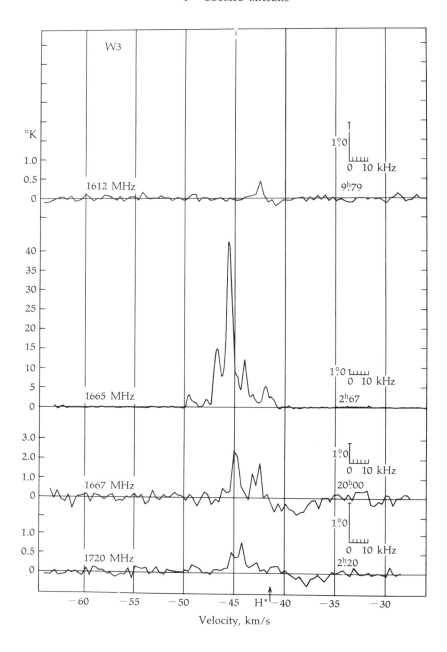

Figure 4.1   Radio-line profiles for the maser emission of the OH molecules in the source located in the nebula W3.

75

whereas the H II zones that normally surround the sources often have an angular size measuring tens of minutes of arc. Nevertheless, it has turned out that not even seconds of arc truly describe the angular size of the OH emission sources.

The most valuable information about the sources of "mysterium" has been furnished by the intercontinental radio interferometers mentioned in the Introduction. These interferometers have such a fantastic resolving power—better than 0".001—that observations with them have revealed the spatial structure of the sources of anomalous emission in the OH lines. The structure is by no means of a trivial character.

To be definite, let us consider one of the best-studied sources, located in the diffuse nebula W3 (Figure 2.4). Comparatively rudimentary interferometer observations suggested that this source is about 1".5 across. But intercontinental interferometry has now shown that in this case perhaps ten extremely compact sources are strewn over a 1".5 area, as illustrated in Figure 4.2. Each compact source emits *one* very narrow line; the frequencies of the lines from the various sources differ somewhat and correspond to the frequencies of the peaks in the spectral profiles of Figure 4.1. The angular size of each of these sources is remarkably small, only a few thousandths of a second of arc! As we know that the nebula W3 is about 2000 pc away, the measured angular sizes give us an estimate for the linear size of the clouds emitting bright lines: no more than 10^{14} cm, hardly ten times the distance from the earth to the sun (the "astronomical unit"). There are red giant stars whose diameter is nearly 10^{14} cm. The whole region containing these clouds is only about 0.01 pc across. That the clouds are moving is apparent from the slight differences in the frequencies of the OH lines radiated by the separate clouds. The frequency differences are caused by the Doppler effect, which indicates that the relative velocities of the clouds are of the order of several kilometers per second. Other sources of anomalous OH line emission exhibit similar structure.

As further observational material accumulated it became clear that the enigmatic emission sources do not form a homogeneous group of objects. At least three types have been identified. The first type has enormously bright OH line components at frequencies of 1665 and 1667 MHz. Sources of this type are associated with H II zones and have a structure like that described above. In the second type of sources, only the component at 1612 MHz has an enhanced intensity. These sources can be attributed with confidence to red and infrared giant stars. Finally, in sources of the third type the 1720-MHz line is enhanced. These sources are generally viewed in projection against radio nebulae representing the remnants of supernova outbursts (Chapter 16). Although the last two types of sources are, of course,

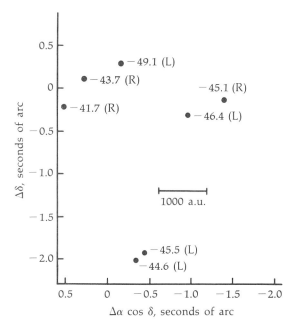

Figure 4.2   Structure of the maser source in the nebula W3. The numbers indicate the radial velocity of the condensations; the letters L and R tell whether their radiation is left- or right-circularly polarized.

very interesting, sources of the first type are especially noteworthy because it is here that we most likely find a relationship to the star formation process.

The combination of very large radio flux and exceedingly small angular size for the clouds means that the sources have a fantastic surface brightness. We can express this brightness in temperature units. If a perfect black body at the frequencies of the OH lines had such a surface brightness, its temperature would be more than 10^{14} °K. Since the width of the lines themselves corresponds to a temperature of only a few tens of degrees Kelvin, the huge "brightness temperature" has nothing to do with the actual kinetic temperature of the radiating material.

In 1969 a group from the University of California, Berkeley, under the direction of Professor Townes, discovered a new type of exceptionally bright "supercompact" sources emitting a radio line of water vapor at 1.35-cm wavelength. This line arises from transitions between the sixth and fifth rotational levels of the ground electronic–vibrational state of the triatomic

H_2O molecule. These sources are usually observed in the same place as compact OH sources of the first type. The brightness temperature of H_2O sources is even higher than for OH sources, attaining a record value of 10^{15} °K! Presumably it is higher still, because for most of these sources the techniques of intercontinental radio interferometry yield only an upper limit on the angular size, which in some instances is as small as 0".0003.

One significant property of the compact sources of "anomalous" H_2O line emission is their variability. Over a time span of several weeks or even days variations occur in the intensity, width, polarization, and even radial velocity of the individual peaks in the line profile. Occasionally these parameters vary on a time scale as short as 5 minutes. From this fact alone and a few simple premises one can infer that the linear size of the sources should be small. They can hardly be larger than the distance that light covers during the period of variability. If a source varies within 5 min, for example, it should be no more than about 10^{13} cm across, or approximately the distance from the earth to the sun.

It is worth noting in this connection that the radiation of some OH sources also varies, but considerably more slowly than for H_2O sources. Even though compact H_2O and compact OH sources may agree in position to 1", the profiles of the 1.35-cm H_2O radio line do not generally show a detailed match of individual peaks. Since H_2O sources have the same type of fine structure as OH sources (tiny condensations, each emitting a separate spectral peak, scattered over a region several seconds of arc across), we may conclude that both clouds emitting only the H_2O line and clouds emitting only OH move within the same region measuring a few hundredths of a parsec. Unlike the OH lines, the H_2O emission is unpolarized.

What sort of emission mechanism might combine these seemingly incompatible characteristics: an extraordinarily high brightness temperature and a low kinetic temperature? Astronomers had not previously needed to devise such a mechanism. More than 10 years before the "mysterium" lines were discovered, however, physicists were making use of quantum generators of coherent radiation—masers in the radio range and lasers in the optical and near-infrared ranges. It is quite understandable, then, that shortly after the surprisingly bright OH lines had been found the sources of these lines were recognized to be none other than *natural cosmic masers.*

The chief property of any maser is the lack of thermal equilibrium between its radiating atoms (or molecules) and the surrounding medium. Let us recall how an ordinary maser operates. To this end we return to the principles of radiation theory set forth by Albert Einstein as long ago as 1915. Although we need not do so, we shall limit ourselves to emission in individual spectral lines. Such emission results from electron transitions between the upper and lower levels of an atomic system. The most important service

Einstein rendered was his demonstration that *two* types of transitions exist. Those of the first type are spontaneous transitions, where an atom for no outward reason—"all by itself," so to speak—goes from a relatively excited state to a less excited one, emitting a photon in the process. This phenomenon simply means that the excited states of atomic systems are unstable. Only the ground state, the deepest one, can be stable in the sense that an atomic system can stay there indefinitely. We discussed this point in some detail in Chapter 2.

However, an atom on an excited upper level can execute a downward transition not just spontaneously but also when photons of the radiation field surrounding the atomic system interact with it. These photons should have the same energy as the photons emitted by the atomic system itself in its transition. Such transitions are said to be stimulated or induced. It is important that the stimulated photon be traveling in the same direction as the stimulating photon.

Suppose that the number density of particles on the upper level is n_2 and on the lower level n_1. Then the number of transitions per unit volume and unit time accompanied by the emission of a photon will be given by

$$Z_2 = n_2(A_{21} + B_{21}u_{21}), \tag{4.1}$$

where A_{21} is the spontaneous-transition probability, u_{21} is the radiation density at the frequency of the spectral line in question, and $B_{21} = (c^3/8\pi h\nu_{12}{}^3)A_{21}$, in which h is the Planck constant, ν_{12} is the frequency of the line, and c is the velocity of light. The quantities A_{21} and B_{21} are called the Einstein coefficients.

On the other hand, atoms or molecules on the lower level will *absorb* photons of the same frequency and undergo a transition to the upper level. The number of such transitions per unit volume and unit time will be

$$Z_1 = n_1B_{12}u_{21}, \tag{4.2}$$

where $B_{12} = B_{21}$ (to within a constant factor, which we shall regard as unity to simplify the discussion). The process 4.2 describes how radiation is absorbed as it passes through matter.

If the emission process represented by Equation 4.1 were not operative, the intensity I_ν of the radiation after traversing a gas layer of thickness l would fall off according to the law

$$I_\nu = I_\nu{}^0 e^{-\kappa_\nu l}, \tag{4.3}$$

where $I_\nu{}^0$ is the intensity before the radiation reaches the gas layer, and the quantity κ_ν, which is proportional to the Einstein coefficient B_{12}, is called the absorption coefficient. But when we allow for stimulated transitions the absorption coefficient will clearly *decrease*, for the transitions will generate *new* photons traveling in the same direction as the radiation incident on the matter. As a result the absorption coefficient will become

$$\kappa_\nu' = \kappa_\nu \left(1 - \frac{n_2}{n_1} \right). \tag{4.4}$$

In thermal equilibrium the ratio n_2/n_1 is given by the Boltzmann formula

$$\frac{n_2}{n_1} = e^{-h\nu_{12}/kT}. \tag{4.5}$$

Whatever the temperature, this ratio will always be less than unity. The allowance for stimulated transitions can here only decrease the absorption coefficient. This effect is particularly strong at the low radio frequencies. For example, the absorption coefficient for the 21-cm line of interstellar hydrogen decreases by hundreds of times because of stimulated transitions!

However, in the absence of thermal equilibrium between the radiation and the medium, a situation can arise in which $n_2 > n_1$. In this event the absorption coefficient will become *negative* (see Equation 4.4). A remarkable phenomenon will then occur: as radiation passes through the medium, instead of diminishing in intensity (as always happens in everyday practice) it will become more intense. The number of photons will grow like an avalanche as they advance through the medium: there will be a swift rise in the number of stimulated photons, overwhelming the inevitable absorption processes. A medium possessing these unusual properties is said to be activated. Formally, on the basis of the Boltzmann formula, we can assign a negative temperature to such a medium.

A negative temperature can never develop in a medium by itself, just from equilibrium thermal processes. In order for the medium to become activated, nonequilibrium processes of some kind must operate and produce an anomalously high excitation of the upper level of the atomic system. These are figuratively called pumping processes. Pumping may, for example, occur when matter is irradiated by a strong monochromatic beam that converts an atomic system from its lower level to some third level higher than the second one. The radiation in the beam must, of course, have a frequency higher than ν_{12}. When the atomic system executes downward transitions from its third level, its second level may become overpopulated. This sort of pumping will raise the atomic system artificially from the first to the second level, thereby imparting a "negative" temperature to it. Needless to say, as soon as the pumping ceases to operate everything will come back into place, the temperature will again be positive, and there will be no amplification of radiation at the frequency ν_{12}.

The method just described for activating a medium is very often applied in practical work with laboratory masers and lasers, but it is not the only possible one. For example, chemical pumping is becoming a more and more important technique. It is based on the principle that in various kinds of chemical processes among atoms or molecules, particles of "working material" (that is, molecules or atoms that can produce maser amplification of a spectral line) can be formed primarily in the second (excited) state.

Maser radiation is highly coherent, since regular phase relations exist between the stimulating and stimulated photons. It may be nearly 100 percent polarized if the activated medium amplifies only radiation with a particular type of polarization. Furthermore, maser radiation may possess an extremely sharp directivity, to a degree that cannot be achieved with any projection devices. This property will arise if only radiation traveling in a very definite direction can be amplified. In principle, though, nearly isotropic masers can also exist.

If the pumping process alone operated in the gas serving as the maser "working material," some negative temperature would be established in it—or, stated more simply, the number of molecules per unit volume in the upper level would exceed the number in the lower level by an amount $\Delta n = n_2 - n_1$. In a real gas, however, processes occur that tend to decrease this overpopulation of the excited level. Paramount among these processes are *collisions* between molecules, which act to establish a distribution between the two levels approaching the Boltzmann law 4.5. Indeed, in a Boltzmann distribution the second level will always have a *smaller* population than the first. Another process serving to reduce the excess population of the higher level is stimulated emission and absorption. If the radiation density is high enough, then according to Equations 4.1 and 4.2 a balance between these processes would make the populations of the two levels equal. Thus in a real gas there are two tendencies opposing each other: pumping tends to establish some definite excess population on the higher level, while collisions and stimulated processes tend to smooth the population out. The operating conditions of real masers depend on the relation between these two tendencies.

Now let us look at these conditions from the quantitative side. Suppose that a pumping process, in the absence of concomitant processes of collision and stimulated emission and absorption, gives the upper level an excess population Δn_0. When the concomitant processes are taken into account the excess population will be

$$\Delta n = \frac{\Delta n_0}{1 + 2(W_c + W_s)/W_p},$$ (4.6)

where W_c, W_s, and W_p are the probabilities per molecule of collisions, stimulated processes, and pumping. For instance, W_p expresses the number of events each second whereby one molecule is excited by pumping to its second level. The probability $W_p = (\Omega/c)B_{12}I$, where Ω is the solid angle of the maser beam.

First take the case where $W_p \gg W_c + W_s$, so that the radiation field of photons of frequency ν_{12} has a comparatively low density. Calculations show that in this case the intensity of the radiation emerging from an "activated" gas layer will be

$$I_\nu = I_\nu^0 e^{\kappa_\nu l} + \frac{\varepsilon_\nu}{\kappa_\nu} e^{\kappa_\nu l}, \tag{4.7}$$

where l, as before, designates the extent of the gas layer within which maser amplification occurs, $\varepsilon_\nu = A_{21} \cdot h\nu_{21} \cdot n_2/4\pi\Delta\nu_D$ is the radiation due to spontaneous transitions emitted per unit volume into unit solid angle in unit frequency interval per unit time, $\Delta\nu_D$ is the width of the amplified line expressed in frequency units (inverse seconds or hertz), I_ν^0 is the intensity of the radiation prior to its passage through the activated gas, and

$$\kappa_\nu = \frac{1}{8\sqrt{\pi}\,\Delta\nu_D} \frac{c^2}{\nu^2} \frac{A_{21}\Delta n}{\nu^2} \exp\left[-\left(\frac{\nu - \nu_0}{\Delta\nu_D}\right)^2 \right] \tag{4.8}$$

is the negative absorption coefficient. Equation 4.7 shows that a maser operating under such conditions (it is said to be unsaturated) will exponentially (very steeply) amplify the ambient radiation striking it from behind as well as its own spontaneous radiation in the ν_{12} line emitted within the gas itself. The absorption coefficient κ_ν depends very sharply on frequency inside the limits of the spectral line; thus because of the exponential character of the amplification the centermost part of the line will be most strongly amplified, and as a result the line will become narrower by a factor of 5 to 6.

If the thickness l of the gas layer is sufficiently great, the radiation will become so intense that stimulated processes will begin to affect the excess population of the second level. There will be changes in the behavior of the maser, especially in its amplification. With $W_p \ll W_s$ and $W_c \ll W_s$ we will now have a saturated maser. Simple calculations show that in this case

$$I_\nu = I_\nu^0 + \frac{\Delta n W_p h\nu l}{\Omega\Delta\nu_D} + \varepsilon_\nu l. \tag{4.9}$$

Equation 4.9 states that the radiant intensity emerging from the output of a saturated maser comprises the ambient background radiation (which is not amplified), stimulated emission, and spontaneous emission. In all cases of practical interest the second term in the expression 4.9 dominates the others. Its significance is very simple: the intensity of the maser radiation will be determined solely by the power of the pumping mechanism. The number of photons of amplified radiation emerging from the maser cannot exceed the number of pumping events throughout the entire maser volume. If absorption by working molecules of higher-frequency photons is responsible for the pumping, then in the case of a saturated maser we can be confident that the number of maser photons will be less than the number of pumping photons (these quantities all refer to unit time).

For an unsaturated maser, intensity variations can easily be interpreted as variability in the background, to which the amplified intensity is proportional (see Equation 4.7). In the case of a saturated maser the intensity variations depend only on its internal properties, such as the pumping power and the length. The intensity of a saturated maser grows linearly with l, far more slowly than for an unsaturated maser. Spectral lines do not become narrower in a saturated maser. Nevertheless, when the amplification is just beginning and l is still comparatively small, every maser will be unsaturated; hence the line width at the output of a saturated maser will in fact be considerably diminished.

We have mentioned that masers may be either highly directive or more or less isotropic. In the latter case the observed angular size of the emission source will appear substantially smaller than the angular size of the volume where the amplification has occurred. This effect is particularly striking for saturated masers, where a globular volume of gas will seem to have a hot spot at its center tens of times smaller in diameter than the actual cloud. Figuratively, one may imagine the radiation of such a spherical maser as resembling a little hedgehog (Figure 4.3a). But if the region where the radiation is amplified is cylindrical in shape, the radiation will emerge mainly from the ends of the cylinder, and will thus be quite directive (Figure 4.3b).

The properties of the compact, extremely bright radio sources emitting in the OH and H_2O lines all indicate that radio astronomers have detected natural cosmic masers. As pointed out above, the radio flux of these sources

Figure 4.3   Schematic diagram of: a) an isotropic maser (a "hedgehog"); b) a cylindrical maser.

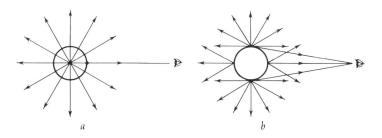

a b

is unusually high. For example, in the H_2O line at 1.35 cm the flux density[1] from the source known as W49 reaches 10^4 Jy, an enormous value. No other sources of cosmic radio waves located outside the solar system send us so much radiation. Even the moon, situated right next to the earth, sends us only about 30 times as much radiation per hertz at this wavelength. Understand that the source W49 is very remote from us. It is about 14,000 pc away, in a completely different part of the Galaxy. This distance is a trillion times that of the moon, and the flux density is inversely proportional to the square of the distance from the source. The radiant power of W49 in the water-vapor line is of the order of 10^{31} erg/s, or just a few hundred times smaller than the bolometric luminosity of the sun. For the radio range, particularly in a single narrow spectral line, this is an extravagantly high value.

We shall now estimate the physical parameters of the first type of maser radiation sources described above (they will be called type I sources). From the angular size of the emission regions (0.''001–0.''01) together with the established distances to the H II zones containing these sources, we may infer that the linear extent of cosmic masers is $l \approx 10^{14}$ cm, just one order of magnitude more than the radius of the earth's orbit. For the brightness temperature to be as high as 10^{13}–10^{15} °K, the radiant intensity must be amplified by a factor of 10^{12} to 10^{14}. Recall that in the radio frequency range the intensity is proportional to the brightness temperature (the Rayleigh–Jeans law). In our rough estimate we shall regard the maser as unsaturated. By Equation 4.7 we then have

$$e^{\kappa_\nu l} \approx 10^{12}\text{–}10^{14},$$

so that $\kappa_\nu l \approx 30$. The expression 4.8 for the negative absorption coefficient, $\kappa_{\nu 12} \approx (\Delta n/\Delta \nu_D)A_{21}c^2/15\nu^2$, contains an Einstein coefficient, $A_{21} \approx 10^{-11}$ s^{-1} in our case. The line width $\Delta \nu_D \approx 10^3$ s^{-1}, whence $\Delta n \approx 1$ cm^{-3}. Let us suppose that $\Delta n/n \approx 0.1$. Then the numerical density of hydroxyl molecules will be $n \approx 10$ cm^{-3}, which is hundreds of millions of times greater than in "normal" interstellar gas clouds (Chapter 2).

For a saturated maser, the more probable case, we obtain a value of n considerably higher still. The total number density of all atoms and molecules in the maser emission region should be at least 10^6–10^7 cm^{-3}. Such regions indeed can no longer be regarded as dense clouds in the interstellar medium. They more closely resemble the rarefied atmospheres of giant stars, and their linear size is actually of the same order. The maser effect narrows the spectral lines by several times, so that the kinetic temperature of the medium in which the radiation is amplified can hardly exceed 2000–

---

[1]It will be recalled from the Introduction that the unit of spectral flux density is the jansky, 10^{-26} W m^{-2} Hz^{-1}.

3000°K and probably is even lower. In their physical properties, then, maser emission regions are comparable to the extended atmospheres of cool giant stars.

The flux density of the maser radiation arriving from the brightest sources is so great that they could, in principle, have been detected even with radio telescopes of the low sensitivity available in 1950 to 1955. It would have been enough to know the frequency of the radiation and to search persistently for sources. But masers were not even invented on the earth until 1954—something to think about when the conversation turns to the role of astronomy in practical affairs and the interconnection between "pure" and "applied" sciences! Now that masers and lasers have become powerful weapons in the present scientific and technological revolution, it is no longer a cause for wonder that in the natural cosmic medium, when radiation and matter are not in thermal equilibrium with each other, conditions can prevail that give rise to maser effects. The problem facing us is to understand how these conditions develop, and above all, what pumping mechanism operates in cosmic masers.

The most straightforward interpretation would be that cosmic masers operating in the OH and H_2O lines are pumped by a radiative mechanism. This is especially the case for hydroxyl molecules, which have the richest infrared and ultraviolet spectra. Presumably in the absence of thermodynamic equilibrium relatively close to extraneous sources of infrared or ultraviolet radiation, the absorption of this radiation in various lines followed by cascade transitions to lower levels might ultimately produce an anomalously high population for the excited levels of these molecules. The first hypothesis regarding the nature of the pumping rested on the idea that the upper Λ-doublet level of the ground rotational level in the OH molecule is pumped when ultraviolet photons corresponding to the resonance electronic transition for this molecule are absorbed. The pumping radiation would then have a wavelength of 3080 Å. This hypothesis was suggested because the first anomalous emission sources to be discovered were type I sources located in H II regions, which, as we know (Chapter 2), contain hot O and B stars. The radiation of these stars in the near-ultraviolet region was thought to be powerful enough to provide the requisite pumping. Alas, the expectations were not warranted!

As a matter of fact, there is every reason to believe that the masers represented by bright cosmic OH sources (as well as H_2O sources) are saturated. Such an inference may be drawn from the spectral profile of the individual peaks: in all cases investigated the profile has a Gaussian shape (that is, the intensity falls off with distance from the center of the peak according to the law $I \propto e^{-(\alpha \Delta \nu)^2}$, where $\Delta \nu$ is the distance from the center). A Gaussian profile is a necessary attribute for the lines of a saturated maser.

If the maser were unsaturated its intensity would decline by another law as $\Delta\nu$ increases. As soon as our maser is saturated we can assert that the number of pumping photons will in no event be smaller than the number of maser radio photons emitted by the source. It should be recognized, however, that each ultraviolet pumping photon has an energy $18/(3 \times 10^{-5})$ $= 6 \times 10^5$ times as great as a radio photon. On the other hand, only a very narrow band in the continuous spectrum of hot stars is utilized for pumping. One conclusion, for example, is that the most powerful maser emission source, W49, would need about 1000 type O stars to provide enough pumping. Yet the optical radiation of this source can be maintained with no more than 10 such hot stars!

Our calculation has presupposed that the radiation of cosmic masers has low directivity, so that the solid angle Ω appearing in Equation 4.9 is of order unity. Of course if we make Ω small enough, say $1/100$, we can remove the energy difficulty described above. But then we will inescapably encounter another difficulty: if $\Omega \ll 1$, there ought to exist at least a hundred times more maser emission sources whose rays are directed so as to miss us. This circumstance would require an excessively large number of hot stars in the Galaxy, which definitely are not observed. Another defect with such a pumping mechanism is the strong absorption of ultraviolet radiation by cosmic dust, which occurs abundantly in sources of cosmic maser emission. Accordingly, the explanation that pumping occurs by ultraviolet radiation from hot stars located close to the OH sources is not tenable.

In 1966, soon after the sources of maser emission in the OH lines were discovered, the author of this book put forward the hypothesis that the pumping might be effected by infrared photons of the OH rotational–vibrational spectrum. The sources of this infrared radiation could be starlike objects with a high luminosity in the long-wave spectral region; they would combine a relatively low surface temperature with a huge linear size. It was perfectly natural to suppose that these infrared objects might be protostars. Indeed, during their free-fall stage protostars should already be powerful sources of infrared radiation. In the next stage of gravitational contraction (the Hayashi stage; see Chapter 5) protostars should again be strong infrared sources, because their surface temperatures will stay at a constant level near 3500°K for a rather long time. We might point out that in 1966 the only OH sources known were identified with H II zones that contain young stars belonging to associations—zones in which presumably the star formation process is continuing "before our eyes" or has just recently concluded. Our hypothesis therefore envisioned OH maser sources as connected with the birth of stars, and the pumping mechanism as the infrared radiation of protostars.

The hypothesis promptly attracted some interest and in the years following it was intensively developed by various authors. At the same time rapid progress was made in collecting observational material, which served to clarify the structure of the sources, to identify them with other objects, and to classify them into the three groups. With regard to the pumping by infrared photons, two completely different processes should be kept in mind. Pumping can first be produced by near-infrared photons with a wavelength of 2.8 μ. Such photons will excite the higher vibrational levels of the OH molecule, and downward transitions can then yield the excess population of the initial level for emission of the 18-cm line. In the second place, far-infrared photons at wavelengths of about 120 and 80 μ can perform the pumping and excite the OH rotational levels.

Theoretical developments have demanded considerable complexity to describe the pumping. In particular, to calculate the pumping by far-infrared rotational photons one must take into account processes whereby the photons are multiply scattered in the medium containing the OH molecules. Careful calculations have shown that the rotational photons by themselves can generate the maser effect only for the Λ-doublet components at 1612 and 1720 MHz. They cannot overpopulate the initial levels of the principal components of the 18-cm line at 1665 and 1667 MHz, the components characteristic of type I sources. But the result for the 1612-MHz line is also of great interest. If we simultaneously admit the presence of a large number of photons in the near-infrared region and take a medium at a high enough kinetic temperature ($\approx$ 2000°K), we will in addition have a comparatively small excess population for the upper levels of the 1665- and 1667-MHz lines. Under such conditions, then, we would expect a very bright 1612-MHz line with considerably fainter 1665- and 1667-MHz lines, whereas the 1720-MHz line should appear in absorption. But this is just what is observed for the type II sources of OH emission identified with infrared stars!

The best-studied source of this type has been identified with the infrared star NML Cygni, discovered not long before. This star is located relatively close to the sun. Such objects are red giant stars of the "late" spectral type M with a very large surplus of infrared radiation in the 2–5 μ wavelength range. Dense envelopes of dust surrounding the stars are responsible for this infrared excess. Such an envelope will absorb the radiation of its central star, become heated to a temperature of 600–800°K, and reemit the radiation in the infrared region. Along with the 1612-MHz maser radiation these stars also emit maser radiation in the 1.35-cm water-vapor line. Stars of this type exhibit several subsidiary components of the 1612-MHz line differing slightly in frequency. Usually these components form two groups in a given star, and the separation of the two groups in the spectrum corresponds to a difference of several tens of kilometers per second in the Doppler velocities. The groups of higher and lower radial velocity are respectively called the red and blue groups. The two groups of lines are probably present because of the rotation of the star.

Objects such as NML Cygni very likely are not regular stars but protostars, although we have no proof as yet. Altogether the problem is far from being simple. One complicating factor is that some red supergiants whose brightness varies sharply and irregularly, such as the famed star Mira (o Ceti), also exhibit the 1.35-cm H_2O maser emission line and the 1612-MHz OH line, but of fairly moderate intensity. These stars have water-vapor lines in their infrared absorption spectra. But Mira-type stars are definitely not young, as their space distribution alone indicates.

This example shows that maser line emission need not necessarily be associated with star formation processes. Hence a very urgent problem confronting astronomy today is establishment of the age of NML Cygni–type objects.

OH maser emission sources of the third type, those with an amplified 1720-MHz line, probably are related genetically to the expanding nebular supernova remnants (see Part III). Thus far, however, these type III maser sources have been very poorly investigated. Dense, fairly cool gas with a high abundance of molecules apparently develops behind the front of the shock wave generated in the interstellar medium by a supernova explosion (Chapter 16).

But let us return now to the OH and H_2O sources of type I found in H II zones. The star formation process is most likely connected with these sources. Both point (starlike) and extended infrared sources are found in the immediate vicinity of the type I maser objects, so that the possibility of pumping by infrared photons cannot yet be excluded. Nevertheless, investigators have shown increasing preference lately for a chemical pumping mechanism in these maser sources.

Some quite general physical properties of maser sources have been estimated above—very roughly, to be sure. We recall that the sources should be rather dense gas clouds with a kinetic temperature of 1000–2000°K and a size comparable to that of red supergiants. The observed 5-min flux variations imply that the brightest sources of the H_2O radio emission line have a maser amplification region hardly more than an astronomical unit (1.5 $\times$ 10^{13} cm) across. Then according to the theory of a saturated maser (Equation 4.9, with $W_p \approx 1$ s^{-1}), the density of working water molecules should be about 10^6 cm^{-3}, and the total density of all molecules (primarily H_2), about 10^{10} cm^{-3}. At such a high density interparticle collisions will be very likely. For example, the ordinary gas-kinetic collision frequency $W_c \approx$

$n_{H_2} \sigma V \approx 1$ s^{-1}, where $\sigma \approx 10^{-15}$ cm^2 is the cross section of a molecule and $V \approx 10^5$ cm/s is its velocity (in estimating n we have assumed that $W_p \approx W_c$). Some of these collisions will serve to excite OH molecules. As a result of this chemical excitation, the initial levels for emission of the OH and H$_2$O radio lines may become overpopulated.

This problem in the kinetics of chemical reactions is fairly complicated, and the question of whether cosmic masers can be chemically pumped has not yet been answered definitively. Various authors have calculated different reactions which in principle might be capable of chemical pumping in cosmic masers. Among these reactions are:

$$OH + H \rightarrow OH^* + H,$$
$$OH + H_2 \rightarrow OH^* + H_2, \quad (4.10)$$
$$H_2O + H + 0.69 \text{ eV} \rightleftarrows OH + H_2.$$

The asterisk designates an excited state of the molecule. Several of the reactions proposed are exothermic, such as the reaction $OH + H_2 \rightarrow H_2O + H + 0.69$ eV. A comparatively high kinetic temperature for the gas would then be a favorable factor. The formation of excited OH and H$_2$O molecules at the shock front is a very promising possibility. Shock waves would be expected to occur in protostars during the terminal phases of the free-fall stage, and also in "old" supernova remnants (see below). Excited OH molecules could also be formed through the collision of water molecules with comparatively energetic hydrogen atoms or ions:

$$H_2O + H + 5.2 \text{ eV} \rightarrow OH^* + 2H. \quad (4.11)$$

This hypothesis faces a most difficult question: Where would such energetic atoms or ions of atomic hydrogen come from? In this case too shock waves might perhaps save the situation. Finally, one should not forget that a great many dust grains are present in regions where maser emission is generated. The grains may serve as catalysts for chemical reactions that yield excited OH and H$_2$O molecules. Moreover, the relatively fast protons that might be formed at shock fronts will simply dislodge the excited OH molecules from the surface layer of the "ice" grains (or, properly speaking, crystals).

We see that the problem of pumping type I cosmic masers may be, and evidently is, a most difficult one for modern astrochemistry. There is some hope, however, that it will be solved before very long.

To conclude this chapter we shall summarize the arguments indicating that the sources of maser emission in the OH and H$_2$O radio lines are associated with regions in which the star formation process is taking place.

1. Many if not all maser sources are connected with bright H II zones. These regions in the interstellar medium are excited to radiate by very hot massive stars of spectral types O and B which, as indicated in Chapter 3, are young objects. Maser sources are certainly not observed, though, in all H II zones. In fact, the age of various H II zones covers quite a wide range, from tens of thousands to millions of years. It seems as if the OH and H_2O maser sources are grouped mainly in young H II zones. A good example of a young H II zone is the great Orion Nebula.

2. Soon after the cosmic masers were discovered, radio sources of an unknown new type were found in the same H II zones in which the masers were observed. They proved to have a thermal spectrum and a very small angular size—a few seconds of arc. It became evident that this radiation was emanating from small, rather dense plasma clouds heated to a temperature of about 10,000°K. That the sources are thermal in nature is graphically demonstrated by the presence of hydrogen recombination radio lines in their spectra (see Chapter 2). These sources have been called compact H II regions. They are approximately 0.1 pc across and their electron density is about 10^4-10^5 cm^{-3}, hundreds of times the average value for bright H II regions. The compact H II regions are ionized, and they radiate only because there should be a hot O or B star inside them. Yet such stars are not observed there, just as the compact H II regions themselves are invisible in optical light.

   Only one conclusion is possible: the layer of light-absorbing dust there must be very thick indeed. Now the density of the surrounding medium is generally lower than inside the compact H II zone, where the temperature is a hundred times higher. Hence external pressure would definitely be unable to arrest the expansion of the compact H II zone and its subsequent dispersal over a time span of some tens of thousands of years. It follows, then, that the compact H II zones and the hot massive stars located inside them represent ultrayoung objects; they were formed "within the memory" of Cro-Magnon man.

   But where did the gas there come from—gas amounting to several solar masses or more? Presumably this gas constitutes the remains of the diffuse medium from which the star was formed. A great deal of dust is present, making such an object completely opaque to optical light rays. The stars located within compact H II regions have therefore received the figurative name "cocoon stars." It is exceptionally interesting that many OH and H_2O maser sources of type I coincide, within the observational error (which is very small, roughly a second of arc), with compact H II regions. This coincidence is indisputable; the only point not yet clear is whether the maser sources are actually inside the compact H II regions or are situated near their periphery. The close association between type I maser sources and compact H II regions undoubtedly proves that the maser sources are young and are directly related to the star formation process.

3. Many type I maser sources have been identified with point infrared sources. In this case the word *point* means that their angular size is smaller than 2″. Such infrared objects occur, in particular, in the Orion Nebula and the nebulae W3 and W49, where the brightest maser sources are located. Careful investigations of the typical point infrared source in the Orion Nebula (it is located next to the source of long-wave infrared radiation, with an angular diameter of about 30″, mentioned earlier) have shown that it cannot possibly be regarded as a "normal" high-luminosity star embedded in a dense gas cloud. The point source in the Orion Nebula has a diameter of 50 astronomical units as computed from its radiation; the point source in W3 is about 600 astronomical units across. The temperatures of the dense, radiating gas–dust clouds that constitute these sources are 550°K and 350°K, respectively. Such objects have a total luminosity thousands of times that of the sun. Thus all the observations together suggest that these objects are none other than *protostars.*

Having taken advantage of the latest technology of quantum radio-physics, astronomers are now fully justified in saying that the protostars being formed from the diffuse interstellar medium are simply clamoring for attention. What the first steps taken by the newborn stars are we shall see in the next chapter.

5

Evolution of Protostars

In Chapter 3 we recounted in detail how dense, cool molecular clouds condense into protostars; gas–dust complexes in the interstellar medium break up because of gravitational instability. Once again we must emphasize that the process is natural and inescapable. In fact, the thermal instability of the interstellar medium, mentioned in Chapter 2, will unavoidably lead to its fragmentation into separate, comparatively dense clouds and an inter-cloud medium.

All by itself, however, the force of gravity is incapable of impelling clouds to contract; they are not dense or large enough for that. But now another factor will come into play: either a shock wave compressing the interstellar medium in a spiral arm (Chapter 2), or the interstellar magnetic field with its characteristic Rayleigh–Taylor instability. In the system of lines of force of the interstellar field, rather deep wells or cavities will be formed in which clouds of interstellar matter will tend to collect (Chapter 3). Huge complexes of gas and dust will develop as a result.

A layer of cold gas will arise in such a complex, because the ultraviolet radiation of stars that ionizes interstellar carbon will be strongly absorbed by the cosmic dust in the dense complex, while neutral carbon atoms will cool the interstellar gas intensively and hold it thermostatically at a very low temperature, about 5–10°K. Since the gas pressure in the cold layer is equal to the external pressure of the warmer gas surrounding it, the density in the layer will be substantially higher, reaching several thousand H atoms and H_2 molecules per cubic centimeter. After the cold layer has become constricted to a thickness of about 1 pc, it will begin to fragment under its own gravitation into even denser separate clouds, which will continue to contract by *their* own gravitation. Associations of protostars will thus be created in the interstellar medium in a perfectly natural way. Each of these protostars will evolve at a rate depending on its mass.

We have already described in Chapter 3 the earliest phase in the evolution of a protostar, the free-fall stage. This stage terminates when the density of the protostar (which has hitherto contracted at a more or less constant temperature) has risen to the point at which the protostar becomes opaque to its own infrared radiation. Afterward the temperature of its central regions will begin to grow rapidly. A large temperature differential will therefore arise between the outer and inner layers. Accordingly, the gravitational energy released in the contraction process ought somehow to be transported outward.

The ensuing evolution of the protostar has been calculated theoretically by the Japanese astrophysicist Chushiro Hayashi, who in 1961 was the first to point out that the energy in a contracting protostar should be transported by convection rather than by radiation, as astronomers had believed. As we shall explain in Chapter 7, convection will set in when other routes for transporting the energy produced in stellar interiors are cut off. In the outermost, photospheric layers of the protostar the mechanical energy of the violent convective motions pervading the whole body of the protostar should be transformed into radiant energy, which will then escape into interstellar space.

On a miniature scale a similar behavior takes place in the outer layers of the solar atmosphere, the chromosphere, whose comparatively high temperature is maintained by the mechanical energy of waves in convective streams rising from the subphotospheric layers of the sun. But in our own star only the outer layers are involved in convection. The conditions in a protostar resemble far more closely the situation in red giants, most of whose volume is caught up in stormy convection, out to the surface itself (see Figure 11.4).

The temperature at which the energy of the convective streams is transformed into radiant energy is determined by a variety of factors, including the chemical composition. We may assume on a purely empirical basis that in the surface layers of a protostar the balance between the influx of mechanical convective energy and the radiation will establish a temperature comparable to that in the photosphere of a red giant, about 3500°K. More accurate calculations yield a somewhat lower value, about 2500°K, for the temperature in the outer layers of a protostar. It is interesting to note that according to the same calculations the surface temperature of a protostar depends on its mass $\mathfrak{M}$ and luminosity L as follows:

$$T_s \propto \mathfrak{M}^{7/31} L^{1/32}. \tag{5.1}$$

Thus T_s is virtually independent of the luminosity of the protostar and depends only weakly on its mass.

We may infer that the surface temperature of a fully convective protostar will remain nearly constant throughout the Hayashi stage of its evolution. As its radius will meanwhile continually diminish (for the protostar will keep on contracting by its own gravitation), the luminosity of the protostar during this stage will steadily *decrease.* For a comparatively short time, when convection is first established throughout the protostar, its luminosity will be at a maximum.

To estimate roughly the value of this maximum luminosity (or flare), we shall adopt the expression 3.8, obtained in Chapter 3, for the radius of the protostar when convection is established in it. Here we are assuming in advance that convection sets in quite rapidly in the protostar, becoming established before the protostar is able to contract appreciably. Then the luminosity of the protostar at the time of the flare will be given by the simple equation

$$L = \left(\frac{T_1}{T_\odot}\right)^4 \left(\frac{R_1}{R_\odot}\right)^2 \approx 10^4 \left(\frac{\mathfrak{M}}{\mathfrak{M}_\odot}\right)^2 L_\odot. \tag{5.2}$$

The duration of the flare can be estimated if we divide the gravitational energy $G\mathfrak{M}^2/R_1$ released in the contraction of the protostar by L. It turns out to be just a few years, which is a short time indeed.

In Chapter 3 we showed that at the close of the free-fall stage as well, the contracting protostar should experience a bright, comparatively brief flare of infrared radiation during which its luminosity will be some thousands of times the sun's bolometric luminosity. This second flare ought to occur quite soon after the first one. The two flares will differ significantly in the spectral distribution of their radiation. At the time of the first flare the radiation should be concentrated toward the long-wave part of the infrared spectrum (at about 20–30 μ), while most of the radiation during the second flare should be emitted in the near-infrared range ($\approx$ 1–2 μ). In view of both the present status of the theory and the level now attainable in observational astronomy, it might even be possible for the two flares of a protostar not to be separated in time but practically to merge with each other.

After the flare that accompanies the end of the stage establishing convection throughout the protostar, the object will continue to contract while its surface temperature stays at almost the same level, as we have said. Hence the luminosity of the protostar will decline in inverse proportion to the square of its radius. At the same time, however, the temperature in its interior will steadily rise. And eventually, when the central temperature has reached several million degrees, the first thermonuclear reactions involving light elements (lithium, beryllium, boron) with a low Coulomb barrier (see Chapter 8) can begin. Even so the protostar will keep on contracting, because the thermonuclear energy production will still be inadequate to heat its interior to a temperature high enough for the gas pressure to balance the force of gravity.

Not until the steady rise of temperature advances to the point where the proton–proton or carbon–nitrogen reactions (Chapter 8) can take place in the interior of the protostar will the gas pressure finally succeed in stabilizing it. The protostar will become a star, and depending on its mass it will occupy a perfectly definite position in the Hertzsprung–Russell diagram. The theory explaining the structure of equilibrium stars formed in this manner will be discussed in Part II.

We have now dealt with the process by which protostars evolve into stars. Needless to say, our account has not been a strict analysis; of necessity it has been semiqualitative in character. A rigorous solution of the problem of star formation from the interstellar medium would hardly even be possible now. One can only work out separate pieces of the theory, continually checking them against the observations.

Figure 5.1 shows schematically how the radius of a protostar whose mass was originally equal to that of the sun depends on time. To indicate scale, the dashed horizontal lines mark the radii of the orbits of some planets in our solar system. At the beginning of the free-fall stage, when a protostar that has only recently been a dense, cool molecular cloud begins to contract by its own gravity, its radius is comparable to that of the orbit of Pluto. The average particle density (primarily hydrogen molecules) is about 10^{12} cm^{-3}. Starting with this density the free-fall stage will last a bit longer than 10 years (see Equation 3.7). After this brief interval the protostar will have contracted to the size of Mercury's orbit—about 100 times as small. Of course the free-fall stage is preceded by a much lengthier phase during which the cloud contracts from an initial density of 10^5 to 10^6 cm^{-3} until it is confined to a region equivalent to Pluto's orbit.

Subsequently, as the protostar becomes opaque to its own radiation, its contraction will slow down abruptly. The Hayashi stage in the life of a protostar embroiled in convection will commence. A flare should occur at the very beginning of this stage, as we have pointed out. Tens of millions of years later the protostar will nearly have stopped contracting, and it will be on the main sequence.

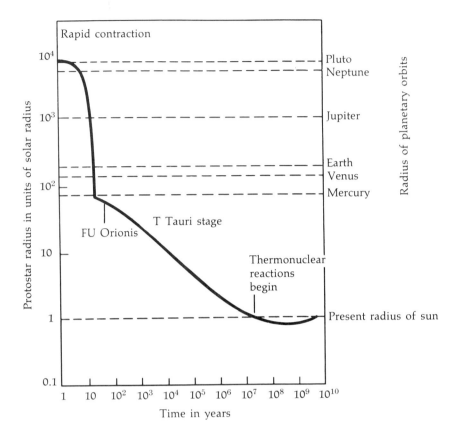

Figure 5.1 Theoretical time dependence of the radius of a protostar (G. H. Herbig).

In Figure 5.2 we show the evolutionary track followed by a protostar in the Hertzsprung–Russell diagram. The free-fall stage, when the protostar was cool and transparent, is designated (schematically, of course) by the dashed curve at the right of the diagram. The maximum of this curve, corresponding to the onset of opacity, matches the first flare of long-wave infrared radiation. After opacity has begun the bolometric luminosity of the protostar diminishes rapidly; however, it rises very rapidly again as the protostar "comes to a boil," with the convective streams escaping outward and transferring their energy to radiation. A second flare occurs, this time in the near-infrared region. At the peak of the second flare the luminosity of the protostar is several times smaller in the diagram than according to our rough estimate 5.2, but naturally we should not be disturbed by that. This short stage in the evolution of the protostar is represented by the wider shaded band. The last part of the evolutionary track, toward the left in the diagram, shows a continuous drop in the luminosity of the contracting protostar, whose surface temperature is maintained at a nearly constant level (the Hayashi stage). Ultimately the track arrives at the main sequence (narrow shaded band), which means that the protostar has become a "normal" star. It is worth emphasizing again that the duration of the individual segments of the evolutionary track is altogether different.

Theoretical astrophysicists in West Germany have recently investigated how a massive, spherical cloud of gas and dust will condense into a star. They have performed numerical calculations for masses of 150, 50, and 20 $\mathfrak{M}_\odot$. These calculations show that when the stars eventually reach the main sequence, their masses will have dropped to 36, 17, and 12 $\mathfrak{M}_\odot$, respectively; thus much of the cloud's original mass will remain uncondensed, forming protostellar envelope structures. One can follow the evolu-

Figure 5.2   Evolutionary track of a protostar in the Hertzsprung–Russell diagram.

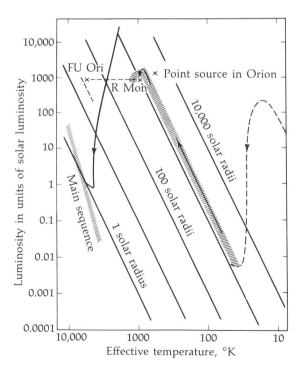

tion of these envelopes theoretically; furthermore, they might well be examined by the techniques of observational astronomy. A new approach to the fundamental problem of stellar cosmogony opens its doors.

An initial radius of about 10^{18} cm has been adopted for the contracting clouds, and they have been regarded as nonrotating and devoid of a magnetic field—assumptions which simplify the problem considerably, of course. Nonetheless, observations suggest that these calculations describe quite accurately the various steps in the evolution of a contracting cloud. Let us summarize the results:

1. A few hundred thousand years after a cloud begins to contract, and shortly after a fairly hot, starlike nucleus develops within the contracting cloud, a dense "cocoon" of gas and dust, opaque to optical light, will form around the nucleus. This cocoon will have an inside radius of $(3-5) \times 10^{13}$ cm and an outside radius of 10^{15} cm. The temperature of its outer layers will be about 500°K, and it would in principle be observable as an infrared source. However, the cool material of the contracting cloud surrounding the cocoon is opaque to infrared rays, so the observer will not see a cocoon within the cloud after all.

2. Strong ($\approx 1000\ L_\odot$) infrared radiation will emanate from the cocoon and exert pressure on the gas–dust medium of the envelope. As a result the outer layers of the cloud will stop contracting in a few tens of thousands of years, and afterward they will begin to expand. Thus an exterior gas–dust envelope or outer cocoon will develop with a radius of roughly 10^{17} cm. Both the inner and the outer cocoon will thenceforth expand. In due course the outer cocoon will become thin enough to transmit the infrared radiation of the hotter, more compact inner cocoon. A distant observer will therefore perceive a warm ($T \approx 500-1000$°K), compact source in the infrared wavelength range, embedded in a cooler ($T \approx 200$°K), more extended source. In certain cases we do observe a situation of this very kind, as in Orion (see above).

3. Up to this point, only a small region within the inner cocoon contains ionized gas. The flux of thermal radio waves associated with this gas is too small to be observable. But as the expansion continues, the thickness of the inner cocoon will become so small that the ionizing ultraviolet radiation of the protostar will begin to pass through it. In only a few thousand years a very compact H II region will be able to form within the outer cocoon, and cool, nonionized gas will surround it. During this phase the observer will see an exceptionally compact H II region surrounded by a more extended infrared source. This combination of sources is also observed fairly often.

4. The compact H II region formed in this way will expand rapidly, and before long it will reach the inner edge of the outer cocoon. The observer will notice an H II region accompanied by an infrared source of comparable size.

5. After the whole outer cocoon has become ionized, a new type of compact H II region will form: its mass will remain constant but the radio emission will weaken rapidly (see Chapter 13 for a fuller discussion). An ionization front will be propagated through the rarefied medium surrounding the protostellar cloud, thereby producing an ordinary extended H II region. Professor Peter G. Mezger in Bonn, who is actively studying the radio astronomy aspects of the star formation process, estimates that one of these ordinary H II clouds would have an average lifetime of about 5×10^5 yr.

This outline for the scenario of star formation has several implications of significance for observational astronomy:

1. During the earliest phase, free fall (lasting about 10^5 yr for O stars), the contracting protostellar cloud will not be observable.

2. For the next 10^4 yr (approximately), the protostar can be observed as an infrared source, but there will be no sign of a compact H II region.

3. After the protostar has turned into a star and has arrived on the main sequence, a compact, expanding H II region will develop, surrounded by an outer, comparatively cool cocoon. This phase will also last about 10^4 yr.

4. In the last phase, traces of a compact H II region (the old outer cocoon), with an extended, relatively faint H II region around it, will persist as long as a million years.

Even though, as we have indicated, the calculations rest on a very schematic model, evidently the main features of the evolution of protostellar clouds into stars are represented correctly. The model is consistent with a large number of observations carried out lately—in particular at Bonn under Mezger's direction. One should also remember that the calculations described above refer to protostellar clouds of high mass. For less massive stars, the fraction of the mass in the protostellar cloud that fails to condense into a star will presumably be small. Hence an outer cocoon might not form at all, and the infrared radiation of the comparatively hot inner cocoon will not be shielded from us.

To what extent do astronomical observations support the scenario we have sketched for the evolution of a protostellar cloud? First of all, the view that groups of stars form in dark molecular clouds of interstellar matter requires empirical underpinning. That H ɪɪ zones around hot, young, massive stars are genetically related to dark molecular clouds has long been recognized: one need only look at photographs of diffuse nebulae with the dark patches and other extended features enclosed inside them (see, for example, Figures 2.2 and 2.3). The new observations add significantly to this picture. For instance, radio emission in the CO molecular line at 2.64-mm wavelength can be detected in nearly all H ɪɪ zones.

This circumstance means that cool molecular gas is located there, a vestige of the original gas–dust cloud from which hot massive stars were formed, spawning H ɪɪ zones. If a protostar is concealed by a dense, opaque cocoon, the cocoon will reemit as infrared rays all the radiation it absorbs from the protostar. Thus by measuring the power of the infrared source one can determine the luminosity of the protostar hidden inside it by absorption. In some cases the power of the compact infrared source may reach a luminosity tens or hundreds of thousands of times that of the sun, indicating the presence of a massive protostar that will soon become a star of spectral type O. It is worth emphasizing that associations of compact H ɪɪ regions (which, as shown above, represent a later phase in the development of protostellar envelopes) and infrared sources are observed fairly often.

The latest radio astronomy research in this field makes wide use of the CO molecular radio line. H ɪɪ zones frequently contain compact regions where the intensity of this line is enhanced. Dense clumps of cool molecular gas surrounded by a hot rarefied medium are consequently located there. Such condensations, with a mass hundreds of times that of the sun, generally occur in association with clusters of young stars.

Since the gravitational contraction time of massive protostars is comparatively short, we would expect to find in their vicinity remnants of the gas–dust cloud that gave rise to them. These remnants would constitute the protostellar envelopes that we have considered in a theoretical sense. If the spectrum of a type A or B star contains not only absorption but also emission lines (such stars are designated as type Ae or Be), we may suspect it of being a T Tauri star (see below) and thereby a protostar. Observations confirm that most such stars are indeed surrounded by compact molecular clouds showing a strong 2.64-mm CO radio line. The observations further imply that these circumstellar clouds are considerably denser and hotter than the ordinary molecular clouds found in the interstellar medium.

The presence of dense circumstellar clouds may also be inferred from observations of the carbon recombination radio line. A star of spectral type B located within a dense cloud has only a small H ɪɪ zone, whereas the star emits so many photons in the wavelength range $912 < \lambda < 1101$ Å (the ionization limit of carbon) that a rather extensive carbon ionization

100

zone can form, with the carbon atoms thousands of times less abundant than hydrogen.

Modern astronomy has direct evidence that in many cases a dense, cool interstellar gas cloud opaque to visible light contains a cluster of very young stars or protostars. A good example is the notable gas–dust cloud in the constellation Ophiuchus, located 160 pc from the sun. In infrared light about 70 stars, invisible optically because of absorption, appear within a region of this dark cloud about 1.5 pc across (Figure 5.3). An analysis of the observations shows that the distribution of these stars with respect to luminosity (called their luminosity function) is the same as in young star clusters. The infrared stars are undoubtedly the brightest members of a cluster embedded in the dense cloud. Thus each star is surrounded by a rather dense envelope that is responsible for additional absorption.

Another interesting fact is that the wavelength dependence of the additional absorption in the envelopes differs from that of the absorption in the cloud as a whole. Accordingly, such properties of the grains making up the protostellar cloud as their size and chemical composition must differ from the average. Figure 5.3 displays curves for the absorption of light in the Ophiuchus cloud. The dots represent the positions of stars observed only at infrared wavelengths. Most of these infrared stars occur within the square shown on a larger scale in Figure 5.4. Two isophotes indicate

Figure 5.3   Infrared stars in the Ophiuchus molecular cloud. The contours map the absorption in magnitudes (Kitt Peak and Meudon Observatories).

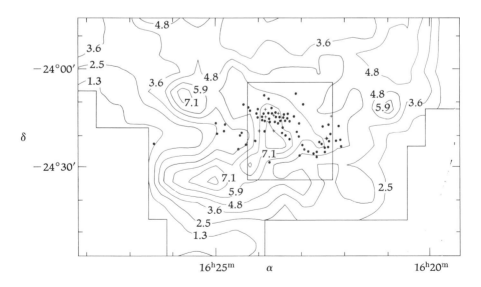

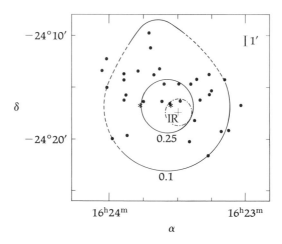

Figure 5.4   Center of infrared star cluster in Ophiuchus.

the brightness distribution of the carbon recombination radio line C157α, while the dot–dash circle marks a source of long-wave ($\lambda \approx 25$ microns) infrared radiation in the area where the density of molecular gas is highest ($\approx 10^6$ cm^{-3}). A certain number of tiny radio sources, probably representing compact H II regions, have been detected in the same area.

    All the observational data we have described accord with the idea that the dark nebula in Ophiuchus contains a protostar cluster whose most massive members will presently become type B stars. This interpretation follows from the comparatively large extent of the carbon ionization zone and the absence of an H II region of similar size. In fact, the theoretical results we have outlined state that the more massive protostars should be surrounded by dense envelopes or cocoons. One would expect that in a hundred thousand years the massive stars formed in this cloud will be resting on the main sequence; having ionized a substantial part of the cloud to the point where it clears up, they will become observable in the optical range. But at the same time one should recognize that many of the steps along the evolutionary path of relatively low-mass stars are still far from adequately understood.

    We turn now to the observations gathered for giant gas–dust complexes, where, as might be expected, the process of star formation from diffuse interstellar matter operates with special intensity. The wealth of radio and infrared data concerning these objects has been interpreted particularly by Mezger's group in West Germany. These astronomers find that star formation takes place somewhat differently in gas–dust complexes situated within the spiral arms of the Galaxy (Figure 5.5) and in complexes between them.

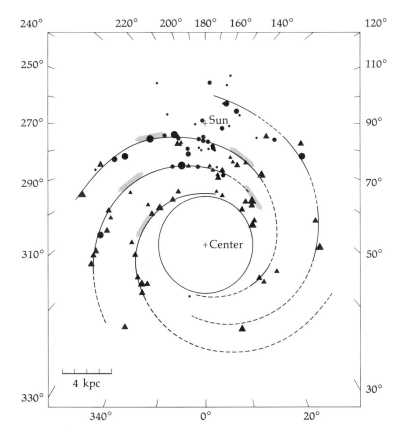

Figure 5.5   H ɪɪ regions in the Galaxy indicate optical radiation; triangles show radio emission (Marseille Observatory).

The main difference is that in the former case the process operates practically all at once, whereas in the latter it may stretch out for many millions of years. This distinction may result from differing circumstances under which compression waves are propagated—the agency that stimulates interstellar clouds to condense through gravitational instability (see Chapter 3). Gas will be compressed almost simultaneously throughout a gas–dust complex located in an arm, but the compression wave will take many millions of years to pass completely through an isolated complex between arms.

Next let us examine in somewhat greater detail the conditions prevailing in the isolated gas–dust complex closest to us, the one in the constellation Orion. Part of this complex has long been familiar as the remarkable Orion Nebula (Figure 2.3). Young stars can be observed in this complex at various phases in their evolution (the Orion OB association), as well as compact H II regions and protostars situated in a dense, opaque cloud of cool gas.

Figure 5.6 maps the brightness distribution in the radio line of the $C^{13}O$ molecular isotope. The cool cloud seems to split the Orion Nebula into two parts (compare Figure 2.3). The molecular gas in the cloud is very dense ($\approx 5 \times 10^4$ particles/cm^3), and the total mass is fully 2000 $\mathfrak{M}_\odot$. Hot O and B stars belonging to the Orion association extend 12° northwest of the molecular cloud, and the age of the stars steadily increases toward the northwest, reaching 10^7 yr. Curiously enough, the CO radio line is not observed within the OB association itself. We may infer that the cool molecular gas from which the stars there were formed has been ionized and scattered by the evolving stars. In the vicinity of the dense molecular cloud we find the celebrated Orion Trapezium, consisting of hot stars formed very recently ($\approxeq 10^5$ yr ago), whereas inside the molecular cloud the development of stars has barely started.

Bright, compact H II regions occur to the south and north of the molecular cloud. Near the two brightness peaks of the CO line, representing the densest parts of the molecular cloud ($n_{H_2} \approx 2 \times 10^6$ cm^{-3}, with a mass of about 200 $\mathfrak{M}_\odot$), sources of long-wave infrared radiation are observed. One of these is the remarkable Kleinmann–Low infrared object, the source we discussed at the end of Chapter 4. We mentioned that within these relatively large ($\approx 1'$ across) sources of long-wave infrared radiation, "point" sources of considerably higher temperature, to judge from their spectrum, have been detected; probably they are associated with protostellar envelopes. In particular, inside the compact Kleinmann–Low infrared nebula there is a star that has just arrived on the main sequence, and we are now able to observe its inner and outer cocoons.

Recently the bright point source inside the Kleinmann–Low nebula was found to be emitting hydrogen lines of the infrared Brackett series—proof that a very small ($r \approx 5 \times 10^{14}$ cm ≈ 30 astronomical units) H II region with an electron density $n_e \approx 3 \times 10^5$ cm^{-3} exists there. This supercompact H II region almost surely represents the part of the inner cocoon facing the star. Other infrared nebulae, probably outer cocoons, have less massive protostars inside. It is now apparent that after a hundred thousand years observers will see at the site of the present dense molecular cloud in Orion one additional feature of the large association located in that part of the sky.

These observations, then, support the picture of a compression wave traveling at a velocity of about 10 km/s through a gas–dust complex some 100 pc across and stimulating the star formation process at the wave front. The factor originally responsible for generating such a wave could be a strong shock wave produced in the interstellar medium at the time of a supernova outburst (Chapter 16).

Figure 5.6   The molecular cloud in the northern part of the Orion Nebula, mapped in the 2.7-mm $C^{13}O$ radio line at the Millimeter Wave Observatory, University of Texas (Kutner et al.)

We now consider how star formation proceeds in the giant complexes of gas and dust found in spiral arms. Let us take the W3 complex (Figure 2.4) as an illustration. Several compact H II regions occur here, each ionized by its own hot massive star or protostar. The thermal radio emission from this giant complex is tens of times more powerful than that from the com-

105

plex in Orion. Figure 5.7 shows a set of radio isophotes for the central part of the W3 complex, derived from observations at 6-cm wavelength with the record angular resolution of 2″. A 5-km antenna array at Cambridge University was used. Crosses designate infrared stars; crosses with dots mark OH and H_2O maser sources; and asterisks indicate stars observed optically. The H II zones depicted in this chart are surrounded by cool, nonionized gas.

At a shorter wavelength, 2 cm, the isophotes displayed in Figure 5.8 have been obtained for a compact H II zone in W3 at the exceptionally high angular resolution of 0″.65, even better than that of optical photographs. The diameter of the region filled with ionized gas at a density of about 10^5 cm^{-3} is only 0.01 pc, and the mass is 4×10^{-3} $\mathfrak{M}_{\odot}$. This ionized gas is embedded in a dark gas–dust cloud, a cocoon, whose radius is about 10 times that of the H II zone inside, as shown by observations of the CO radio line in this region. The + symbols in Figure 5.8 indicate OH maser sources within the compact H II zone. Another set of isophotes at $\lambda = 6$ cm, obtained for a larger region at the lower resolution of 4″, is shown in Figure 5.9. Along with the compact H II zone A of Figure 5.8, we find four other, less bright, compact H II regions containing less massive protostars inside.

Much the same picture is observed in all the gas–dust complexes that have been investigated. In all cases we encounter compact H II, CO, and infrared sources in a characteristic combination. This evidence fully con-

Figure 5.7 Part of the gas–dust complex W3 at λ_4 6 cm (Harris and Wynn-Williams).

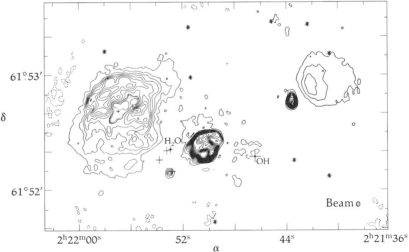

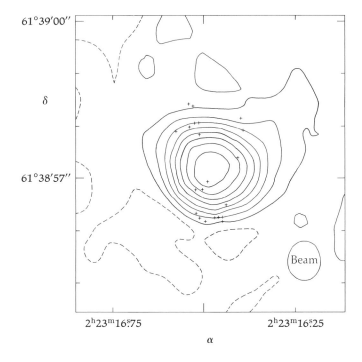

Figure 5.8   OH maser sources inside a compact H II zone in W3 at
$\lambda = 2$ cm.

firms the picture outlined above whereby protostars condense from a
medium of gas and dust.

It remains to say a few words about the place of the OH and H_2O maser
sources in this general scheme of star formation. Some mention of this matter
was made toward the close of Chapter 4, where we pointed out the close
association between OH maser sources of type I and compact H II zones.
Figure 5.8 provides a good example of this relationship. It has recently been
established that OH masers coincide with compact H II zones to 1″ accuracy.
Upon investigating the relationship, one finds that when the size of an
expanding compact H II zone reaches about 0.1 pc, there will no longer be
any OH maser sources nearby. In view of the expansion velocity of about
10 km/s for compact H II zones, one may conclude that cosmic OH masers
are no older than 10^4 yr. When a size of 0.1 pc is attained, the density of the
molecular gas in the protostellar envelope will be roughly 10^5 cm^{-3}; hence
it is natural to conclude that OH masers work at densities of order 10^6 cm^{-3}
and temperatures of about 100°K, and that they are situated outside the
ionization front.

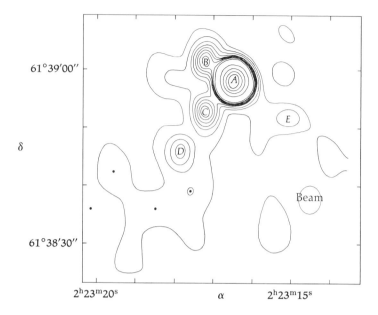

61°39'00"

δ

61°38'30"

$2^h23^m20^s$ α $2^h23^m15^s$

Figure 5.9   Compact H II zones in W3 at λ = 6 cm, as observed with the Westerbork radio telescope.

Unlike OH masers, H_2O masers do not coincide with compact H II zones. Apparently these water masers are associated with an earlier phase in the evolution of the protostellar cloud, when the compact H II zone has not yet formed. The gas density in the region where water masers are generated is evidently about 10^9 cm^{-3}, and the temperature is about 1000°K, corresponding to the inner part of the inner cocoon. Might an H_2O maser be the earliest sign that a protostar is forming out of a condensing protostellar cloud of gas and dust?

A dominant part of the radiation of protostars, then, should be concentrated in the far- and near-infrared regions of the spectrum. Although infrared astronomy is just taking its first steps, its progress is already most impressive. In particular, some of the infrared stars discovered in the past few years (such as the star in Orion mentioned in Chapter 4, the star R Monocerotis, and other objects), whose surface temperature is 300–700°K, appear to be protostars at their flare stage. They are marked by crosses in Figure 5.2. One cross marks the rather peculiar star FU Orionis. The radius of this star is 25 times the solar radius. In 1936 it flared up suddenly, increasing its brightness by tens of times, and ever since the brightness has remained almost unchanged. Its surface temperature, however, is about 5000°K, which is decidedly higher than the temperature expected at the time of a protostellar flare. Another curious object, the star V1057 Cygni, might be similar to FU Orionis. Before its flare this star was of the T Tauri type (see below). An interesting observation was made during the flare, when a bright 1720-MHz OH maser line was recorded there. But unlike FU Orionis, V1057 Cygni has begun to decline in brightness fairly rapidly. In general, much work still needs to be done in searching for the flare phase of protostars. The problem is a difficult one, though, since this phase of protostellar evolution is very transient, and few such objects may be anticipated.

Incomparably more observational material is available for protostars that are in their convective contraction stage. Well over 30 years ago astronomers were aware of a very interesting class of stars, called the T Tauri type after the name of a typical representative. As a rule they are cool stars whose brightness varies rapidly and irregularly. From all indications their atmospheres are experiencing violent convection. One characteristic property of T Tauri stars is the presence in their spectra of lithium absorption lines; evidently lithium is hundreds of times as common there as in the solar atmosphere. This circumstance may signify that the first nuclear reactions, in which "burning" of light elements takes place, have not yet begun in the interiors of these stars. T Tauri stars are always observed in *groups* known as T-associations. Such associations include an aggregate of dense clouds of the gas–dust interstellar medium in which the T Tauri stars are literally embedded. Often, but not always, the T-associations coincide with O-associations, groups of unquestionably young and massive hot stars. In the Hertzsprung–Russell diagram the T Tauri stars lie *above* the main sequence. This placement is fully understandable if we regard them as protostars in their convective contraction stage: the more massive protostars evolving into O and B stars will reach the main sequence relatively soon, while the less massive protostars observed as T Tauri objects will evolve considerably more slowly.

The spectra of T Tauri stars often exhibit emission lines of hydrogen, ionized calcium, and certain other elements. An analysis of the conditions under which these lines are formed shows that the temperature increases with height in the outer atmospheric layers of T Tauri stars. This is similar to the situation in the upper layers of the solar atmosphere, where the temperature rises with height because solar material is heated by mechanical energy of motion. Everything points to the fact that T Tauri stars are involved in rapid convective motions: their outer layers are actually boiling.

Another interesting property of the spectra of T Tauri stars is the presence of absorption line components shifted toward the blue. Such a shift implies a continuous ejection of matter from the stellar surface, amounting to about $10^{-7} \, \mathfrak{M}_\odot$/yr. It follows that these stars will lose a substantial part of their original mass while they are perched on the main sequence.

Their shedding of mass may also be attributed to the strong turbulent motions pervading the stars. The kinetic energy flux carried off by the gas clouds that T Tauri stars eject comprises an appreciable fraction (10 to 20 percent) of their radiation flux. All this evidence gives serious grounds for believing that T Tauri stars represent the Hayashi stage of protostellar evolution.

Our discussion above refers to the evolution of protostars whose mass is smaller than the solar mass. More massive protostars evolve in their own distinctive fashion toward the concluding stage. Even before they have reached the main sequence the convective transfer of energy is replaced by radiative transfer. This circumstance results from a faster temperature rise in the interior of such stars, which in particular causes the opacity of their material to diminish (see Part II). The change in the mode of energy transfer turns the evolutionary track of the protostar rather abruptly leftward in the diagram. Thus as it continues to contract the protostar will maintain nearly the same luminosity, and its temperature therefore will progressively rise. Figure 5.10 shows some evolutionary tracks calculated theoretically for protostars of several masses; this behavior is strikingly apparent. In particular, it becomes evident why T Tauri stars include not only cool objects with temperatures around 3500°K but also considerably hotter ones.

We are certainly interested in understanding the earliest stages in the evolution of the sun, when it was still a protostar. In this case too, even before arrival on the main sequence, the energy in the central part of the protosun began to be transported by means of radiation. As trial calculations have shown, the radius of the sun at that time was approximately twice its present value and the luminosity was 1.5 times as great. During the protosun's subsequent contraction its luminosity decreased to half its brightness today. By then the interior of the sun was almost wholly in a state of radiative equilibrium. Thenceforth the luminosity of the sun gradually rose again to the value it has now.

As Figure 5.10 demonstrates, when massive protostars are in the closing phase of their evolution and their luminosity is holding almost constant, they have all the characteristics of giant stars. We may accordingly suppose that some of the giant stars in young star clusters are actually protostars. One should keep in mind, however, that massive protostars rush through the final horizontal branch of their evolutionary track very rapidly, in just a few thousand years. Hence they should be rather scarce.

Let us close this chapter with a look at the interesting question of the age of various protostars belonging to the same association. The theory of protostellar evolution that we have sketched briefly above has now reached a level at which the age of the protostars can be estimated from the characteristics we observe. And what we find is that protostars of comparatively

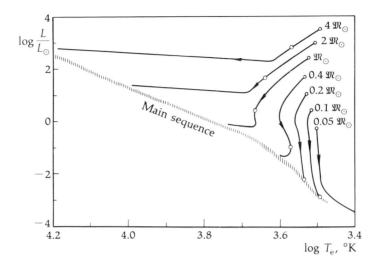

Figure 5.10 Theoretical evolutionary tracks for protostars differing in mass (C. Hayashi).

low mass are always significantly older than the more massive protostars as well as O and B stars in the same association. How are we to explain this rule? One might think instead that in the process of gravitational condensation the more massive stars ought to be formed first from the gas–dust medium.

At the very beginning of the fragmentation process the mean density of the gas–dust complex would have been lower, and the theory of gravitational instability states that a low mean density corresponds to a breakup of the complex into more massive fragments. In fact, we may write Equation 3.4 in the form

$$\mathfrak{M} \geq \frac{1}{2}\left(\frac{A}{\mu G}\right)^{3/2}\frac{T^{3/2}}{\rho^{1/2}}. \tag{5.3}$$

For a given temperature, then, the lower the mean density the higher will be the mass of the fragments evolving into protostars. Actually, we must also take into account the continuous rise in the kinetic temperature in the gas–dust complex, which occurs because of the gradual depletion of carbon in the complex as it adheres to dust grains. This is a very slow process, lasting at least 10 million years. We saw in Chapter 3 that carbon serves as the main temperature control in cold, dense gas–dust complexes; hence if its abundance drops the gas temperature will inevitably rise.

111

Now according to the expression 5.3 the mass of the resulting fragments that are to become protostars depends rather strongly on the temperature but weakly on the density, so that the protostars will grow more massive with time. If, for example, the initial temperature of the cold gas in the dense gas–dust complex was $T \approx 6°K$, most of the stars formed would have had a mass smaller than the solar mass. For the development of O and B stars with a mass tens of times the solar mass, the gas temperature ought to rise to 40–50°K, and correspondingly the abundance of carbon serving to cool the gas ought to diminish by a factor of about 10. One should recognize also that the radiation of the newly formed protostars would have gradually heated the interstellar gas. We have thereby interpreted what would seem to be a paradox, that the massive stars in a given complex are formed later.

After a substantial part of the mass of the primordial gas has turned into stars, the interstellar magnetic field, which had been maintaing the gas–dust complexes by its pressure, will of course have no effect on the stars and young protostars. They will begin to fall toward the galactic plane through the influence of the gravitational attraction of the Galaxy. Thus young stellar associations should always be approaching the galactic plane. This behavior, predicted theoretically by Professor Pikel'ner, has been confirmed by analysis of observational data, so we have a graphic demonstration of the origin of stars from the interstellar medium. Note that we are referring here to young associations. Old associations will have been able to cross the galactic plane and then oscillate about it with a period of 30 to 50 million years.

In the last quarter of the twentieth century, then, astronomy finds itself in a position to follow all the steps in the momentous process by which stars form from the interstellar medium. Unlike the astronomers of decades past, who sought to develop cosmogonic hypotheses from very weak foundations, astronomers today can cope with the difficult problems of the evolution of cosmic objects by relying on the firm ground of observational evidence. Of special value are the facts gleaned through progress in the "astronomy of the invisible"—radio and infrared astronomy. We are surely not unmindful of our debt to the great classical figures of cosmogony, Laplace, Henri Poincaré, and particularly Sir James Jeans, whose creative research on the theory of gravitational stability runs like a bright thread through all of modern cosmogony. The author merely wishes to say that every scientific problem is solved in good time, and that the time for solving the classic problems of cosmogony has not arrived until now. Much work is still ahead of us, of course; we have suggested that at some length. But the fundamental idea that stars originate from the diffuse interstellar medium has been supported by a wealth of factual material, and today it may be regarded as a durable achievement of science.

112

Part II

Stars Radiate

So I travelled, . . . in great strides of a thousand years or more, drawn on by the mystery of the earth's fate, watching with a strange fascination the sun grow larger and duller in the westward sky . . . . At last, more than thirty million years hence, the huge red-hot dome of the sun had come to obscure nearly a tenth part of the darkling heavens.

H. G. WELLS, *The Time Machine*

(A brilliant insight into the red giant phase as a closing chapter in the evolution of the sun, first expressed in June 1894, before the laws of radiation had been discovered.)

6

A Star: A Globe of Gas in Equilibrium

It might seem superfluous to say that the large majority of stars maintain their properties unchanged for great lengths of time. Surely this proposition is evident at least for the six decades in which astronomers from many countries have been sedulously measuring the brightness, color, and spectrum of a multitude of stars. Although some stars do undergo changes (such stars are called variables; see the end of Chapter 1), their variations are either strictly or imperfectly periodic in character. Systematic variations in the brightness, color, or spectrum of a star are encountered only in very rare cases. For example, although changes in the periods of pulsating stars (Cepheids) have indeed been detected, they are so slight that millions of years or more would have to pass before the pulsation period is substantially altered. Yet we know (Chapter 1) that the luminosity of a Cepheid depends on its period. We may therefore conclude that for at least several million years there is little change in the most important property of these stars, the power of their radiant energy. From this illustration we see that whereas observations have been in progress for only a few dozen years, an utterly negligible span on a cosmic scale, we may infer that the properties of Cepheids hold constant for incomparably longer time intervals.

But another opportunity is available to us for estimating the length of time that the radiant power of stars remains almost unchanged. Geological data show that over the past 2 or 3 billion years or more, the temperature at the earth's surface has stayed constant to a few tens of degrees centigrade. This fact is apparent simply from the continuous evolution of life on the earth. And it tells us that over an enormous time span the sun could never have been radiating three times more strongly, or three times more weakly, than now. Evidently, during its long history our luminary has indeed experienced periods when its radiation was appreciably (though not too greatly) different from the present level, but these eras were quite transitory. We have in mind the ice ages, which will be mentioned in Chapter 9. On the average, however, the radiant power of the sun has been remarkably constant for the past few billion years.

Now the sun is a fairly typical star. We know (Chapter 1) that it is a yellow dwarf of spectral type G2. At least several billion such stars occur in our galaxy. It is perfectly logical to conclude that most of the other main sequence stars whose spectral types differ from the sun's should also be very long-lived objects.

The overwhelming majority of stars, then, change very little with time. This is not to say that they can persist indefinitely in unaltered form. On the contrary, we shall see below that although the age of stars is very great, it is limited. Moreover, the age differs greatly for different stars, depending primarily on their mass. But even the most short-lived stars keep almost the same properties for millions of years. What can we infer from this fact?

Even the simplest analysis of the spectra of stars indicates that their outer layers should be in a gaseous state. Otherwise the spectra could certainly never exhibit the sharp absorption lines typical of gaseous material. By a more penetrating analysis of stellar spectra we can gain a much better idea of the properties of the material in the outer layers of stars (the stellar atmospheres), which send radiation to us.

The study of stellar spectra allows us to assert with every confidence that stellar atmospheres represent ionized gas—plasma—heated to thousands or tens of thousands of degrees. Spectrum analysis can establish the chemical composition of stellar atmospheres, which generally is much the same as that of the sun. Finally, by investigating spectra we can also ascertain the density of stellar atmospheres, a quantity that varies over an extremely wide range for different stars. So the outer layers of a star are gaseous.

But these layers contain only a tiny fraction of the mass of the whole star. Although the inside of a star is much too opaque for us to observe by direct optical techniques, we can now state with certainty that the inner layers of stars are likewise in a gaseous state. This claim is by no means obvious. For instance, if we divide the mass of the sun, 2×10^{33} g, by its volume of 1.4×10^{33} cm³, we find that the average density (or specific gravity) of solar material is about 1.4 g/cm³, higher than the density of water. Of course in the central part of the sun the density should be considerably greater than the average. And most dwarf stars have a higher average density than the sun. The question naturally arises, how can our claim that the interiors of the sun and stars are in a gaseous state be reconciled with these high densities? Our answer is that, as we shall shortly demonstrate, the temperature of stellar interiors is so great (far higher than in the surface layers) as to preclude the existence there of matter in a solid or liquid state.

Thus stars are huge globes of gas. It is important to understand that such a gaseous ball is welded together by the universal attractive force of gravitation. Every volume element of the star is subject to the gravitational attraction of all the other elements of the star. Gravitation prevents the various parts of the gas comprising the star from dispersing into the space around it. If there were no force of this kind, the gas making up the star would first spread out, forming something akin to a dense nebula, and then would ultimately diffuse into the vastness of interstellar space surrounding the former star. Let us estimate very roughly how long it would take for a star to spread out in this fashion until it has expanded by, say, a factor of 10. Suppose that the star expands at the thermal velocity of hydrogen atoms (its main constituent) having the same temperature as the outer layers of the star—about 10,000°K. This velocity is approximately 10 km/s, or 10^6 cm/s. Since the radius of a typical star might be a million kilometers, or 10^{11} cm, a spreading process with a tenfold increase in the size of the star would take hardly any time at all: $t = 10 \times 10^{11}/10^6 = 10^6$ s $\approx$ 10 days!

In other words, were it not for the force of gravity stars would disperse into surrounding space in a negligible time (on an astronomical scale), amounting to days for dwarf stars or years for giants. Without universal gravitation, then, there could be no stars. Operating continuously, this force tends to bring together the various elements of a star. In fact, by its very nature gravity tends to bring all the particles of a star together without any limit—to gather up the whole star into a point, as it were. But if universal gravitation were the *only* force acting on the particles that comprise it, the star would at once start to collapse with catastrophic speed.

Let us see how much time would pass before this contraction made significant progress. If gravitation were unopposed by any other force, material would move in toward the center of the star by the laws of freely falling bodies. Consider an element of matter somewhere beneath the surface of the star at a distance R out from its center. This element is subject to a gravity acceleration $g = G\mathfrak{M}/R^2$, where $G = 6.67 \times 10^{-8}$ dyn cm²/g² is the gravitational constant and $\mathfrak{M}$ is the mass inside a sphere of radius R. As the fall toward the center proceeds, both $\mathfrak{M}$ and R will decrease, so that g will also change; but our estimate will not be too far wrong if we suppose that $\mathfrak{M}$ and R remain constant. If we apply to our problem the elementary equation of mechanics relating the path length R traversed in free fall to the gravity g, we will have the first part of Equation 3.6, already derived in Chapter 3:

$$ t = \left(\frac{2R^3}{G\mathfrak{M}}\right)^{1/2} \approx 38 \text{ min,} \qquad (6.1) $$

where t is the falling time and we have taken $R \approx R_{\odot}$, $\mathfrak{M} \approx \mathfrak{M}_{\odot}$. Thus without some force to counteract the gravitation, the outer layers of the star would literally cave in, and the star would catastrophically collapse in a fraction of an hour!

What is the force that operates continuously throughout the star to oppose the force of gravity? Notice that in every elementary volume of the star this force should be directed opposite to the attractive force, but with the same amplitude. Otherwise local violations of equilibrium would occur, leading—in the very short time just estimated—to major changes in the structure of the star.

The force counteracting gravity is the pressure of the gas.[1] Gas pressure continually seeks to expand the star, to disperse it over as large a volume as possible. We have already calculated how rapidly a star would disperse if its separate parts were not held together by the force of gravity. Thus the simple fact that stars are globes of gas which survive practically without change (neither contracting nor expanding) for at least millions of years enables us to conclude that every element of stellar material is in equilibrium under oppositely directed forces of gravitation and gas pressure. This type of equilibrium is called hydrostatic equilibrium; it is a widespread phenomenon in nature. In particular, the earth's atmosphere is in hydrostatic equilibrium under the combined action of the gravitational attractive force of the earth and the pressure of the gases in the atmosphere. If there were no such pressure the earth's atmosphere would very swiftly "fall down" onto the surface of our planet.

A point worth emphasizing is that hydrostatic equilibrium is fulfilled with great accuracy in stellar atmospheres. The slightest deviation from it would at once induce forces tending to alter the mass distribution in the star, redistributing the material in such a way as to restore the equilibrium. We are always speaking here of ordinary, "normal" stars. In exceptional cases, which we shall discuss later, a violation of the equilibrium between the force of gravity and the gas pressure can lead to grave, even catastrophic consequences for a star. For the time being, all we need say is that the life history of any star truly represents a titanic struggle between the force of

---

[1] Properly speaking, the differential in the gas pressure at various depths inside the star.

gravity, which strives to compress it without limit, and the force of gas pressure, which seeks to disperse it away into surrounding interstellar space. The struggle goes on for many millions and billions of years. Throughout these immensely long time spans the forces remain equal. But eventually, as we shall see, gravitation wins out. Quite a drama is the evolution of any star. Soon we shall describe in some detail the separate acts in this drama, corresponding to definite phases of stellar evolution.

Near the center of a normal star, the weight of the matter within a column whose base has an area of 1 cm^2 and whose height is the radius of the star will be equal to the gas pressure at the base of the column. On the other hand, the weight of the column is equal to the force with which it is attracted toward the center of the star. Now we shall perform a very simplified calculation which nonetheless fully reflects the essential features of the problem. Suppose that our column of matter has a mass $\mathfrak{M}_1 = \bar{\rho}R$, where $\bar{\rho}$ is the mean density of the star, and let us take $R/2$ as the "effective" distance between the center of the star and the column as a whole. Then the condition of hydrostatic equilibrium may be written as

$$P = \frac{G\mathfrak{M}\bar{\rho}R}{(R/2)^2} = \frac{4G\mathfrak{M}\bar{\rho}}{R}, \tag{6.2}$$

where $\mathfrak{M}$ is the mass of the entire star.

From Equation 6.2 we can estimate the gas pressure P in the central part of a star such as our sun. Introducing the numerical values of the quantities on the right-hand side, we find that $P = 10^{16}$ dyn/cm^2, or 10 billion atmospheres! So great a pressure is unheard of in our experience. The highest steady pressure attained in laboratories on the earth is of the order of a few million atmospheres.[2]

In elementary physics we learn that the pressure of a gas depends on its density ρ and temperature T. The equation relating all these quantities,

$$P = \frac{A}{\mu}\rho T, \tag{6.3}$$

is sometimes named after the French engineer B. P. É. Clapeyron, who in 1834 was the first to formulate it in this way; here the constant $A = 8.32 \times 10^7$ cm^2 s^{-2} deg^{-1}. In the central regions of normal stars the density ρ_c is of course higher than the mean density $\bar{\rho}$, but not by so very much. Accordingly, Equation 6.3 at once shows us that a high density in stellar interiors would not of itself explain the enormous gas pressure required to satisfy the

---

[2]Note, however, that when a powerful laser beam is focused on a target (which, of course, will instantaneously be vaporized), the evaporating atoms can produce a recoil pressure for 10^{-9} second that reaches 10^{12} atmospheres.

hydrostatic equilibrium condition. The primary demand is for a very high temperature of the gas.

The law 6.3 also contains the mean molecular weight μ. Hydrogen is the principal chemical element in stellar atmospheres, and there is no reason to believe that in the interiors of most stars, at any rate, the chemical composition should differ very greatly from that observed in the outer layers. At the same time, since we do expect a high temperature in the central zone of a star, the hydrogen there should be almost fully ionized, or "split" into protons and electrons. Electrons have a negligible mass compared to protons, but the number of electrons will be equal to the number of protons, so the mean molecular weight of the mixture should be close to $\frac{1}{2}$. Then Equations 6.2 and 6.3 tell us that the temperature in the central regions of stars should be, to order of magnitude,

$$T_c \approx \frac{2\bar{\rho}}{\rho_c} \frac{\mathfrak{M}G}{AR}. \tag{6.4}$$

The ratio $\bar{\rho}/\rho_c$ is a fraction, several percent in typical cases, but dependent on the structure of the stellar interior (see Chapter 11). Simple expressions such as Equation 6.4 show that the temperature in the central regions of the sun should be roughly 10 million degrees. More accurate calculations yield a value differing by only 20 to 30 percent. Thus the temperature near the center of a star is very high indeed, perhaps a thousand times as high as at the surface.

Now let us consider what the properties of matter should be when it is heated to so high a temperature. First of all, despite its high density, such matter should be in a gaseous state. We have already mentioned that; now we can substantiate our claim. The temperature in stellar interiors is so high that, even though the density is also high, the properties of the material there will be almost indistinguishable from the properties of a *perfect gas*—one whose constituent particles (atoms, electrons, ions) interact by collisions only. It is for a perfect gas that the law 6.3 holds, the relationship we are using to estimate the temperature in the central regions of stars.

At a temperature of the order of 10^7 °K and for the corresponding density, all atoms should be ionized. In fact, the mean kinetic energy of a single gas particle, $\frac{1}{2}m\bar{v^2} = \frac{3}{2}kT$, will be about 2×10^{-9} erg or 10^3 electron volts. Every collision of an electron with an atom can therefore ionize the atom, as the binding energy of the electrons in an atom (the ionization potential) is generally less than 10^3 eV. Only the deepest electron shells in heavy atoms will remain intact and be kept by their atoms.

The state of ionization of the material in stellar interiors determines its mean molecular weight—a value, as we have already had occasion to show, which plays a significant role. If stellar material consisted only of fully ionized hydrogen, as we have assumed above, the mean molecular weight μ would be $\frac{1}{2}$. If fully ionized helium were the sole constituent, then $\mu = \frac{4}{3}$, because when a helium atom with its atomic weight 4 is ionized, three particles are formed: the helium nucleus and the two electrons. Finally, if a stellar interior contained nothing but heavy elements (carbon, oxygen, iron, and so on), and if all the atoms were fully ionized, then the mean molecular weight would be close to 2, because for such elements the atomic weight is approximately twice the number of electrons in the atom.

Actually, the material in stellar interiors is a mixture in some proportion of hydrogen, helium, and heavier elements. The relative abundance of these principal components (not by number of atoms, but with respect to mass) is customarily designated by the letters X, Y, and Z, which specify the chemical composition of a star. For typical stars more or less similar to the sun, $X = 0.73$, $Y = 0.25$, and $Z = 0.02$. The ratio $Y/X = 0.34$ means that there are about 12 hydrogen atoms for every helium atom. The relative number of heavy atoms is very much smaller; for example, there are about a thousand times fewer oxygen than hydrogen atoms. Nevertheless, heavy elements make a substantial contribution to the structure of the inner regions of stars, because they strongly affect the opacity of stellar material.

We may now apply the simple expression

$$\bar{\mu} = \frac{1}{2X + \frac{3}{4}Y + \frac{1}{2}Z} \tag{6.5}$$

to determine the mean molecular weight. The role of Z in this estimate for $\bar{\mu}$ is negligible, but the values of both X and Y are decisive. For stars in the middle of the main sequence (and, in particular, the sun), $\bar{\mu} = 0.6$. The value of $\bar{\mu}$ is practically the same for most stars, so we may write a simple equation for the central temperatures of various stars, expressing their mass and radius in terms of the solar mass and radius:

$$T = T_{\odot} \frac{\mathfrak{M}}{\mathfrak{M}_{\odot}} \left(\frac{R}{R_{\odot}}\right)^{-1}, \tag{6.6}$$

where $T_{\odot}$ is the temperature in the central part of the sun. We have given above $T_{\odot} = 1.0 \times 10^7 \, °K$ as a rough estimate; accurate calculations yield $T_{\odot} = 1.4 \times 10^7 \, °K$. Equation 6.6 implies, for instance, that the central temperature of the hot (at the surface!), massive stars of spectral type B is at most two to three times higher than the temperature in the solar interior, while for red dwarfs it is no more than two to three times lower.

An important fact to understand is that a temperature of order $10^7 \, °K$ prevails not only very close to the center of a star but throughout a large volume around the center. Inasmuch as the density of stellar material increases toward the center, one finds that the bulk of the mass within stars is at a temperature which in any event exceeds about 5×10^6 °K. If we further recall that most of the mass of the universe resides in stars, the inference suggests itself that, as a rule, matter in the universe is hot and dense. But we should add that we are speaking of the universe as it is today. In the remote past and future the state of matter in the universe was and will be altogether different. We have alluded to this point in the Introduction.

7

How Do Stars Radiate?

An enormous amount of radiation should pervade the interior of a star, given its temperature of some 10 million degrees and its high density. Photons of this radiation will continually interact with the matter there, being absorbed and reemitted by it. As a result of these processes the radiation field will take on an equilibrium character (strictly speaking, it will be almost in equilibrium—see below); that is, it will be described by the familiar Planck equation with a parameter T equal to the temperature of the medium. For example, the energy density of the radiation at frequency ν within a unit frequency range is expressed by

$$u_\nu = \frac{8\pi h}{c^3} \frac{\nu^3}{e^{h\nu/kT} - 1},$$

(7.1)

while the total radiation density is given by the Stefan–Boltzmann law

$$u = \frac{4\sigma}{c} T^4,$$

(7.2)

which we have met before as Equation 1.6.

Another important quantity describing the radiation field is its intensity, generally designated by the symbol I_ν. The intensity at frequency ν is defined as the amount of energy crossing an area of 1 cm^2 each second in a unit frequency range and entering a solid angle of 1 sr (steradian) in a specified direction perpendicular to the area crossed. If the amount of energy is the same in all directions, it will be related to the radiant energy density by the simple expression

$$I_\nu = \frac{c}{4\pi} u_\nu. \tag{7.3}$$

Similarly, the total intensity I is expressed in terms of the total radiant energy density u by

$$I = \frac{c}{4\pi} u. \tag{7.4}$$

Finally, the radiation flux, denoted by H, is of special significance for the theory of the internal structure of stars. We can define this important quantity as the total amount of energy passing through a fictitious sphere drawn around the center of the star:

$$L = 4\pi r^2 H. \tag{7.5}$$

If energy is generated only in the innermost part of the star, the quantity L will remain constant, independent of what radius r we select. Taking r equal to the radius R of the star, we recognize the meaning of L: evidently it is just the luminosity of the star. The flux H, on the other hand, falls off as r^{-2} as one progresses outward through the star.

If the radiant intensity were strictly the same in every direction (an isotropic radiation field), the flux H would be equal to zero.[1] This fact is easily understood if we note that in an isotropic field the amount of radiation flowing outward through a sphere of any radius will be equal to the amount of radiant energy flowing inward through the same sphere. Under the conditions prevailing in stellar interiors the radiation field is almost isotropic; that is to say, the total intensity I far exceeds the flux H. We can see this directly. According to Equations 7.4 and 7.2, if $T = 10^7$ °K then $I = 2 \times 10^{23}$ erg cm^{-2} s^{-1} sr^{-1}, while the amount of radiation traveling in one direction or the other ("upward" or "downward") will be slightly greater: $F = \pi I = 6 \times 10^{23}$ erg cm^{-2} s^{-1}. On the other hand, in the case of the sun the radiation flux in the central region, say at a distance of 10^5 km (or one-seventh of the solar radius) out from the center, will be $H = L/4\pi r^2$ $= (4 \times 10^{33}$ erg s$^{-1})/(1.3 \times 10^{21}$ cm$^2) = 3 \times 10^{12}$ erg cm^{-2} s^{-1}, a value hundreds of billions of times smaller. The reason is that in the solar interior

---

[1] It is for this reason that, curious though it may seem, the flux of primordial background radiation in the universe is equal to zero.

the outward (upward) radiation flux is almost exactly equal to the inward (downward) flux. *Almost* equal: that is the vital point. For this tiny differential in the intensity of the radiation field governs the whole pattern of radiation of a star. That is why we have stipulated at the beginning of this chapter that the radiation field is *almost* in equilibrium. A strictly equilibrium field would release no radiation flux at all! We would emphasize again that the departures of the actual radiation field in stellar interiors from Planck radiation are wholly insignificant, as is apparent from the smallness of the ratio $H/F = 5 \times 10^{-12}$.

At $T = 10^7$ °K the maximum energy in the black-body Planck spectrum is emitted in the x-ray range. This conclusion follows from the Wien law, a standard relationship in elementary radiation theory:

$$\lambda_m T = 0.288 \text{ cm deg,} \tag{7.6}$$

where λ_m is the wavelength of the maximum in the Planck function. For $T = 10^7$ °K we have $\lambda_m = 3 \times 10^{-8}$ cm $= 3$ Å, a typical x-ray wavelength.

Now the amount of radiant energy confined in the interior of the sun (or any other star) depends strongly on the distribution of temperature with depth, since u is proportional to T^4. From the accurate theory of stellar interiors one can establish such a dependence, and one finds that our sun has a supply of radiant energy of more than 10^{47} erg. If there were nothing to restrain them, these hard x-ray photons would leave the sun in a couple of seconds, and this monstrous flare would undoubtedly incinerate all life on the surface of the earth. That does not happen because the radiation is literally locked up inside the sun. The great thickness of overlying solar material serves as a reliable buffer. Photons of radiation are continually and very frequently being absorbed by atoms, ions, and electrons of the solar plasma, and they can only leak out at an extremely gradual pace. In the course of this diffusion process the photons undergo a major change in their chief property—their energy. Although in the interior of a star, as we have seen, their energy corresponds to the x-ray region of the spectrum, the photons escaping from the stellar surface have become severely "emaciated," as it were; by that time their energy has shifted primarily to the optical range.

A fundamental question arises here: What determines the luminosity of a star, the power of its radiation? Why does a star with immense energy resources expend them so economically, losing to radiation a quite definite but very small part of its supply? We have estimated the supply of radiant energy in the interior of a star. Remember that this energy, as it interacts with matter, is continually being absorbed and then restored in the same amount. The *thermal* energy of the particles of matter in the stellar interior serves as a reservoir of available radiative energy.

It is not hard to estimate the amount of thermal energy stored in a star. For definiteness let us take the sun. Assuming for simplicity that it consists exclusively of hydrogen, and knowing its mass, we readily find that there are about 2×10^{57} particles (protons and electrons) in the sun. At a temperature $T \approx 10^7$ °K the mean energy per particle will be $\frac{3}{2}kT = 2 \times 10^{-9}$ erg, so that the sun has quite a substantial supply of thermal energy: $W_t \approx 5 \times 10^{48}$ erg. At its observed radiant power $L_\odot = 4 \times 10^{33}$ erg/s the sun's supply would last 1.3×10^{15} s or about 40 million years.

Why, then, does the sun have the very luminosity we now observe? Or, stated differently, why does a gaseous sphere in a state of hydrostatic equilibrium with a mass equal to the sun's mass have a perfectly definite radius and a perfectly definite temperature at the surface from which radiation escapes outside? Indeed, we can represent the luminosity of any star, including the sun, by the simple expression

$$L = 4\pi R^2 \sigma T_e^{\,4}, \tag{7.7}$$

where T_e is the temperature of the solar surface.[2] In principle, a sun of the same mass and radius could have a surface temperature of, say, 20,000°K, and then its luminosity would be a hundred times greater. But that is not the case; and, of course, it is no mere accident.

We have spoken of the thermal energy supply in a star. Along with its thermal energy a star also has available to it a considerable reserve of other types of energy. Above all there is its gravitational energy. This energy comprises the energy of gravitational attraction of all the particles in the star for one another. Naturally it represents potential energy of the star, and it has a minus sign. It is equal numerically to the work that must be done in order to overcome the force of gravity and pull all the parts of the star asunder to an infinitely great distance from its center. We can evaluate this energy by finding the energy of gravitational interaction of the star with itself:

$$W_g \approx \frac{G\mathfrak{M}^2}{R} \approx 4 \times 10^{48} \text{ erg}. \tag{7.8}$$

An accurate calculation using simple methods of higher mathematics yields approximately twice this value, and the following relation, known in mechanics as the "virial theorem," is strictly fulfilled:

$$2W_t = -W_g. \tag{7.9}$$

---

[2]Since the radiation emerges from layers with somewhat different depth in the stellar atmosphere and hence slightly different temperatures, T_e refers to an "effective temperature."

Now let us consider a star not in a steady equilibrium state but in a phase of slow contraction (as with a protostar; see Chapter 5). During the contraction process the gravitational energy of the star will gradually decrease (remember that it is negative). However, Equation 7.9 indicates that only half the gravitational energy released will be transformed into heat, that is, will be consumed in heating the material. The other half of the energy released will necessarily leave the star in the form of radiation. It follows thats if gravitational contraction is the source of a star's radiant energy, then the amount of energy it radiates over the course of its evolution will be equal to its thermal energy supply.

Setting aside for the present the important question of why a star should have a perfectly definite luminosity, we must at once stress that if we consider the star's energy source to be the liberation of gravitational energy as it contracts (the opinion held at the close of the nineteenth century), we will be confronted with some very serious difficulties. The trouble is not the fact that in order to achieve the luminosity now observed the sun would have to shrink its radius by about 20 meters annually: the techniques of modern observational astronomy would not be able to detect so slight a change in the size of the sun. Rather, we will be in difficulty because the sun's gravitational energy supply would suffice only for some 40 million years of radiation—provided, of course, the sun had been radiating at about the same rate in the past as now. Although in the mid-nineteenth century, when the famed British physicist William Thomson (later Lord Kelvin) debated the gravitational hypothesis for the maintenance of solar radiation, knowledge about the age of the earth and the sun was very hazy, the same is not true today. Geological evidence permits us to assert with every confidence that the sun is at least several billion years old, a hundred times the Kelvin scale for its life span.

We thereby arrive at the very important conclusion that neither thermal nor gravitational energy is capable of sustaining such prolonged radiation for the sun or indeed the great majority of other stars. Many years ago a third energy source was suggested for the radiation of the sun and stars, and it is of decisive significance for our problem. We refer to nuclear energy (see Chapter 3). In the next chapter we shall discuss in fuller detail the specific nuclear reactions taking place in stellar interiors.

The sun has a supply of nuclear energy $W_n = 0.008 \mathfrak{X} c^2 \mathfrak{M} \approx 10^{52}$ erg that exceeds the sum of its gravitational and thermal energy by more than a factor of 1000. This is the state of affairs for stars generally. With so much energy the sun can keep on radiating for 100 billion years! It might not radiate for such a vast time span at the same level as now, but in any event the sun and stars clearly have ı more than adequate nuclear fuel reserve.

It is worth emphasizing that the nuclear reactions occurring in the interior of the sun and stars are thermonuclear. That is, although high velocity (and hence quite energetic) particles are reacting, they are nonetheless thermal. Gas particles heated to some given temperature have a Maxwellian velocity distribution. At temperatures in the 10^7 °K range the mean energy of the thermal motions of the particles will be close to 10^3 eV. This amount of energy is too small to overcome the Coulomb repulsive force when two nuclei collide, which would allow a nucleus to be struck and thus induce a nuclear transformation. Tens of times as much energy, at the very least, is needed for that to happen. But in a Maxwellian velocity distribution there will always be some particles whose energy is much higher than the average. They will be scarce, to be sure, but only when such particles collide with other nuclei can nuclear transformations take place and energy be released. The number of these anomalously fast but still thermal nuclei depends in an extremely sensitive way on the temperature of the matter.

One might think that under such a circumstance nuclear reactions, accompanied by the liberation of energy, could swiftly raise the temperature of stellar material, whereupon the rate of the reactions would in turn increase sharply, and the star could exhaust its nuclear fuel supply in a relatively short time by becoming more luminous. As a matter of fact, energy cannot accumulate in a star; if it did, the gas pressure would rise abruptly and the star would simply explode like an overheated steam boiler. Hence all the nuclear energy released in the stellar interior ought to escape from the star: this is the process that determines the star's luminosity. But in any event, whatever the thermonuclear reactions may be they cannot operate in the star at any arbitrary rate. As soon as stellar material becomes heated locally, if only to a negligible extent, it will expand because of the rising pressure. As a result, in view of the perfect-gas law 6.3, cooling will take place. The nuclear reaction rate will diminish at once, and the material will thereby revert to its original state. This restoration of hydrostatic equilibrium that has been violated by local heating occurs very rapidly, as we have seen earlier.

Thus the nuclear reaction rate tends to adjust itself to the temperature distribution inside the star. Paradoxical as it may seem, the value of a star's luminosity does *not* depend on the nuclear reactions taking place in the interior. The significance of the nuclear reactions is that they serve to maintain an established temperature regime at the level set by the structure of the star, enabling the star to sustain its luminosity for cosmological time intervals. Accordingly, an ordinary star such as the sun is a splendidly controlled machine that can go on operating under stable conditions for an immense length of time.

128

Now we can return to the fundamental question posed earlier in this chapter: If a star's luminosity does not depend on its energy sources, then what does determine it? To answer this question we must first of all understand how energy is transported from the central part of a star outward to the periphery. There are three principal modes of energy transfer:

1. Heat conduction.   2. Convection.   3. Radiation.

In most stars, including the sun, energy transfer by heat conduction turns out to be utterly inefficient compared to the other mechanisms. The interiors of white dwarfs are an exceptional case; we shall describe them in Chapter 10. Convection occurs whenever thermal energy is transported along with matter. For example, heated gas contiguous to a hot surface may expand; its density will thereby diminish and it will recede from the heated body, simply "floating upward." Cooler gas sinking down to replace it will likewise become heated and rise upward, and so on. If conditions are suitable the process may be quite a violent one. Its role in the innermost regions of comparatively massive stars, as well as in their outer subphotospheric layers, can be very substantial, as we shall see presently. However, the principal mode of energy transfer in stellar interiors is radiation.

We have explained that the radiation field in stellar interiors is *almost* isotropic. Imagine a small volume of stellar material somewhere inside a star; then the intensity of radiation coming from below (outward from the center of the star) will be ever so slightly higher than the intensity from the opposite direction. It is for just this reason that a radiation flux exists within the star. What determines this flux, that is, the difference in intensity between radiation coming from above and below? Suppose for a moment that the material in the stellar interior is nearly transparent. Then the radiation traveling upward through our elementary volume will have come from a place far away, somewhere in the central region of the star. Since the temperature is high there, the intensity will be very strong. On the other hand, the intensity of the radiation traveling downward will correspond to the relatively low temperature of the outer layers of the star. In this imaginary case the intensity difference between upward and downward radiation would be very great, representing a huge radiation flux.

Now consider the opposite extreme: a star whose material is very opaque. Then we can look outward from our elementary volume only to a distance of order $1/\kappa\rho$, where κ is the absorption coefficient per unit mass.[3]

---

[3]The mass absorption coefficient κ is defined as follows. Suppose we have a layer of material with a very small thickness l and a density ρ. After radiation has traversed this layer, its intensity I will have fallen by the amount $\kappa l \rho I$.

In the interior of the sun the value of $1/\kappa\rho$ is about 1 mm. At first glance it may seem strange that gas could be so opaque; for we, immersed in the earth's atmosphere, can see things tens of miles away. The great opacity of the gaseous material in stellar interiors results from its high density and especially its high temperature, which causes the gas to be ionized. Certainly the difference in temperature over a distance of 1 mm should be completely negligible. It can be estimated roughly by regarding the temperature as declining uniformly from the center of the sun to its surface. We then find that the temperature differential along 1 mm of the solar radius should be of the order of 10^{-5} degree. Accordingly, the intensity difference between upward and downward radiation will also be insignificant, and the radiation flux will be negligible compared to the intensity, as we have said before.

We arrive, then, at the important conclusion that the opacity of stellar material determines the radiation flux passing through it, and hence the luminosity of the star. The greater the opacity of the material, the less will be the radiation flux. Furthermore, the radiation flux should also depend on how rapidly the temperature of the star varies with depth. Suppose that we have a hot globe of gas whose temperature is perfectly constant. Obviously in this case the radiation flux would be zero, however great or small the absorption of the radiation may be. Indeed, for any absorption coefficient κ the intensity of radiation coming from above would equal the intensity of radiation from below, since the temperatures are strictly equal.

We can now fully understand the significance of the exact equation relating the luminosity of a star to its basic properties:

$$L_r = 64\pi r^2 \frac{\sigma}{3} \frac{T^3}{\kappa\rho} \left(-\frac{dT}{dr} \right). \tag{7.10}$$

In this equation all the quantities are to be evaluated at a given distance r from the center of the star. The temperature gradient dT/dr represents the amount by which the temperature changes as one moves 1 cm farther out from the center; if the temperature were strictly constant, dT/dr would be equal to zero. Equation 7.10 bears out the statement we have just made: a star will have a high radiation flux (and thus a high luminosity) if the opacity of its material is low and if the temperature falls off rapidly from the center outward.

This equation is useful above all for estimating the luminosity L of a star if we know its basic parameters. But before we mention numerical values let us first transform Equation 7.10. Applying Equations 6.3 and 6.2, we express $T(r)$ in terms of $\mathfrak{M}$ (as in Equation 6.4); we note that $\bar{\rho} = 3\mathfrak{M}/4\pi R^3$, consider a typical interior density $\rho(r) = 5\bar{\rho}$, and replace the gradient $-dT/dr$ by $T(r)/R$. In this manner we obtain the very rough approximation

$$L = 23 \frac{\sigma G^4}{A^4} \frac{\mu^4}{\kappa} \mathfrak{M}^3. \tag{7.11}$$

The significant feature of Equation 7.11 is that the luminosity no longer depends on the radius of the star. Although there is a strong dependence on the mean molecular weight of the material in the stellar interior, for most stars the quantity μ itself is nearly the same. The opacity κ of stellar material depends primarily on the heavy elements present there. In fact, the conditions in stellar interiors are such that hydrogen and helium are fully ionized, and in that state they can absorb very little radiation. For a photon of radiation to be absorbed its energy must be completely expended in detaching an electron from a nucleus—that is, in ionization. With the hydrogen and helium atoms fully ionized, there simply is nothing to detach.[4] The situation is different for the heavy elements. As we have seen, they still retain some of their electrons even in the innermost layers, so they can absorb radiation quite efficiently. Thus although their relative abundance in stellar interiors is low, heavy elements play a disproportionately great role because they are mainly responsible for determining the opacity of stellar material.

Theory yields a simple law, named for the Dutch physicist Hendrik A. Kramers, relating the absorption coefficient to the properties of the material:

$$\kappa \propto \frac{\rho}{T^{7/2}}. \tag{7.12}$$

This relation is quite approximate in character. It nonetheless tells us that we will not make a very serious error if we consider that the value of κ is of the same general order for stars of different types. Accurate calculations show that $\kappa \approx 1$ for hot, massive stars, and is tens of times greater for red dwarfs. In any event, we may conclude from Equation 7.11 that the luminosity of a normal star (one in equilibrium on the main sequence) depends chiefly on its mass. If we introduce the numerical values of the constants into Equation 7.11, it will take the form

$$\frac{L}{L_\odot} \approx 10^3 \frac{\mu^4}{\kappa} \left(\frac{\mathfrak{M}}{\mathfrak{M}_\odot} \right)^3. \tag{7.13}$$

From Equation 7.13 we can obtain a rough idea of the absolute luminosity of a star if we know its mass. In the case of the sun, for example, suppose that we take an absorption coefficient $\kappa \approx 20$ cm²/g and a mean molecular weight $\mu = 0.6$ (see Chapter 6); then we find $L/L_\odot = 6.5$. We should not be too disturbed that $L/L_\odot$ turns out to be different from unity:

---

[4]A special mechanism ("free–free transitions") does exist whereby radiation can be absorbed by a fully ionized gas, but in stars such as the sun this mechanism is unimportant.

our model is extremely crude. When accurate calculations are made, taking into account the distribution of temperature with depth in the sun, the value of $L/L_\odot$ is indeed found to be nearly unity.

The principal meaning of Equation 7.13 is that it demonstrates how the luminosity of a main sequence star depends on its mass. An equation of the type 7.13 is thus generally called a mass–luminosity relation. Once again we emphasize that so vital a parameter of a star as its radius does not appear in this relation. Nor is there any hint that the luminosity of a star depends on the power of its internal energy sources. This last fact is of fundamental significance. As we have explained above, a star of given mass is capable of controlling the power of its energy sources, which in a sense become adjusted to the structure and opacity of the star.

Sir Arthur Stanley Eddington, the illustrious British astronomer to whom we owe the foundations of modern theories for the internal structure of stars, was the first to derive a mass–luminosity relation. He established the relation theoretically; it was confirmed only later, on the basis of extensive observational evidence. This relation, which as we have seen can in effect be obtained from very simple premises, on the whole represents the observations quite well. There are some deviations for very large and very small stellar masses (blue giants, red dwarfs). Further theoretical refinements, however, have removed the discrepancies.

We have stated in Equation 7.10 a relationship between the radiation flux and the temperature gradient. It rests on the assumption that energy is transported away from the stellar interior only by radiation. The condition of radiative equilibrium is satisfied in the interior of such a star. This condition ensures that every volume element of the star will absorb just as much energy as it emits. But an equilibrium of this kind will not always be stable. Let us illustrate with a simple example. Take a small volume element inside the star and imagine that it travels upward a short distance toward the surface. Since both the temperature and pressure of the gas surrounding it will diminish with distance away from the center of the star, our volume element should expand as it moves along. We may consider that no energy is exchanged between our volume element and the surrounding medium during this motion. In other words, the expansion of the volume element as it travels upward may be considered adiabatic. The expansion will proceed in such a way that the internal pressure in the element is always equal to the external pressure of the surrounding medium. If after the displacement we look again at our particular volume of gas, we will find that it either returns to its original position or continues to travel upward. On what does its direction of travel depend?

Figure 7.1 schematically depicts our problem of the elementary gas volume. Quantities describing the volume and the surrounding medium

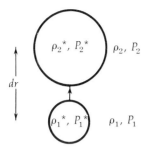

ρ_2^*, P_2^* ρ_2, P_2

dr

ρ_1^*, P_1^* ρ_1, P_1

Figure 7.1   Schematic diagram of gas convection in the interior of a star (M. Schwarzschild).

in their initial state are assigned a subscript 1; for the final state, after the displacement, we use a subscript 2. An asterisk designates quantities that refer to the volume element. Since the initial characteristics of the volume are entirely indistinguishable from the surrounding medium, we will have the equations

$$\rho_1^* = \rho_1, \qquad P_1^* = P_1, \tag{7.14}$$

where ρ and P denote the density and pressure. After the volume element has moved upward (has "undergone a perturbation") with its internal pressure balanced by the pressure of the surrounding medium, its density should differ from the density of that medium. The reason is that as our volume element rises and expands, its density will change in a special fashion called the adiabatic law. As a result of this process we will have

$$P_2^* = P_2, \qquad \rho_2^* = \rho_1 \left(\frac{P_2}{P_1}\right)^{1/\gamma}, \tag{7.15}$$

where $\gamma = c_p/c_v$ is the ratio of the specific heats at constant pressure and at constant volume. For the perfect gas comprising the substance of "normal" stars, $c_p/c_v = \frac{5}{3}$.

And now let us take stock of the situation. After the volume element has drifted upward, the pressure of the surrounding medium acting on it will be equal to the internal pressure, as before, but the gravitational force exerted on the unit volume will have changed, because the density has become different. It is now clear that if the new density is greater than the density of the surrounding medium, the volume element will begin to sink downward until it returns to its original position. But if the density after the adiabatic expansion process has become less than the density of the surrounding medium, the volume element will continue its upward motion, buoyed by an Archimedean force. In the first case the state of the medium will be stable. Any motion of gas that may develop by chance in the medium will be suppressed, and the mass element that had begun to drift away will at once resume its original place. In the second case, however, the state of the medium will be unstable. The slightest perturbation (against which there can never be any insurance) will be amplified more and more. Disordered upward and downward gas motions will develop in the medium. The moving masses of gas will carry along the thermal energy they contain. A state of convection will set in.

Convection is very often observed under conditions on the earth: recall, for example, how water heats up in a teakettle on the stove. Energy transfer by convection differs qualitatively from the radiative energy transfer we have hitherto been discussing. In the radiative case, as we have seen, the amount of energy transported in the radiation flux is restricted by the opacity of the stellar material. If the opacity is very high, for instance, then for a given drop in temperature the amount of energy transported can be indefinitely small. That is not so if energy is transported by convection. The principle of the mechanism itself implies that the amount of energy transported by convection cannot be restricted by any of the properties of the medium.

In stellar interiors energy transfer usually takes place by means of radiation, because the medium is stable against disturbances of its motionless state (see above). But in stellar interiors layers and even whole large regions exist wherein the condition of stability obtained above is not satisfied. In these instances the bulk of the energy is transported by convection. Usually convection occurs when radiative energy transfer is blocked for some reason. This might happen, for example, if the opacity is too high.

We have derived the fundamental mass–luminosity relation on the premise that the energy transfer in stars is accomplished by radiation alone. The question now arises, if a star is also experiencing convective energy transfer, will this mass–luminosity relation break down? No, as it turns out! Fully convective stars, in which energy would be transported only by convection throughout the range from the center to the surface, do not actually exist in nature. Real stars may contain either comparatively thin layers or large regions near the center where convection plays the predominant role. But it is enough to have just one layer inside a star through which energy is transported by radiation in order for the opacity of the layer to have the most radical effect on the throughput of energy released in the stellar interior. The presence of convective regions in the interior does, however, change the numerical value of the coefficients in Equation 7.13. This circumstance is one reason why the solar luminosity given by Equation 7.13 is several times the observed value.

Thus because of the specific instability we have described, large-scale gas motions take place in the convective layers of stars. The hotter masses of gas rise upward while the cooler masses sink. An intensive process of mixing takes place in the material. Calculations demonstrate, however, that the temperature difference between moving gas elements and the surrounding medium is entirely negligible, only about $1°K$; and this is true despite the fact that the temperature of matter in stellar interiors is of the order of 10 million degrees! The explanation is that convection itself tends to balance out the temperature of the layers. Rising and sinking gas elements also have an insignificant velocity, on the average: their speed is only a few tens of meters per second. It is worthwhile to compare this speed with the thermal velocities of ionized hydrogen atoms in the stellar interior, which are of the order of several hundred kilometers per second. Since the gas participating in convection moves at a rate tens of thousands of times lower than the thermal velocities of stellar material, the pressure induced by convective flow is almost a billion times lower than the ordinary gas pressure. Accordingly, convection can have no effect whatever on the hydrostatic equilibrium of the material in stellar interiors, which is established by the balance between the forces of gas pressure and gravitation.

One should not think of convection as some orderly process in which regions of rising gas alternate regularly with sinking regions. The behavior of convective motion is not laminar but turbulent; it is quite chaotic in character, varying irregularly in time and space. This random fashion in which the gas elements move results in a complete mixing of material. As a result the chemical composition in a region of a star pervaded by convective motions should be homogeneous. This last circumstance is of great significance for many problems of stellar evolution. For example, if nuclear reactions in the hottest part of a central convective zone alter the chemical composition there (hydrogen might be depleted, with some of it turning into helium), then in a short time this change would be propagated throughout the whole convection zone. Therefore the nuclear reaction zone—the central region of a star—may continually receive fresh nuclear fuel, a process that is of decisive importance for the evolution of the star.[5] On the other hand, situations may perfectly well arise in which no convection exists in the central, hottest part of a star; then as evolution progresses the chemical composition of that zone will change radically. We shall return to this subject in Chapter 12.

---

[5] In such an event hydrogen would "burn" only inside the convection zone, whereas the outer layers of the star, containing the bulk of its mass, would not mix with the convective core.

135

8

Nuclear Sources of Stellar Energy

Thermonuclear reactions, as we have mentioned in Chapters 3 and 7, are the energy sources that support the radiation of the sun and stars for vast cosmological time spans, billions of years if the mass of the star is not too great. We shall now take a more searching look at this important topic.

The principles underlying the theory of the internal structure of stars were laid down by Eddington at a time when the energy sources had not yet been identified. We have already seen that many important results concerning the equilibrium conditions in stars, the temperature and pressure in their interior, the dependence of luminosity on mass, and the opacity and chemical composition (which determines the mean molecular weight) of stellar material can be obtained without even knowing the nature of the stellar energy sources. Nevertheless, some insight into these energy sources is absolutely essential if we are to explain the long periods that stars persist in a nearly unchanged state. An understanding of the stellar energy sources is still more vital for the problem of stellar evolution—the regular variation in the basic characteristics (luminosity and radius) of a star on a cosmological time scale. Not until the nature of stellar energy sources had become apparent was it possible to comprehend the significance of the Hertzsprung–Russell diagram, which represents fundamental laws of stellar astronomy.

The identity of stellar energy sources came into question almost immediately after the discovery of the energy conservation law, for it then became clear that the radiation of stars is produced by energy transformations of some kind and cannot continue eternally. It is no accident that the first hypothesis about stellar energy sources is due to the German physicist J. Robert Mayer, who in 1842 formulated a principle for the conservation of energy. Mayer suggested that the source of the sun's radiation is a continuous infall of meteors toward its surface. Simple calculations, however, showed that such a source could not possibly furnish enough energy to account for the observed luminosity of the sun. Lord Kelvin and the German scientist Hermann von Helmholtz sought to explain the prolonged radiation of the sun by a slow contraction process accompanied by the release of gravitational energy. While this hypothesis is very important, especially for modern astronomy, it has proved incapable of explaining the persistence of the sun's radiation for billions of years. In Helmholtz and Kelvin's day, though, a reasonable estimate of the age of the sun was not yet available. Only rather recently has it become evident that the sun and its whole planetary system are about 5 billion years old.

Around the turn of the twentieth century one of the greatest discoveries in the history of mankind was made: radioactivity. The completely new world of atomic nuclei was thereby discovered. A decade was hardly long enough, however, for the physics of the atomic nucleus to be placed on a firm scientific basis. By the 1920s it had become clear that the energy source of the sun and stars was to be found in nuclear transformations. But Eddington himself believed that it would be premature to identify the specific nuclear processes occurring in real stellar interiors and liberating the requisite amount of energy. The imperfect state of knowledge at that time concerning the nature of stellar energy sources is evident simply from the fact that Sir James Jeans, the illustrious British physicist and astronomer at the beginning of our century, thought that a suitable source could be none other than radioactivity. This process, of course, is also a nuclear one, but as can easily be shown it is wholly inadequate to explain the radiation of the sun and stars. We need only recognize that such an energy source depends not at all on the external conditions, for radioactivity is a *spontaneous* process. Such a source therefore would have no power to adjust itself to the changing structure of the star. In other words, the radiation of the star could not be regulated. The whole behavior of stellar radiation in such an event would be in sharp conflict with the observations. The first to understand this point was the prominent astronomer Ernst J. Öpik, originally from Estonia, who concluded shortly before World War II that the energy source of the sun and stars could only be thermonuclear synthesis reactions.

It was not until 1939 that the famed German-American physicist Hans A. Bethe gave a quantitative theory for nuclear sources of stellar energy. Just what are these reactions? In Chapter 7 we pointed out that the reactions taking place in stellar interiors ought to be thermonuclear. Let us examine the situation somewhat more fully. Nuclear reactions, accompanied by transformations of nuclei and the release of energy, occur when suitable particles collide. Such particles can first of all be nuclei themselves. But nuclear reactions can also take place when nuclei collide with neutrons. However, free neutrons (those not bound in nuclei) are unstable particles, so we would expect to find only an insignificant number of them in stellar interiors.[1] On the other hand, since hydrogen is the most abundant element in stellar interiors and is fully ionized, collisions of nuclei with protons will take place particularly often.

If in such a collision the proton is to penetrate the nucleus with which it is colliding, it must approach to within nearly 10^{-13} cm of the nucleus. This is the distance at which specific forces of attraction come into play to cement together the nucleus and the incoming stranger, the proton. But in order to come so close to the nucleus, the proton must overcome the very considerable force of electrostatic repulsion (the Coulomb barrier), because the nucleus, like the proton, is positively charged. To surmount this electrostatic force the proton needs a kinetic energy exceeding the potential energy of electrostatic interaction:

$$E = \frac{Ze^2}{r} \approx 2 \times 10^{-6} \text{ erg} \approx 1 \text{ MeV} \tag{8.1}$$

(1 MeV $= 10^6$ electron volts $= 1.60 \times 10^{-6}$ erg).

As mentioned in Chapter 7, however, the mean kinetic energy of thermal protons in the solar interior is only about 1 keV, or 1000 times smaller. Stellar interiors thus contain hardly any protons with enough energy for nuclear reactions. It might seem that in a situation of this kind no nuclear reactions at all could take place. But this is not so: according to the laws of quantum mechanics, protons whose energy is substantially lower than 1 MeV might nevertheless, with some low probability, be able to overcome the Coulomb repulsion force and strike a nucleus. Although this probability diminishes rapidly as the proton energy decreases, it never quite reaches zero. On the other hand, as the energy under consideration approaches the mean thermal energy, the number of protons having such energy increases. Hence there should be a "compromise" energy at which the low probability of protons penetrating a nucleus is compensated by their great

---

[1]In certain rare cases, when a catastrophic process causes a star to explode, reactions involving neutrons can evidently take on great importance.

abundance. One finds that under the conditions prevailing in stellar interiors this energy is about 20 keV. Scarcely 10^{-8} of all protons have this much energy; yet that number turns out to be enough for nuclear reactions to proceed at a rate sufficient to liberate energy corresponding exactly to the luminosity of a star.

We have concentrated our attention on reactions with protons not merely because they are the most abundant constituent of the material in stellar interiors. In encounters between heavier nuclei, which have a much higher charge than the elementary proton charge, the Coulomb repulsion forces become correspondingly greater; even at $T \approx 10^7$ °K the nuclei will have practically no chance of penetrating each other. Only at far higher temperatures (which indeed occur inside stars in certain cases) can nuclear reactions involving just heavy elements take place.

Early in Chapter 3 we indicated that the dominant feature of nuclear reactions inside the sun and stars is the combining, through a series of intermediate steps, of four hydrogen nuclei into one helium nucleus (an α-particle), with the excess mass being released in the form of energy that heats the medium in which the reactions are operating. In stellar interiors there are two modes of converting hydrogen into helium, differing in the sequence of nuclear reactions. The first mode is usually called the proton–proton chain; the second, the carbon–nitrogen cycle.

First let us consider the proton–proton chain. This series of reactions begins with a collision between two protons which yields a nucleus of heavy hydrogen (deuterium). Even in stellar interiors such collisions occur very rarely. Ordinarily, collisions between protons are elastic: after colliding the particles simply fly off in different directions. In order for two protons to merge into a single deuterium nucleus (a deuteron), two independent conditions must be satisfied. First, one of the colliding protons must have a kinetic energy 20 times the mean energy of thermal motion at the temperature of the stellar interior. As we have said, only 10^{-8} of all protons have the relatively high energy needed to overcome the Coulomb barrier. Second, during the collision process one of the two protons must succeed in turning into a neutron, emitting a positron and a neutrino as it does so. For only a proton plus a neutron can form a deuteron! Now a collision lasts just 10^{-21} second—a time span of the same order as the classical radius of a proton divided by its velocity. Taking these facts into account, one finds that any particular proton will have a real chance of becoming part of a deuteron only after some tens of billions of years. But there are so many protons inside a star that reactions of this kind will indeed take place in the number necessary.

The fate of the newly formed deuterons is altogether different. They are greedy; in just a few seconds a deuteron will snap up some nearby proton and turn into the helium isotope He^3. After this event, three nuclear reaction sequences (branches) are possible. Most often the helium isotope will interact with another nucleus like itself to produce a nucleus of ordinary helium together with two protons. Since the He^3 isotope is very scarce, a few million years will be needed for two of them to combine.

Table 8.1 Proton–Proton Chain

Reaction	Energy released MeV	Reaction time* log t, s
$H^1 + H^1 \rightarrow D^2 + e^+ + \nu$	1.44	17.6
$D^2 + H^1 \rightarrow He^3 + \gamma$	5.49	0.7
$He^3 + He^3 \rightarrow He^4 + 2H^1$	12.86	13.5

*Values log t = 0, 5, 10, 15 correspond to $t \approx$ 1 s, 1 day, 300 yr, 30 million years, respectively.

In Table 8.1 we collect this series of reactions and the energy released in each of them. The letter ν designates a neutrino, and γ is a gamma-ray photon. Not all the energy liberated by this reaction chain is transmitted to the star, because some of the energy is carried away by the neutrino. After correcting for this circumstance one finds that the total energy released in the formation of a single α-particle is 26.2 MeV or 4.2×10^{-5} erg.

The second branch of the proton–proton chain starts when a He^3 nucleus combines with an ordinary He^4 nucleus, yielding a nucleus of the beryllium isotope Be^7. The Be^7 nucleus can in turn capture a proton to form a nucleus of the boron isotope B^8, or it can capture an electron and become a lithium nucleus. If the radioactive B^8 isotope is formed, it will experience beta-decay: $B^8 \rightarrow Be^8 + e^+ + \nu$. Incidentally, attempts have been made to detect the neutrinos generated in this reaction by means of unique and expensive equipment. We shall describe this experiment in some detail in the next chapter. The radioactive Be^8 isotope is highly unstable, and it splits quickly into two α-particles. The other alternative, representing the third branch of the proton–proton chain, includes the following links: after capturing an electron the Be^7 nucleus turns into Li^7, which then captures a proton to become the unstable Be^8 isotope, which decays as in the second branch into two α-particles.

We emphasize that in the great majority of cases the proton–proton chain follows the first sequence, but the role of the other branches is by no means small, as we learn in particular from the celebrated neutrino experiment discussed in Chapter 9.

Table 8.2   Carbon–Nitrogen Cycle

Reaction	Energy released MeV	Reaction time log t, s
$C^{12} + H^1 \rightarrow N^{13} + \gamma$	1.94	14.6
$N^{13} \rightarrow C^{13} + e^+ + \nu$	2.22	2.6
$C^{13} + H^1 \rightarrow N^{14} + \gamma$	7.55	13.9
$N^{14} + H^1 \rightarrow O^{15} + \gamma$	7.29	16.0
$O^{15} \rightarrow N^{15} + e^+ + \nu$	2.76	1.9
$N^{15} + H^1 \rightarrow C^{12} + He^4$	4.96	12.5

Now we turn to the carbon–nitrogen cycle. This cycle consists of the six reactions listed in Table 8.2. First a proton collides with a carbon nucleus C^{12}, converting it to the radioactive nitrogen isotope N^{13}. A photon is emitted in this reaction. The N^{13} nucleus undergoes β-decay, emitting a positron and a neutrino to become the carbon isotope C^{13}. This particle collides with a proton and, with the emission of a photon, turns into an ordinary nitrogen nucleus, N^{14}. Then the nitrogen nucleus collides with a proton, forming the radioactive oxygen isotope O^{15} and another photon. Next the oxygen isotope β-decays into the nitrogen isotope N^{15}. Finally, upon acquiring a proton by collision, the N^{15} nucleus splits into ordinary carbon and helium nuclei. The whole reaction chain represents a successive weighting of a carbon nucleus by attachment of protons followed by β-decay processes. The last link in the chain serves to restore the carbon nucleus and to produce a new helium nucleus out of the four protons which had adhered to the C^{12} nucleus and the isotopes formed from it during the process. Notice that the number of C^{12} nuclei in the material in which these reactions occur remains unchanged. Carbon here acts as a catalyst of the reaction cycle.

The second column of Table 8.2 gives the energy liberated at each step of the carbon–nitrogen cycle. Some of this energy is released in the

141

form of neutrinos generated by the decay of the radioactive N^{13} and O^{15} isotopes. Neutrinos can escape freely from the stellar interior, so their energy is not used to heat the material of the star. For example, when O^{15} decays the energy of the resulting neutrino averages about 1 MeV. Altogether 25.1 MeV of energy (not counting the neutrinos) is released to the star through the formation of a single helium nucleus by the carbon–nitrogen cycle, and the neutrinos carry away another 5 percent of the energy.

The rates of the various links in the carbon–nitrogen reaction chain are given by the last column of Table 8.2. For β-decay processes the rate simply corresponds to the half-life. It is considerably more difficult to determine the reaction rate when a nucleus becomes heavier by acquiring a proton. One must know both the probability that a proton will penetrate the Coulomb barrier and the probability of the corresponding nuclear interaction, because the penetration of a proton into the nucleus will not of itself ensure the nuclear transmutation in which we are interested. Nuclear reaction probabilities may be derived from laboratory experiments or calculated theoretically. A reliable determination of these quantities calls for years of laborious effort by nuclear physicists, both theoreticians and experimentalists. The values in the third column represent the lifetimes of the various nuclei in the central region of a star with a temperature of $1.3 \times 10^7 \, °K$ and a hydrogen density of $100 \, g/cm^3$. For example, in order for a C^{12} nucleus to capture a proton under such conditions and turn into a radioactive nitrogen isotope, it must "wait" 13 million years! Thus for each nucleus participating in the cycle the reaction takes place extremely slowly; but the whole point is that enough nuclei are available.

As we have already stressed, the rate of thermonuclear reactions depends in a sensitive manner on the temperature. This behavior is quite understandable: even slight changes in temperature have a strong effect on the density of the energetic protons needed for the reactions—protons whose energy is 20 times the mean thermal energy. In the case of the proton–proton chain, an approximate expression for the rate of energy release per gram of material is

$$\varepsilon \approx 10^{-5} \, \rho X^2 T_6^4 \, \text{erg g}^{-1} \, \text{s}^{-1}, \tag{8.2}$$

where X is the fraction of hydrogen by mass (Chapter 6) and T_6 is the temperature in millions of degrees. This equation holds for the comparatively narrow but important temperature range $T_6 = 11$–16. At lower temperatures, $T_6 = 6$–10, another expression may be used:

$$\varepsilon \approx 10^{-6} \, \rho X^2 T_6^5 \, \text{erg g}^{-1} \, \text{s}^{-1}. \tag{8.3}$$

The main source of energy in the sun, whose central zone has a temperature close to 14 million degrees, is the proton–proton chain. For more massive and thus hotter stars the carbon–nitrogen cycle becomes important, and its temperature dependence is much stronger still. In the temperature range $T_6 = 24$–36, for instance,

$$\varepsilon \approx 3.5 \times 10^{-17} \, \rho X Z T_6^{\,15} \text{ erg g}^{-1} \text{ s}^{-1}. \tag{8.4}$$

It is clear why this equation contains as a factor the relative abundance Z of heavy elements: carbon and nitrogen are among them. These are the elements whose nuclei serve as the catalysts in the carbon–nitrogen cycle. As a rule the combined abundance of these two elements by mass is about one-seventh the abundance Z of all the heavy elements, and this ratio has been incorporated into the numerical factor in Equation 8.4.

Slowly but surely the nuclear reactions continually operating in the central regions of stars change the chemical composition of stellar interiors. The primary trend of this chemical evolution is the conversion of hydrogen into helium. In addition, the carbon–nitrogen cycle alters the relative abundance of the various isotopes of carbon and nitrogen until a certain equilibrium state is established. When equilibrium prevails, the number of reactions per unit time generating some given isotope will be equal to the number of reactions that destroy it. However, a very long time may be needed to establish such an equilibrium. And until equilibrium is established the relative abundances of different isotopes are free to vary over a remarkably wide range.

Here are the equilibrium abundances of some isotopes[2] at a temperature of 13×10^6 °K:

$$\frac{[C^{12}]}{[C^{13}]} = 4.3; \qquad \frac{[N^{14}]}{[N^{15}]} = 2800; \qquad \frac{[N^{14} + N^{15}]}{[C^{12} + C^{13}]} = 20. \tag{8.5}$$

These do not depend on the density of the material, for the rates of *all* reactions are proportional to the density. The first two isotopic ratios 8.5 are also independent of the temperature. The computed equilibrium abundances are subject to an error of tens of percent because the corresponding reaction probabilities are uncertain. In the earth's crust the ratios $[C^{12}]/[C^{13}] = 89$ and $[N^{14}]/[N^{15}] = 270$.

For the proton–proton chain an equilibrium state will not be reached until the enormous time span of 1.4×10^{10} yr has elapsed. Calculations performed for $T = 13 \times 10^6$ °K yield the abundance ratios

$$\frac{[D^2]}{[H^1]} = 3 \times 10^{-17}, \qquad \frac{[He^3]}{[H^1]} \approx 10^{-4}. \tag{8.6}$$

---

[2]The brackets here designate the mass of particles of a given kind per gram of stellar material.

At the rather lower temperature $T = 8 \times 10^6$ °K, the $[He^3]/[H^1]$ ratio is almost 100 times higher, or about 10^{-2}. Thus the He^3 isotope formed in the interior of comparatively cool dwarf stars is very abundant.

Apart from the proton–proton and carbon–nitrogen sequences, other nuclear reactions can also be significant under certain conditions. There are, for example, some interesting reactions of protons with the nuclei of light elements (deuterium, lithium, beryllium, and boron): $Li^6 + H^1 \rightarrow He^3 + He^4$, $Li^7 + H^1 \rightarrow 2H^4$, $B^{10} + 2H^1 \rightarrow 3He^4 + e^+$, and several others. Since the target nucleus with which the proton is colliding has a smaller charge, the Coulomb repulsion is not so strong as in the case of collisions with carbon and nitrogen nuclei. These reactions therefore proceed at a comparatively high rate, even at a temperature of just 10^6 °K. Unlike carbon and nitrogen nuclei, however, the nuclei of light elements are not restored in the course of further reactions, but are irretrievably consumed. This is why the abundance of light elements on the sun and in the stars is so very low. They were long ago "burned up" in the earliest phases of the stars' existence. When the temperature inside a protostar contracting by gravity reaches about 10^6 °K, the first nuclear reactions to operate there are reactions involving light nuclei.

The fact that faint spectral lines of lithium and beryllium are observed in the atmospheres of the sun and other stars calls for an explanation. It might testify to a lack of mixing between the outermost layers of the sun, say, and layers deep enough for the temperature to exceed 2×10^6 °K and for these elements to be burned up. An entirely different interpretation is possible, however. It has now been demonstrated that in active regions (where flares occur) on the sun, charged particles are accelerated to very high energies. Such particles, upon colliding with the nuclei of the atoms comprising the solar atmosphere, give rise to various nuclear reactions.

At the time of the great solar flares in August 1972, the γ-ray detector carried by the specialized American satellite OSO-7 (*Orbiting Solar Observatory*) recorded two lines in that part of the spectrum. One line, corresponding to photons of 0.511-MeV energy, has been identified with the radiation produced when electrons and positrons annihilate each other; the second line, at 2.22 MeV, is emitted when deuterium is formed from protons and neutrons. These important experiments thus show that nuclear reactions are taking place in active regions on the sun and, of course, on other stars too. Such reactions offer the only explanation for the abnormally high lithium abundance in the atmospheres of certain stars and for the occurrence of technetium lines in stars of the rare spectral type S. The most durable technetium isotope has a half-life of about 2×10^5 yr, so the element is missing on the earth. Only nuclear reactions in the surface layers of stars can account for the presence of technetium lines in type S stellar spectra.

If for any reason the temperature in the interior of a star becomes exceptionally high (of the order of 10^8 °K), as may happen after almost all the hydrogen has been burned, a completely new reaction will then serve as the source of nuclear energy. This reaction is called the triple-alpha process. At such high temperatures reactions between α-particles can proceed fairly rapidly, for it becomes easier to surmount the Coulomb barrier. In this case the Coulomb barrier corresponds to an energy of several million electron volts. When collisions occur the barrier can be penetrated efficiently by α-particles whose energy is of the order of 100 keV. The energy of thermal particle motion at temperatures in the 10^8 °K range is about 10 keV. Under such conditions two α-particles can collide to form the radioactive beryllium isotope Be^8. This isotope will very rapidly decay again into two α-particles. But it can happen that a Be^8 nucleus which has not yet managed to decay will collide with a third α-particle—one with high enough energy, of course, to penetrate the Coulomb barrier. The reaction $He^4 + Be^8 \rightarrow C^{12} + \gamma$ will then take place, producing the stable carbon isotope C^{12} along with 7.3 MeV of energy.

Even though the equilibrium abundance of the Be^8 isotope is exceedingly low (at a temperature of 10^8 °K there is only one Be^8 nucleus for every 10^{10} α-particles), the rate of this 3α reaction process is nevertheless adequate to release a substantial amount of energy in the interior of very hot stars. The temperature dependence of the rate of energy release is remarkably strong. For example, at temperatures in the range 100–200 million degrees the 3α process generates energy at a rate

$$\varepsilon \approx 10^{-8}\, \rho^2 Y^3 T_8^{30} \text{ erg g}^{-1} \text{ s}^{-1} \tag{8.7}$$

where, as before, Y is the fractional helium abundance by mass in the stellar interior and T_8 is the temperature in units of 10^8 °K. If nearly all the hydrogen has been burned the value of Y will be quite close to 1. From an energy standpoint hydrogen burning is a more effective process, because in that case 1 g of fuel supplies 10 times as much energy as in the 3α process.

On a logarithmic scale Figure 8.1 shows how the rate of energy production depends on temperature for the three most important reaction processes that can occur in stellar interiors: the proton–proton and carbon–nitrogen sequences and the triple collision of α-particles which we have just discussed. The arrows indicate which nuclear process is most significant for certain stars.

To sum up this chapter, we would say that progress in nuclear physics has given a full explanation of the nature of stellar energy sources.

It is often thought that the world first learned of the bountiful realm of atomic nuclei with A. Henri Becquerel's landmark discovery of radioactivity. And of course there can hardly be any dispute about this point. But through-

145

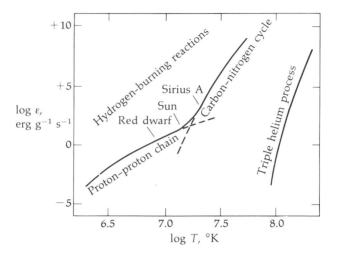

Figure 8.1 Temperature dependence of the rate of nuclear energy release for three reaction processes. From Martin Schwarzschild, *Structure and Evolution of the Stars* (Princeton, N.J.: Princeton University Press, 1958).

out human existence we have bathed in the rays of the sun. To say that the sun is the source of life on earth has long since become a banality. Yet solar rays represent transformed nuclear energy. This means that were it not for the occurrence of nuclear energy in nature there would be no life on the earth. While owing everything to the atomic nucleus, for long millennia people did not even suspect it existed. Still, just to look at something does not mean a discovery has been made. And we would never want to encroach on the fame of the noted French physicist.

Nuclear processes, as we have seen in this chapter, play a fundamental role in the calm, protracted evolution of stars on the main sequence. But nuclear energy is also a decisive factor in the swift, unstable, eruptive processes that mark the turning points of stellar evolution. We shall examine this subject in Part III. Finally, even with respect to so very ordinary and quiet a star as our sun, nuclear reactions would appear to open up for us an opportunity to explain phenomena that are quite remote from nuclear physics. This is our topic in the next chapter.

9

The Problem of Solar
Neutrino Radiation

Until fairly recently one of the most challenging problems in astronomy, the internal structure and evolution of stars, had seemingly succumbed to the joint efforts of theoretical astrophysicists and observational astronomers. As we have emphasized repeatedly, this problem cannot possibly be solved unless the conclusions of theory are continually verified by astronomical observations. Precise observations of the brightness and color of stars belonging to clusters have provided an especially valuable check on the theory (see Chapter 12). It has been and still is believed that the validity of the theory of the internal structure and evolution of stars is borne out by its capability for explaining various subtle features of the Hertzsprung–Russell diagrams for star clusters of differing ages. Nevertheless, a vague feeling unquestionably remains that something is not quite satisfactory. Ideally, we would like a chance to learn about the basic properties of stellar interiors by observing them *directly*.

Not long ago the very thought of peering into stellar interiors would have seemed an altogether fantastic possibility. The enormous thickness of stellar matter makes it opaque to all types of electromagnetic radiation, including even the most energetic γ rays. Photons generated in the central regions of stars by the nuclear reactions occurring there take millions of years to filter through to the surface layers and escape out into interstellar space. Meanwhile the photons interact with stellar material and experience a great many absorptions and reemissions, thereby undergoing major transformations. Originally their frequencies would have been in the x-ray range, but by the time the photons emerge from the stellar surface they will have become far "softer," with frequencies in the optical range and the adjacent infrared and ultraviolet ranges. In other words, their properties will no longer have anything to do with the properties of the medium in which they were once formed. It might seem that we would have no hope of receiving any direct information from the stellar interior. But the marvelous development of physics in our century has brought, in a wholly unexpected way, an opportunity at least in principle for finding a solution to this previously intractable problem.

In 1931 the Austro-Swiss theoretical physicist Wolfgang Pauli, starting from a firm conviction that the conservation laws are satisfied for elementary particles and analyzing the β-decay phenomenon which at that time was not fully understood, put forward the bold hypothesis that a new elementary particle exists. This particle, called the neutrino, would have very remarkable properties. Being electrically neutral, it would have an infinitesimal rest mass, quite likely zero. As a result neutrinos would possess the altogether extraordinary capability of penetrating immense thicknesses of matter. It was calculated that a neutrino beam of 1-MeV energy could pass without appreciable absorption through a steel slab a hundred times thicker than the distance from the sun to the nearest stars! Such particles could go straight through any star as if it were empty space. The weakness of the power of neutrinos to interact with matter is remarkable, but this same coin has a reverse side. Twenty-five years had to pass after Pauli's brilliant theoretical prediction before this unusual particle was detected in a laboratory experiment and thereby emerged from the realm of speculation.

Since that discovery there has been considerable progress in neutrino physics. Like every respectable elementary particle, the neutrino possesses a twin, an antiparticle called the antineutrino. The eminent Italo-Soviet physicist Academician Bruno M. Pontecorvo theoretically predicted the existence of two kinds of neutrino, the "electron neutrino" ν_e and the "muon neutrino" ν_μ. Very soon this prediction was splendidly confirmed by experiment. Pontecorvo was also the first to point out the value of neutrinos for probing stellar interiors, especially that of the sun.

The theory of thermonuclear reactions taking place in the central regions of the sun, the basic principles of which were outlined in Chapter 8, affords quite a reliable estimate of the solar neutrino flux arriving at the earth. In fact, as we have stressed above, the primary nuclear reaction occurring in the interior of our star is the combination of four protons into one α-particle. Two neutrinos are emitted in this process. Each such combination releases to the sun about 26 MeV of energy, which escapes into interstellar space in the form of radiation. Hence the total number of neutrinos produced every second in the solar interior is $N = 2L_\odot/(26\,\text{MeV}) \approx 2 \times 10^{38}\,\text{s}^{-1}$, and their flux density at the earth is $N/4\pi r^2 \approx 7 \times 10^{10}\,\text{cm}^{-2}\,\text{s}^{-1}$. This is a huge quantity. We are literally immersed in a flux of solar neutrinos.

However, the negligibly low interaction probability of solar neutrinos with matter makes it extremely difficult to detect them experimentally. Such an experiment was first proposed in 1946 by Pontecorvo. The detection of neutrinos would be based on the reaction

$$\nu_{\text{solar}} + \text{Cl}^{37} \rightarrow \text{A}^{37} + e^-, \tag{9.1}$$

where Cl^{37} is a stable chlorine isotope and A^{37} is a radioactive argon isotope. This reaction is called inverse β-decay. Although the probability that a neutrino will be absorbed by a chlorine isotope is small indeed, thus far there is no other practical way to detect solar neutrinos.

As a working material rich enough in the Cl^{37} isotope, the transparent liquid perchloroethylene (carbon tetrachloride), whose chemical formula is C_2Cl_4, has been used since 1955. This inexpensive substance is commonly used as a fluid for routine cleaning of surfaces. The first neutrino detection experiments by this technique were aimed not at the sun but at nuclear reactors generating a great many neutrinos. These experiments, devised by the noted American physical chemist Raymond Davis, Jr., were conducted in an effort to distinguish between neutrinos and antineutrinos. Antineutrinos would not be absorbed by the Cl^{37} isotope. Davis used a tank of C_2Cl_4 with the rather small capacity of 3900 liters. The experiment involved estimating the number of nuclei of the radioactive A^{37} isotope formed in the tank of C_2Cl_4. Such an estimate can be made by modern radio-chemical methods.

Even though the underlying purpose of the experiment had nothing to do with astronomy, nevertheless as a byproduct Davis set the first upper limit on the solar neutrino flux, a very crude estimate at that time. The sensitivity of Davis's first experiment was about a thousand times lower than the solar neutrino flux expected in the energy range absorbed by the Cl^{37} isotope.

This last stipulation is very important. We have estimated above the total expected flux of solar neutrinos. But a C_2Cl_4 detector cannot by any means absorb all solar neutrinos with the same efficiency. Moreover, the energy spectrum of solar neutrinos depends very sensitively on the physical conditions in the solar interior—the temperature, density, and chemical composition. In other words, the model adopted for the solar interior strongly affects the energy spectrum of solar neutrinos, and thereby the rate of formation of radioactive A^{37} nuclei in the perchloroethylene.

From 1955 onward Davis and his collaborators have persistently endeavored to improve the sensitivity of their C_2Cl_4 detectors Through their efforts the detector is now 30,000 times as sensitive! In its present form the neutrino detector has become a grandiose facility (Figure 9.1). A giant tank filled with liquid C_2Cl_4 has a volume of about 400 m^3, which is comparable to the volume of a standard 25-m swimming pool. The apparatus is placed at the bottom of a deep old mine pierced through rock in South Dakota. The mine is 1.5 km deep, which is equivalent to shielding by a layer of water about 4.5 km thick. The detector is buried far underground to minimize interference that would produce radioactive argon isotopes without neutrinos having been absorbed by chlorine nuclei. This interference comes from the penetrating component of cosmic rays. Mu mesons belonging to that component interact with matter, generating fast protons that collide with chlorine nuclei to form the radioactive A^{37} isotope.

Today the sensitivity of the neutrino detector is limited mainly by the amplitude of the cosmic-ray background, which is responsible for the "parasitic" A^{37} nuclei. Some idea of the sensitivity of this giant installation can be gained from the fact that irradiation by solar neutrinos can produce only a few dozen radioactive A^{37} nuclei at the same time in the whole vast C_2Cl_4 tank. (In this connection note that the half-life of the A^{37} isotope is about 35 days.) It has been possible to detect this tiny amount of A^{37} by blowing helium through the tank and then chemically separating the argon isotope from the helium. The entire procedure is of course encumbered by serious experimental difficulties, which Davis and his colleagues have successfully overcome.

Surely, then, it is a real paradox that the Davis experiments have yielded a negative result. At the present time all one can say is that the number of solar neutrinos absorbed each second per absorbing chlorine atom is less than 1.0×10^{-36} (a special name, "solar neutrino unit" or SNU, has been given to the number 10^{-36} absorptions). But if the model that has been generally accepted for the solar interior is accurate, there ought to be about 10 times as many absorptions.

This disparity between anticipated result and observational data is surprisingly large. Some of the discrepancy should, of course, be attributed to imperfections in the theories, both the purely physical and the astronomical. The problem of calculating the absorption probability of solar neutrinos by chlorine is a purely physical one. However, this theoretical probability is supported by the results of direct laboratory experiments, so there is no reason to question it. The errors here can hardly be more than 10 percent. But the accuracy of the customary model for the interior regions of the sun is a more serious affair. As we have explained, the model determines the energy spectrum of solar neutrinos, and thereby the number of radioactive

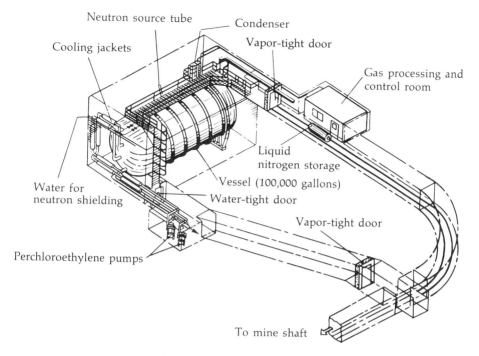

Neutron source tube

Condenser

Cooling jackets

Vapor-tight door

Gas processing and control room

Water for neutron shielding

Liquid nitrogen storage

Vessel (100,000 gallons)

Water-tight door

Vapor-tight door

Perchloroethylene pumps

To mine shaft

Figure 9.1 Neutrino radiation detector (R. Davis, Jr.).

argon nuclei formed in the C_2Cl_4 tank. For example, the rate of neutrino production through β-decay of B^8 (formed in one of the branches of the proton–proton chain—see Chapter 8) depends very strongly on the temperature T, approximately as T^{13}. In fact, the C_2Cl_4 detector records mainly neutrinos generated in the decay of B^8, because these have the highest energy (14 MeV). Yet such neutrinos comprise only a very small fraction of the combined flux of neutrinos at all energies—a quantity almost independent of the solar model.

In principle, theory has now developed to the point where quite an accurate model can be constructed for any main sequence star if we know its mass and its chemical composition at all depths in the interior. For the sun we know the mass very precisely, but there is considerable uncertainty as to the distribution of its chemical composition. The composition depends on the way that matter is mixed inside the sun. Most likely the relative helium abundance is higher in the core of the sun than in the outer layers. The difference between the helium abundance in the central regions and near the periphery also depends on the age of the sun, which may be taken as 4.7×10^9 yr. Laboratory data for the rates of various nuclear reactions occurring in the solar interior are also important for constructing model representations of the sun. For instance, the estimated lifetime of free neutrons was revised in 1967, and laboratory rates of certain nuclear reactions significant in astrophysics have been improved; as a result there has been some change in the rate adopted for the proton–proton reaction chain, the most important thermonuclear process in the solar interior.

Models of the sun proposed in recent years yield a wide variety of values for the number of neutrinos expected to be absorbed in Davis's experiments—from 30 to 6 SNU. But even the lowest theoretical value is still several times the upper limit observed.

Does this strange result from the solar neutrino experiments mean that our ideas about the internal structure and evolution of stars are wrong and need to be fundamentally revised? Thus far there is no serious basis for so far-reaching a conclusion. But the problem of interpreting the outcome of Davis's experiments remains.

To begin with, not all the possibilities of devising solar models have been exhausted. The low value of the neutrino flux recorded by the C_2Cl_4 detector (which reacts, as we have said, primarily to neutrinos formed by the radioactive β-decay of B^8 in a side branch of the proton–proton chain) might, in principle, be explained by supposing that the relative abundance of heavy elements in the solar interior is at least 20 times lower than the abundance observed at the surface. If the heavy element abundance is low the material in the solar interior would become more transparent, the temperature would drop, and hence the flux of neutrinos produced in the decay of B^8 would diminish. A difficulty arises at once, however: on this assumption the primordial helium abundance in the material from which the sun was formed ought to have been several times lower than the helium abundance now observed in the interstellar medium. It is not easy to see how the interior of the sun could have become so deficient in heavy elements compared to its surface. Nevertheless, it is certain that attempts to explain Davis's experiments by various modifications of the solar model will continue, and who knows but what they may eventually be successful.

Another approach toward interpreting the negative result of the experiments for detecting solar neutrinos would be to revise our basic concepts of the nature of the neutrino. It has been suggested, for example, that the neutrino may be an unstable particle. This hypothesis would require that the neutrino have a tiny yet finite rest mass. If we were to assume that the half-life of the neutrino is shorter than a few hundred seconds, then clearly the neutrinos formed in the solar interior would simply fail to reach the earth. Such a hypothesis would, however, demand a radical change in current views of the properties of elementary particles. That is too high a price

152

to pay in order to understand the negative result of Davis's experiments. Hypotheses of this general kind are simply not very realistic.

The problem has been approached in a completely different fashion in a hypothesis advanced by the American physicist William A. Fowler in 1972. Fowler suggests that several million years ago a comparatively rapid, abrupt mixing of material occurred in the inner layers of the sun. Thus the solar interior has been in an unusual, transition state for the past few million years. Several million years from now the physical conditions inside the sun will revert to their former state, as before the abrupt mixing. Without pausing here to inquire into what factors might be responsible for a catastrophic mixing of this kind, let us consider what the consequences would be for the problem of solar neutrinos.

The whole point of Fowler's hypothesis is that the flux of neutrinos from the sun is governed by the instantaneous state of the solar interior. Accordingly, if the temperature in the interior should change for any reason, the flux of neutrinos escaping from the sun would immediately be affected. The sun's flux of photon radiation behaves altogether differently. We have pointed out several times that the photons produced in the central regions of the sun require millions of years to make their way outward and emerge into interstellar space. In principle, then, the following situation could arise: the temperature near the center of the sun suddenly drops, the solar neutrino flux at once falls off sharply, and yet the luminosity of the sun remains unchanged.

Fowler's idea strikes us as a highly fruitful one. It has been developed further by Dilhan Ezer and Alastair G. W. Cameron. If the energy generated by nuclear reactions at the center of the sun should increase sharply for some reason, the solar core would expand swiftly and its temperature would drop. This drop in the temperature of the solar interior would serve to diminish the rate of all thermonuclear reactions. After the excess energy has left the core of the sun, the core would return to its original state and the flux of solar neutrinos would be restored.

How might the rate of energy release in the central part of the sun increase so sharply? It turns out that the infinitesimal amount of the rare helium isotope He^3 present in the solar interior is of great significance here. Ordinarily the concentration of this isotope in the interior is maintained by a dynamic equilibrium between the nuclear reactions leading to its formation and those involved in its destruction. Now as we have pointed out in Chapter 8, the He^3 constituent is vital for the proton–proton chain that operates in the solar interior and generates almost all the sun's luminosity. One finds that as the temperature rises the equilibrium He^3 abundance falls. It follows directly that the equilibrium He^3 density ought to increase outward from the center of the sun. Beginning at a certain distance, however, the He^3 density will no longer be able to rise: the temperature will have become too low for an equilibrium density to have been established during the 5 billion years the sun has existed. Calculations show that the He^3 isotope will reach its maximum density at a distance equal to 0.6 of the sun's radius. Now suppose that for some reason a sudden mixing takes place in the solar interior. This event will raise the He^3 density in the central part of the sun considerably, because material will come down from regions where the He^3 density has been higher. Since the density of the He^3 isotope determines the rate of the proton–proton chain, the energy release will grow sharply and we will have the situation indicated above.

Sudden mixing in the solar interior might be caused by a gradual buildup of some type of instability which, upon reaching a definite limit, is thrown off. For example, the phenomenon might be connected with a meridional circulation of matter in the interior, which would tend to transport angular momentum from the peripheral layers of the sun toward its center. As a result the central regions of the sun would begin to rotate considerably faster than the periphery. A situation of this kind would induce an instability that could be thrown off by mixing. The Japanese aerodynamicist Takeo Sakurai has examined this hypothesis. An important feature of this mechanism for abrupt mixing is its periodicity: after the accumulated instability has been shed it will begin to build up again, since the meridional circulation in the solar interior will continue. Ezer and Cameron estimate that the time interval between these rapid mixing processes in the solar interior should be of the order of 10^8 yr, so that over the evolution of our star such processes would have occurred a few dozen times.

The phase during which sudden mixing causes the temperature of the solar interior to fall below its "normal" value lasts roughly 10^7 yr; hence for about 10 percent of its evolution time the solar interior would be in this minimum state. It would seem that we are particularly fortunate to be living in such an era of the sun's evolution. This remark, as we shall see below, may have a far deeper meaning than appears at first glance.

Ezer and Cameron have performed numerical calculations for the time variation of the solar neutrino flux during the course of such mixing. Figure 9.2 shows their results. This graph indicates that prior to mixing, the "normal" sun would have emitted a neutrino flux corresponding to about 10 SNU in Davis's perchloroethylene detector. In the middle of the mixing phase the flux would drop to a value just slightly below the experimental limit.

However, Ezer and Cameron have not merely developed Fowler's idea in order to make it more specific. They have gone much further. The expansion of the sun's core would inevitably be reflected in its luminosity, that is,

154

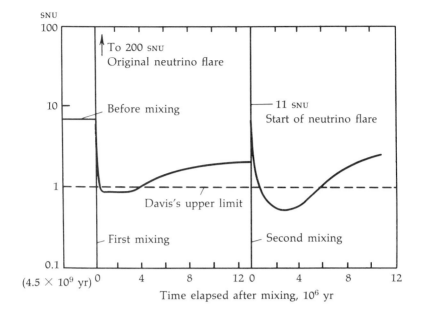

Figure 9.2 Hypothetical time dependence of the solar luminosity
(D. Ezer and A. G. W. Cameron).

in the flux of *photon* radiation from the sun. Moreover, the sun's radius would shrink somewhat. While the surface temperature of the sun would remain almost constant, its luminosity would decrease significantly during the mixing phase. The results of the calculations are displayed in Figure 9.3. Notice that there would be very substantial fluctuations in luminosity. A perfectly natural question arises: Might these cyclic dips in the sun's luminosity manifest themselves in the geological history of the earth?

If Fowler's proposed interpretation of the negative result of the solar neutrino detection experiments is correct, the current level of solar radiation would be well below the "normal" level. Figure 9.3 shows a drop from the "normal" luminosity of the sun corresponding to a decrease in the earth's equilibrium temperature by a factor $(L_1/L_\odot)^{1/4}$, where L_1 is the "normal" luminosity and $L_\odot$ is the present value. Thus the present temperature of our planet should be some 30°C lower than during "normal" periods, when the power of the sun's radiation is close to L_1. Admittedly, the existence of a prominent cloud layer and atmospheric circulation on the earth would considerably smooth out the difference between the earth's mean temperature at "normal" times and today. Even allowing for this circumstance, we would

155

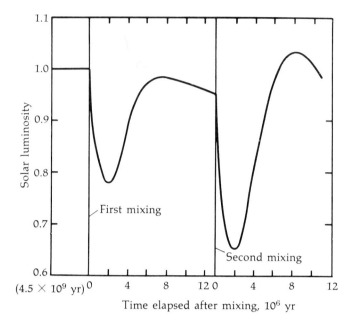

Figure 9.3 Hypothetical time dependence of the solar luminosity (D. Ezer and A. G. W. Cameron).

still have a difference of 10 to 15°C. And this implies that the earth is now going through an *ice age!*

But that is just what is happening! According to geological evidence a glacial period has been in progress on our planet for the past 2 million years. The earth is relatively warm at the moment only because we are living in a brief interglacial period just some 15,000 years long.[1]

Geologists have demonstrated fairly recently that the glaciation of the earth has always been global in character, proceeding simultaneously in the Northern and Southern Hemispheres. It follows that the ice ages must be caused by some cosmic factor. If we are today beginning to comprehend that even terrestrial meteorology is controlled by solar activity, can there be any doubt that the great glaciations of the earth result from much more pronounced changes in the level of solar radiation? We speak of glaciations

---

[1]It is interesting to note that the preceding interglacial period, which occurred on the earth about 100,000 years ago, ended very quickly, even suddenly on a geological time scale. In just a century or so, warmth-loving forms of animal life departed from the waters of Greenland and Newfoundland. There are some grounds for believing that our own interglacial period will come to a close just as rapidly.

in the plural, for it has long been recognized that in the remote geological past (for example, in Archaean time) there were also great glaciations. Such events have been shown to have occurred periodically on our planet every 200 to 300 million years, with ice ages lasting about 10 million years. Ezer and Cameron's development of Fowler's idea predicts just such a behavior.

One cannot help but be struck by the remarkable way in which natural phenomena are interrelated. In a surprising and entirely unexpected fashion the problems of neutrino astronomy can be tied to a very fundamental problem in geology—one which until recently had remained unsolved, despite many attempts.

It is worth reflecting, too, on the fact that the ice age has been the "cradle" of humanity. *Australopithecus* would hardly have culminated the long process of human evolution were it not for the ice age. Even if future developments in science should provide a different explanation for the negative results of Davis's experiments, the ultimate cause of glacial periods that we have described might continue to be the correct one and to impress us with its elegance.

10

How Are White Dwarfs Organized?

When in Chapter 1 we discussed the physical properties of various stars plotted in the Hertzsprung–Russell diagram, we mentioned the objects called white dwarfs. A typical representative of this class of stars is the famous companion of Sirius, known as Sirius B. As we have indicated, these odd stars by no means constitute a rare category of pathological freaks in our Galaxy. On the contrary, white dwarfs form a very numerous group of stars. The Galaxy should contain at least a few billion of them, perhaps as many as 10 billion, or some 10 percent of all the stars in our giant stellar system.

White dwarfs, therefore, ought to have been formed by some kind of regular process which a substantial fraction of stars have experienced. Indeed, our understanding of the world of stars would be very incomplete without an explanation of the nature of white dwarfs and the manner in which they originate. This chapter will not enter into the problems concerning the formation of white dwarfs; we are deferring that topic to Chapter 13. Our aim here is to try to comprehend the nature of these remarkable objects.

The chief peculiarities of white dwarfs are:

1. Their mass is not much different from that of the sun, yet they have a radius a hundred times as small. Their size is comparable to that of the earth.

2. Accordingly, their material is enormously dense, averaging up to 10^6–10^7 g/cm^3, or hundreds of tons pressed into a cubic inch!

3. They are very faint, with a luminosity hundreds or even thousands of times lower than that of the sun.

A first attempt to analyze the conditions in the interior of a white dwarf at once encounters a very serious difficulty. Equation 6.4 establishes a relation among the mass of a star, its radius, and its central temperature. Since the temperature should be inversely proportional to the radius of the star, one might think that the central temperature of a white dwarf would reach a huge value—many hundreds of millions of degrees. At such temperatures an exorbitant amount of nuclear energy should be released there. Even if we suppose that all the hydrogen in a white dwarf has burned up, the triple helium reaction would be highly efficient. The energy liberated in the nuclear reactions would leak out to the surface and escape into interstellar space as radiation, which should be exceptionally strong. Yet the luminosity of white dwarfs is quite insignificant—several orders of magnitude weaker than that of "ordinary" stars having the same mass. What is the matter here?

Let us look into this paradox. First of all, such a severe discrepancy between expected and observed brightness means that Equation 6.4 is simply not applicable to white dwarfs. Remember what basic assumptions we made in deriving that equation. To begin with, the star was considered to be in a state of equilibrium under the action of two forces, gravitation and gas pressure. There can be no doubt that white dwarfs are indeed in a state of hydrostatic equilibrium, which we have discussed in detail in Chapter 6. If they were not they would cease to exist in a short time, dispersing into interstellar space if the pressure exceeded the gravitation, or collapsing almost to a point if the gravitation were not compensated by the gas pressure. Nor can there be any question that the law of gravitation is universally valid; gravity operates everywhere and depends on no other properties but the amount of matter. There remains, then, just one possibility: we may challenge the temperature dependence of the gas pressure that we obtained by adopting the familiar law 6.3.

This law holds for a perfect gas. In Chapter 6 we saw that a perfect gas affords a fairly accurate description of the material inside ordinary stars. Hence it is a natural conclusion that the very dense matter in the interior of white dwarfs cannot be a perfect gas.

In fact, it is quite reasonable to question whether the matter is a gas at all. Could a white dwarf be a liquid or a solid body? We can easily show that this is not the case. Liquids and solids consist of densely packed atoms coming into contact at their electron shells, which are not so extremely small in size—about 10^{-8} cm. Atomic nuclei, wherein nearly all the mass of atoms resides, cannot approach closer to each other than this distance. It follows at once that the mean density of liquid or solid matter cannot be much more than about 20 g/cm^3. Since the mean density of the matter in white dwarfs may be tens of thousands of times higher, the nuclei there must be considerably closer together than 10^{-8} cm. The electron shells of the atoms are "crushed," so to speak, and the nuclei are stripped of their electrons. In this sense we may regard the matter in white dwarf interiors as a very dense plasma. But a plasma is above all a gas, a state of matter in which the distance between the constituent particles is much greater than the size of the particles themselves. In our case the nuclei are at least some 10^{-10} cm apart, while the size of the nuclei is far smaller still, of order 10^{-12} cm.

Thus the material inside white dwarfs is a very dense ionized gas. But because of its enormous density its physical properties differ strikingly from those of a perfect gas. This profound distinction should not be confused with the modestly different properties of real gases, about which much is said in physics courses.

The special properties of ionized gas at ultrahigh densities result from its degeneracy. This phenomenon can be explained only in terms of quantum mechanics; the concept is alien to classical physics. Just what is degeneracy? To answer this question we should first consider briefly how electrons move within an atom, as described by the laws of quantum mechanics. The state of any electron in an atomic system is specified by a set of quantum numbers. These are the principal quantum number n, which determines the energy of the electron in the atom; the quantum number l, which expresses the orbital angular momentum of the electron; the quantum number m, which gives the projection of the angular momentum on some physically defined direction (such as the direction of the magnetic field); and finally the quantum number s, which specifies the intrinsic angular momentum (spin) of the electron.

The fundamental law of quantum mechanics is the *Pauli principle,* which forbids any two electrons in a given quantum system (such as a complex atom) from having all their quantum numbers the same. Let us illustrate this principle by the simple semiclassical Bohr model of an atom. The three quantum numbers other than the spin s determine the orbit of an electron in the atom. Applied to this atomic model, the Pauli principle forbids more than two electrons from following the same quantum orbit. If there are in fact two electrons in such an orbit, they should have oppositely aligned

160

spins. That is, although such electrons may have three coincident quantum numbers, the numbers characterizing their spins should differ.

The Pauli principle is of immense significance for all of atomic physics. In particular, only on the basis of this principle can one interpret all the features of the Mendeleev periodic system of elements. The Pauli principle holds universally and is applicable to all quantum systems comprising a large number of identical particles. Ordinary metals at room temperature are an example of such a system. In metals the outer electrons are not bound to their "own" nuclei but are collectivized; they move in the complex electric field of the ion lattice of the metal. To a rough, semiclassical approximation we may regard the electrons as moving along certain trajectories—very complicated ones, to be sure. And of course the Pauli principle also ought to be satisfied for these trajectories. No more than two electrons, differing in spin, can travel along any of the complicated trajectories. We should emphasize that according to the laws of quantum mechanics, even though the number of possible trajectories is very large, it is nevertheless finite. Thus an electron certainly is not permitted merely to follow *any* geometrical orbit.

Actually, of course, our description is highly simplified. We have spoken of "trajectories" just for intuitive clarity. In place of the classical picture of motion along trajectories, quantum mechanics deals only with the *state* of an electron, as described by several perfectly definite (quantum) parameters. In every possible state the electron has some definite energy. In terms of our model of motion along trajectories, the Pauli principle may also be formulated as follows: no more than two electrons with the same velocity (that is, the same energy) can travel along any given "permitted" trajectory.

For complex atoms with many electrons, the Pauli principle explains why their electrons have not been poured into the deepest orbits—those with a minimum amount of energy. In other words, it gives us a key to understanding atomic structure. The situation is just the same for electrons in a metal as in the case of the material in white dwarf interiors. If the same number of electrons and atomic nuclei filled a sufficiently large volume, there would be room enough for all of them. But now suppose that this volume is restricted. Then only a small fraction of the electrons would occupy all the trajectories along which they are permitted to move, the number of which is necessarily limited. The other electrons would have to travel along the same trajectories that are already occupied. But because of the Pauli principle they would be compelled to move along these trajectories at higher velocities, so they must have a higher energy. This behavior is exactly analogous to that in a multielectron atom, where the Pauli principle forces the extra electrons to move along their orbits with a greater amount of energy.

161

In a piece of metal or in any volume element inside a white dwarf, the number of electrons exceeds the number of allowed trajectories of motion. Affairs are different in an ordinary gas, as for example in the interiors of main sequence stars. The number of electrons there is always fewer than the number of allowed trajectories. Hence the electrons can travel along different trajectories at different velocities without impeding one another. In this case the Pauli principle is not reflected in the motion of the electrons: such a gas will have a Maxwellian velocity distribution and will obey the standard laws of elementary physics for the gaseous state of matter, including the perfect-gas law. If ordinary gas is strongly compressed, the number of possible electron trajectories will become much smaller, and eventually a condition will arise in which every trajectory must have more than two electrons. Because of the Pauli principle these electrons must have different velocities, in excess of some critical value.

If we now cool this compressed gas, the velocities of the electrons will by no means diminish, for otherwise the Pauli principle would clearly be violated. Even near absolute zero the velocities of the electrons in such a gas would remain high. Gas possessing such distinctive properties is said to be degenerate. The behavior of a degenerate gas is fully explained by the circumstance that its particles (electrons in our case) occupy all possible trajectories and travel along them at velocities as high as necessary. Unlike the degenerate-gas case, the velocities of particles moving in an ordinary gas become very low as the temperature drops, and there is a corresponding decrease in the pressure. How does the pressure behave in a degenerate gas?

To understand this point let us recall what we mean by gas pressure. It is the momentum that the gas particles impart every second to some surrounding wall with which they collide. Clearly, then, the pressure of a degenerate gas should be very high, since its constituent particles have such high velocities. Even at very low temperatures the pressure of the degenerate gas should remain high, for the velocities of its particles hardly decrease at all as the temperature falls. We would expect the pressure of a degenerate gas to depend little on its temperature, as the velocity of its constituent particles is determined mainly by the Pauli principle.

In addition to electrons, white dwarf interiors should contain stripped nuclei as well as preserved inner electron shells of highly ionized atoms. Here one finds that the number of allowed trajectories is always greater than the number of particles. Therefore these particles comprise not a degenerate but a normal gas. Their velocities are determined by the temperature of the material in the white dwarfs and are always much lower than the electron velocities governed by the Pauli principle. Hence the pressure in white dwarf interiors is due entirely to the degenerate electron gas. It follows that the equilibrium of a white dwarf is almost independent of its temperature.

Quantum-mechanical calculations show that the pressure of a degenerate electron gas, expressed in atmospheres, is given by the equation

$$P = K\rho^{5/3}, \tag{10.1}$$

where the constant $K = 3 \times 10^6$ and the density ρ is expressed, as usual, in grams per cubic centimeter. For a degenerate gas Equation 10.1 replaces the perfect-gas law and serves as the equation of state. One noteworthy property of this equation is that the temperature is missing from it. Furthermore, unlike the perfect-gas law 6.3, in which the pressure is proportional to the first power of the density, the dependence of pressure on density is stronger here. This circumstance is readily understood. The pressure is proportional both to the number of particles per unit volume and to their velocity. The former quantity is of course proportional to the density, while in a degenerate gas the particle velocity increases with the density, because according to the Pauli principle at higher densities there will be a greater number of extra particles compelled to move at high velocities.

In order for Equation 10.1 to be applicable, the thermal velocities of the electrons must be small compared to the velocities due to the degeneracy. At very high temperatures Equation 10.1 should be replaced by the perfect-gas law 6.3. If Equation 10.1 gives a higher pressure for a gas of density ρ than does Equation 6.3, the gas will be degenerate. We thereby have the degeneracy condition

$$K\rho^{5/3} > \frac{A}{\mu}\rho T$$

or

$$\rho > \left(\frac{A}{\mu K}T\right)^{3/2}, \tag{10.2}$$

where μ is the mean molecular weight.

What is the value of μ in the interiors of white dwarfs? First of all, hardly any hydrogen should be present there; at such enormous densities and for high enough temperatures hydrogen would long since have been burned in nuclear reactions. The main element in white dwarf interiors should be helium. Since helium has an atomic weight of 4 and yields two electrons when ionized (remember also that the particles responsible for the pressure are only the electrons), the mean molecular weight ought to be very close to 2. Numerically, the degeneracy condition 10.2 takes the form

$$\rho > \left(\frac{T}{75,000°K}\right)^{3/2} \text{ g/cm}^3. \tag{10.3}$$

If, for example, the temperature $T = 300°K$ (room temperature), then $\rho > 2.5 \times 10^{-4}$ g/cm³. This is a very low density; hence we may at once conclude that the electrons in metals should be degenerate (actually in that instance the constants K and μ have different values, but the principle remains the same). If the temperature T is representative of that in stellar interiors (about 10 million degrees), then $\rho > 1000$ g/cm^3. Two conclusions follow immediately:

1. In the interiors of ordinary stars, where the density while high is certainly lower than 1000 g/cm^3, the gas is not degenerate. Accordingly, the conventional laws of the gaseous state, applied consistently in Chapter 6, are justified.

2. In white dwarfs the mean density, and even more so the central density, is much higher than 1000 g/cm^3. Hence the usual laws of the gaseous state are not applicable there.

To understand white dwarfs we must know the properties of degenerate gas described by its equation of state 10.1. This equation first of all implies that the structure of white dwarfs is practically independent of their

Figure 10.1   Relation between the mass and radius of white dwarfs.

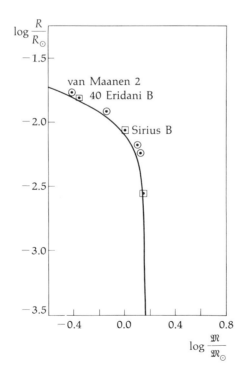

temperature. But since the luminosity of these objects is determined by their temperature (for example, the temperature governs the rate of thermonuclear reactions), we may infer that the structure of white dwarfs is also independent of their luminosity. In principle, a white dwarf can exist (that is, it can survive in an equilibrium configuration) even at temperatures close to absolute zero. We therefore conclude that unlike ordinary stars, white dwarfs do not conform to a mass–luminosity relation.

These singular stars, however, do obey their own mass–radius relation. Just as balls of equal mass made out of some particular metal should all have equal diameters, white dwarfs of the same mass should also have the same size. This statement clearly is incorrect for other stars: giant stars and main sequence stars may have equal masses but very different diameters. White dwarfs differ from other stars in this respect because temperature plays hardly any role in their hydrostatic equilibrium, which in turn determines their structure.

Since this is the case, there should be some universal relation between the mass of a white dwarf and its radius. It would take us too far afield to derive this important relationship, which is by no means an elementary one. The relation itself is displayed on a logarithmic scale in Figure 10.1. In this diagram the circles and squares mark the position of several white dwarfs whose masses and radii are known. The mass–radius relation shown in Figure 10.1 has two curious properties. In the first place, the more massive a white dwarf is, the *smaller* its radius. Here is a way in which white dwarfs behave oppositely from solid balls of metal. Second, the mass of a white dwarf can be no higher than a certain limit.[1] Theory predicts that white dwarfs cannot exist in nature if their mass exceeds 1.43 solar masses.[2] If the mass of a white dwarf approached this critical value from the low side, its radius would shrink to zero. In effect, beginning at a certain mass the pressure of the degenerate gas can no longer balance the force of gravity and the star will collapse catastrophically.

This result, of exceptional significance for the whole problem of stellar evolution, merits more careful examination. As the mass of a white dwarf increases its central density will continue to rise. The electron gas will become progressively more degenerate. This situation means that an increasing number of particles will travel along the same allowed trajectory. They will be closely crowded there, and in order not to violate the Pauli

---

[1]This fact was first pointed out in 1928 by the Soviet theoretical physicist Yakov I. Frenkel'. Two years before that the British physicist Sir Ralph H. Fowler became the first to apply the theory of degenerate gas for explaining the nature of white dwarfs.

[2]Corrected for the neutronization of matter at high density (Chapter 22), this limit drops to 1.2 $\mathfrak{M}_\odot$.

principle the electrons will move at higher and higher velocities. These velocities will eventually become quite close to the speed of light. Matter will enter a new state, known as relativistic degeneracy. Such a gas has a different equation of state, no longer described by Equation 10.1. That equation will be replaced by a law of the form

$$P \propto \rho^{4/3}. \tag{10.4}$$

To assess the new situation let us take $\rho \approx \mathfrak{M}/R^3$, as in Chapter 6. Then in a relativistically degenerate star $P \propto \mathfrak{M}^{4/3}/R^4$, while the force resisting gravitation and equal to the pressure drop will obey the law

$$\frac{P}{R} \propto \frac{\mathfrak{M}^{4/3}}{R^5}. \tag{10.5}$$

But the gravitational force $\rho G \mathfrak{M}/R^2 \approx \mathfrak{M}^2/R^5$. Notice that both forces— gravity and the pressure drop—depend in the same manner on the size of the star, as R^{-5}, but they depend differently on the mass. We would therefore expect some perfectly definite value of the stellar mass to exist at which the two forces are in balance. If the mass should exceed some critical value, the force of gravity would always predominate over the force due to the pressure drop, and the star would undergo catastrophic collapse.

Now suppose that the mass is below the critical value. Then the pressure force will exceed the gravitational force and the star will begin to expand. As it expands the relativistic degeneracy will be replaced by the ordinary nonrelativistic degeneracy. In this event the equation of state $P \propto \rho^{5/3}$ implies that $P/R \propto \mathfrak{M}^{5/3}/R^6$, so that the force opposing gravity will depend more strongly on R. Thus at a certain value of the radius the star will cease to expand.

The qualitative analysis we have given shows that a mass–radius relation must exist for white dwarfs, with the radius decreasing toward higher masses. It also establishes the existence of a limiting mass, an inescapable consequence of the onset of relativistic degeneracy. How far can a star whose mass exceeds $1.2\ \mathfrak{M}_\odot$ contract? This is an absorbing problem that has aroused great interest in recent years; we shall return to it in Chapter 24.

The material inside white dwarfs is noteworthy for its high transparency and heat conductivity. Its transparency once again is explained by the Pauli principle. Light is absorbed by matter when electrons experience a change in state through transitions from one orbit to another. But as the great majority of trajectories or orbits in a degenerate gas are occupied, such transitions are very difficult. Only a very few particularly fast electrons in the plasma of a white dwarf can absorb photons of radiation. The high heat conductivity of degenerate gas is exemplified by ordinary metals.

Because of the high transparency and conductivity, significant temperature differentials cannot arise in the material of a white dwarf. Nearly the whole change in temperature from the surface of a white dwarf in toward its center occurs in a very thin outer layer of matter which is in a nondegenerate state. Within this layer, whose thickness is roughly 1 percent of the radius of the star, the temperature jumps from a few thousand degrees at the surface to about 10 million degrees, but then it remains almost the same all the way to the center.

White dwarfs do radiate, even though only weakly. What is the source of this radiant energy? As we have pointed out, hydrogen, the main nuclear fuel, is virtually absent from the interiors of white dwarfs. It has nearly all been burned during the evolutionary phases of the star preceding the white dwarf stage. But spectroscopic observations clearly demonstrate that the outermost layers of white dwarfs contain some hydrogen. Either it has not managed to burn up or, more likely, it has fallen onto the surface from the interstellar medium. The energy source of white dwarfs might possibly be hydrogen nuclear reactions taking place in a very thin spherical layer at the interface between the atmosphere and the dense degenerate matter in the interior.

Furthermore, white dwarfs can maintain a fairly high surface temperature by ordinary heat conduction. As a result, lacking internal energy sources white dwarfs will tend to cool, emitting radiation at the expense of their heat reserves. But these reserves are very substantial. Since the motion of electrons in white dwarf material represents a degeneracy phenomenon, the heat in the stellar interior will be stored in nuclei and ionized atoms. Assuming that white dwarf material consists primarily of helium (atomic weight 4), we can readily find the amount of thermal energy stored in such a star:

$$E_t = \frac{3}{2}kT\frac{\mathfrak{M}}{4m_H},\tag{10.6}$$

where m_H is the mass of the hydrogen atom and k is the Boltzmann constant. We can estimate the cooling time of a white dwarf if we divide E_t by the luminosity L. It turns out to be many hundreds of millions of years.

For a number of white dwarfs Figure 10.2 illustrates how the luminosity depends empirically on the surface temperature. The straight lines represent geometrical loci of constant radius, expressed in terms of the solar radius. The empirical points seem to follow these lines quite closely, indicating that observed white dwarfs are in different stages of cooling.

Recently the absorption lines in the spectra of two white dwarfs have been found to be sharply split by the Zeeman effect. The amount of splitting

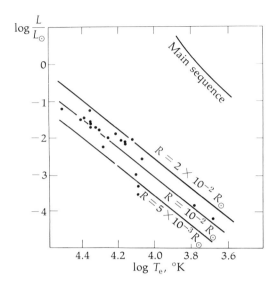

Figure 10.2   Empirical dependence of the luminosity of white dwarfs on their temperature (from data by D. L. Harris).

implies that the magnetic field strength at the surface of these stars reaches the huge value of around 10^7 gauss. This tremendously strong magnetic field evidently results from the conditions under which the white dwarfs were formed. For example, if we suppose that a star contracts without an appreciable loss of mass, we may expect that the magnetic flux (the product of the star's surface area by its magnetic field strength) will maintain a constant value. Accordingly, as the star contracts its field strength will grow in inverse proportion to the square of its radius. In this way the field may become hundreds of thousands of times stronger.[3] Such a mechanism for strengthening the magnetic field is especially important for neutron stars, our subject in Chapter 22.

---

[3]Because of the presence of a strong magnetic field, the radiation of white dwarfs should exhibit a slight circular polarization. By studying the time dependence of this polarization one can, in principle, determine the rotation periods of white dwarfs. In the few cases in which these highly delicate observations have been carried out, comparatively lengthy periods of axial rotation have been measured, of the order of a day. This result should be significant for the problem of stellar evolution.

11

Model Stars

In Chapter 6 we described the basic properties of stellar interiors (temperature, density, and pressure) by using a method for obtaining rough estimates of the quantities that appear in the equations expressing the equilibrium state of stars. These estimates afford a good idea of the physical conditions in the central regions of stars, but they are of course completely inadequate for understanding how various stars resemble and differ from one another.

For example, in order to decide the important question of which nuclear reaction process (proton–proton or carbon–nitrogen) is responsible for the radiation of a given star, we need fuller information on the conditions in its interior. And there remains a fundamental problem that we have not yet considered: what is the physical meaning of the Hertzsprung–Russell diagram? This problem, as we shall see below, is intimately related to the issue of stellar evolution. Although the nuclear energy supply in the interior of a star is very large, it is nevertheless finite. Sooner or later, depending on the mass of the star, it will come to an end. What will happen to the star then? How will its properties change?

To understand the relationship between various kinds of stars and the explanation of the distinguishing features we observe, it is necessary to have a thorough knowledge of the instantaneous state of different stars, like a snapshot of their internal structure. In the same way as a real physical process, the very gradual process whereby a star evolves (due to depletion of its nuclear fuel supply) may be thought of as a sequence of quasi-steady equilibrium states. Such configurations, obtained theoretically by calculation, are called models of a star.

By a stellar model we mean a collection of tables or graphs representing idealized distributions of the density, temperature, pressure, and chemical composition of material at various depths in the star, expressed as a fraction of its radius. Notice that a model of this kind is far from being identical with any actual star. Even so, a properly calculated model, correctly incorporating the basic physical laws that govern the structure of a star, can and should provide an essentially faithful picture of the properties of matter in stellar interiors. It would be unfair to claim that the calculation of model stars inherently contains an element of arbitrariness. Quite the contrary: rigid checks are continuously made in the course of the calculations themselves. And ultimately, after it has been completed, the calculation should be fully consistent with the observed properties of the stars being modeled. For instance, if we are calculating a model for a main sequence star, the theoretical model should fit the mass–luminosity relation.

If we could observe the internal regions of stars directly, we would not have to build models for them. In the case of nebulae, which we can look through, the structure can be deduced at once from optical and radio observations. But alas, stellar interiors are hidden from us by a vast thickness of material, and there is almost no chance to perceive what is going on there. We have included the word *almost* because there still remains one hope for direct observation of stellar interiors, as explained in Chapter 9. Building model stars is a compulsory task, for otherwise we would be unable to draw quantitative conclusions about the main trends of development for a large part of the matter in the universe.

How are model stars calculated? Such calculations are founded above all on the physical laws determining the equilibrium configuration of a star. We discussed these laws in Chapters 6 and 7. First, there is the condition of hydrostatic equilibrium, which must be satisfied for every volume element inside the star (see Equation 6.2). Then there is the condition of radiative equilibrium, which describes the transfer of radiation from the interior of the star to its surface (see Equation 7.10). We must also take into account the way in which the opacity of stellar material depends on the temperature and density, as well as the dependence of the pressure on temperature and density—that is, the equation of state. For the material in normal stars the equation of state is the perfect-gas law 6.3; for white dwarfs it is given by Equation 10.1. Further, one should allow for the very strong temperature dependence of the rate of nuclear energy release (see Equations 8.2 to 8.4). Such fundamental parameters of model stars as their mass, luminosity, and radius are specified at the outset.

The system of equations describing the state of a star is so complicated that models cannot be calculated analytically, that is, by a straightforward even if very cumbersome formula. Success can be achieved only by the *numerical* method of solving these differential equations. The model star is assumed to be spherically symmetric, so that all the properties of any of its volume elements (temperature, density, and so on) will depend merely on the distance of that element from the center of the star.

What is behind this concept of a numerical calculation method? Imagine that our star is made up of a great many concentric spherical layers. Inside each layer, provided it is taken to be thin enough, all the parameters may be regarded as having constant values. Let us specify the pressure and the density at the center of the star. The conditions of hydrostatic equilibrium will then allow us to find the pressure at the surface of the first (innermost) sphere. From the perfect-gas law we calculate the temperature at the center. Knowing how the rate of nuclear energy release depends on temperature, and applying the equation of radiative energy transfer (Equation 7.10), we can now determine the temperature at the surface of the spherical zone, and thereby the density from the perfect-gas law. Such a procedure (a rather intricate one, as we shall see) enables us to stipulate the temperature, density, and pressure at the center of the star and obtain the same basic parameters at some relatively small distance out from the center. Afterward we repeat the procedure by the same method and evaluate the parameters of the stellar material at the surface of a second sphere whose radius is twice that of the first.

Thus step by step we derive a profile of the whole star, a set of values describing the main properties of its material at various distances from the center. In order for the model calculation to meet with success, the thickness of the imaginary spherical layers into which we subdivide the star must be sufficiently small. On the other hand, of course, it is impractical to make them too thin, for the labor of the calculations would become excessive. In practice the number of layers is generally a few hundred, sometimes even a few thousand.

The mass of the theoretical model is obtained by adding up the partial masses contained in the elementary spherical shells. By taking into account the production of thermonuclear energy in the various layers, one can find at the end of the calculation the theoretical luminosity of the model star.

These model calculations were once performed manually on desk calculating machines. But for the past two decades models have been calculated primarily on electronic computers. The consequent great increase in efficiency has not only facilitated the numerical work but has also permitted a wide variation of the parameters entering into the calculation, so that the values yielding the most reasonable and consistent models may be selected. In particular, a consistency criterion for a model of any main sequence star, implying certain values for the radius, mass, and luminosity of the star, is the fulfillment of the mass–luminosity law.

But how can the calculations result in models that manifestly do not simulate real stars? To a large extent this may happen because we are so uncertain about the chemical composition in the interior of the star whose model we are calculating. The calculations have to be performed by trial and error, with assumptions about the chemical composition being rejected if they lead to absurd results.

There is another rather specific reason for discrepancy between the fundamental computed parameters (mass, luminosity, and radius) of a model star and the observed parameters of the corresponding real star. Under certain conditions the character of the energy transfer process in the stellar interior may change. For example, radiative energy transfer may be replaced by convection. Various factors may be responsible for such behavior. If, for instance, the temperature begins to rise fairly sharply as one penetrates down into the star, then radiation will no longer suffice to carry away all the energy generated in the stellar interior. Instability will set in and convection will become the predominant energy transfer mechanism. (We mentioned this point in Chapter 8.) Accordingly, during the step-by-step calculations one must follow the process attentively and watch the behavior of the energy transfer mechanism as one builds the model star.

One must also keep in mind that the chemical composition of a star, as described by the parameters X, Y, Z (Chapter 6), does not remain constant throughout the star but may undergo systematic and indeed radical changes in different zones (see Chapter 7). In comparatively old stars, for example, X may be substantially smaller in the central part of the star than at the periphery because of hydrogen burning. Model stars that allow for this circumstance are said to be inhomogeneous. Such models are especially interesting because they offer a good representation of reality.

In illustrating how a model star is constructed we have considered a process of building outward from the center. It is also possible, and often more convenient, to calculate models from the surface in toward the center. We would then specify the radius and luminosity (or temperature) of the star. Naturally, when the calculation is finished the combined mass of the spherical layers should equal the mass of the star. If the calculation method is incorrect, the mass of the model star may be exhausted long before the calculation has reached the center. The author once encountered this curious phenomenon in the work of some students who were beginning to specialize in the field of internal stellar structure.

Thanks to extensive efforts by theoretical astrophysicists working in this specialty, many stellar models are now available. These models describe stars occupying various places in the Hertzsprung–Russell diagram. We shall next discuss the basic properties of models corresponding to the stellar population of our Galaxy. Models of stars on the main sequence in the Hertzsprung–Russell diagram are of special interest. Stars near the top of this sequence (hot, massive objects of high luminosity) turn out to differ markedly in structure from the red dwarfs occupying the lower right-hand end.

Figure 11.1 schematically illustrates the structure of a hot, massive star. This model has been calculated for a star with a mass 10 times that of the sun, a radius 3.6 times as great, and a bolometric luminosity 3000 times the solar luminosity. The model star therefore is of spectral type B0, and its surface temperature is about 25,000°K. In the central part of the star, as the calculations demonstrate, energy is transported by convection. The convection zone extends outward to about 25 percent of the star's radius, and it contains about 25 percent of the total mass of the star.

One characteristic feature here is the rather high density concentration toward the center. At the center itself the density is approximately 25 times the average density of the star and is close to 7 g/cm^3. The central temperature is fairly high, about 27×10^6 °K, or approximately twice that of the sun. Qualitatively the simple expression 6.4 suggests a higher central temperature than for the sun, although the model calculation gives a different value for the temperature. The model does not depend very strongly on the chemical composition adopted for the star ($X = 0.90$, $Y = 0.09$, $Z = 0.01$).

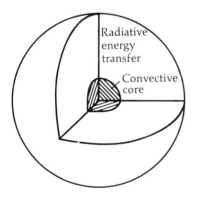

Figure 11.1 A model of a massive main sequence star.

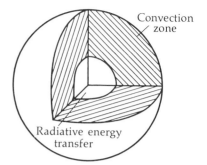

Figure 11.2   A model of a red dwarf.

It is interesting to see how the structure of stars of this type will change if we decrease the mass. Calculations demonstrate that the central temperature will drop, in accord with Equation 6.4, and furthermore that the relative size of the convective core will shrink. The main energy source in such stars is the carbon–nitrogen cycle. As we have seen (Equation 8.4), the rate of this process grows very sharply with temperature. Radiative transfer alone is no longer adequate to remove the enormous amount of energy released in this reaction cycle. Convection becomes necessary to take care of the transport of energy. This circumstance explains why such stars have more or less extended convective cores in their central zones.

The diagram in Figure 11.1 represents the structure of a typical star on the upper part of the main sequence. Figure 11.2 schematically shows a model for a red dwarf whose mass is 0.6 solar units, luminosity 0.56, and radius 0.64. We are dealing, then, with a model dwarf star of spectral type K. Notice that the structure of such a star differs fundamentally from the structure of hot, massive stars on the upper main sequence. To begin with, in the central part of a dwarf star we find no convection zone at all. On the other hand, in the outer layers of such stars energy is transported primarily by convection. For the model depicted in Figure 11.2, the convection zone occupies the outer part of the star, and about 10 percent of all the mass is concentrated there.

We can understand why the stars on the lower part of the main sequence differ structurally from the hot and massive stars by noting the comparatively low temperature in dwarf star interiors. In a cool star the opacity of the material is greater, and it becomes difficult for radiation to transport the energy produced in the central region of the star. Convection comes to the rescue. The concentration of mass toward the center is not so high in dwarf stars as in hot giants; the central density is now only 20 times the mean value, but in absolute terms the central density is far higher, about 60 g/cm^3. In agreement with Equation 6.4, the model calculation yields a relatively low temperature at the center of our cool dwarf, about 9×10^6 °K. At this temperature only the proton–proton chain is responsible for the energy generation.

The sun is a rather typical star belonging to the middle part of the main sequence. A model for the internal structure of the sun naturally is of special interest to us. In recent years several solar models have been calculated, differing from one another in the numerical values adopted for certain parameters used in the calculation (particularly the chemical composition).

One other important feature is involved in calculating models for the sun. Stellar models for the upper and lower parts of the main sequence, such as those illustrated in Figures 11.1 and 11.2, are marked by their homogeneity. This circumstance means that the chemical composition of the star is assumed to remain constant throughout the whole star. It is perfectly reasonable to make such an assumption for dwarf stars of low mass and comparatively low luminosity—stars that will stay on the main sequence for a longer time than the present age of the Galaxy. Only a small part of the original supply of hydrogen fuel will thus far be exhausted in the central zones of these stars. In the case of massive stars on the upper main sequence, the model of Figure 11.1 represents a relatively young star of this kind.

But for the sun things are different. We know the age of the sun: about 5 billion years. Over this vast time span we would expect the amount of hydrogen in the central part of our sun to have diminished somewhat and an appreciable part of the sun's original hydrogen fuel reserve to have been used up. For all that, the sun will go on shining for a very long time to come. Here, then, lies the notorious uncertainty we face in calculating a solar model—it should be *inhomogeneous.* What proportion of the sun's hydrogen has burned, and within what volume? In fact, one may test different values for both the volume and the percentage of burned hydrogen, and that is what is done in model calculations.

Curiously enough, the central temperature of the sun turns out to be almost independent of the special properties of various models. It is about 14×10^6 °K, the value we have adopted in Chapter 8. Evidently the principal thermonuclear reaction process in the solar interior is the proton–proton chain, although the carbon–nitrogen cycle also makes a small contribution. For the model illustrated in Figure 11.3, the hydrogen fraction X is assumed to be 0.50 in the central region and to rise gradually until, at a distance one-fourth of the radius out from the center, X reaches about 0.75; after that it holds constant all the way to the surface.

Just as with red dwarfs, this solar model has no convective core. There is indeed an outer convection zone, but it is considerably smaller than for red dwarfs; it contains only about 2 percent of the sun's mass. The central density of the sun is quite high, more so than for model stars on either the

175

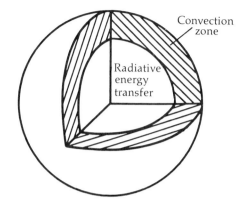

Figure 11.3   A model of the sun.

upper or the lower part of the main sequence. It is fully 135 g/cm³, or nearly 100 times the mean density. This high concentration of mass toward the center can be explained in a natural way by the partial burning of hydrogen in the innermost regions of our star. Such an effect, as we shall see, is manifested most strongly in red giants. Scientific developments in our day are opening up a wholly unforeseen opportunity for refining models of the sun, as explained in Chapter 9.

Subdwarfs are distinctive for their very low abundance of heavy elements; we mentioned this fact in Chapter 1. Hence in calculating models for such stars the value of Z should be set equal to zero. Since the heavy-element abundance is of decisive significance for the opacity of stellar material, the gas in subdwarfs should be very transparent, even though the temperature is comparatively low. Radiation will therefore be quite efficient in transporting energy and there will be no need for convection to develop. The central temperature of such stars depends rather sensitively on the adopted helium abundance, whose value is not well established by observation.

But it may be red giants that have the most interesting structure. Figure 11.4 shows a model for a typical red giant whose mass, radius, and luminosity are 1.3, 21, and 225 times the solar values, respectively.

At the very center of this giant star is a tiny core with an exceptionally high temperature, 40×10^6 °K. There is hardly any hydrogen in this core; it has already been completely burned and converted into helium. But the temperature in the core still is not high enough for the triple-alpha process to operate (Chapter 8). For want of energy sources the temperature inside the core is constant. Such a core is therefore said to be isothermal. Despite its very small size (its radius is about 10^{-3} that of the star), the isothermal core contains no less than one-fourth of all the star's mass. We may conclude at once that the isothermal core has an extremely high density, around 3×10^5 g/cm^3. Accordingly, the electron gas in the core must be degenerate (Chapter 10). In all its properties the matter in the isothermal core of a red giant is indistinguishable from the matter in a white dwarf. The two objects resemble each other not only in their mean density but also in their chemical composition and in the absence of nuclear reactions. We therefore have every reason to claim that a red giant has a white dwarf at its center! This result is of great significance for the problem of the origin of white dwarfs, as we shall find in the next chapter.

The degenerate isothermal core of our red giant is surrounded by a very thin envelope in which the thermonuclear reactions of the carbon–nitrogen cycle take place. The thickness of this shell is somewhat smaller than the radius of the isothermal core. Within this thin layer the temperature of the material drops abruptly from 40 to 25 million degrees. The matter in this envelope is several thousand times less dense than at the center of the iso-thermal core. Surrounding the shell in which the nuclear reactions operate we find another comparatively thin layer (extending to about 10 percent of the radius of the star) through which the energy released in the inner nuclear-reaction shell is transported by radiation. But the bulk of the outer envelope of the red giant—which contains almost 70 percent of the mass of the star and covers approximately the last 90 percent of the radius out toward

Figure 11.4 A model of a red giant.

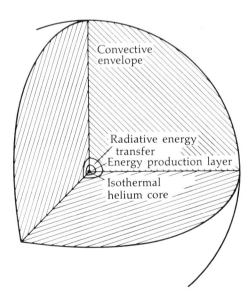

the surface—is caught up in a state of convection. The factor responsible for the formation of such an extensive convection zone is the high opacity of the material, just as in red dwarfs. Thus the structure of red giants is characterized by great nonuniformity.

By contrast to the very complex structure of red giants, the structure of white dwarfs is notable for its simplicity. We have already described this structure in Chapter 10. In a word, a white dwarf is a globe of very dense gas whose electrons are degenerate, surrounded by a comparatively thin envelope of ordinary gas (Figure 11.5). The paradox in the situation, though, is that such diverse objects as red giants and white dwarfs would seem to be genetically related to each other. We shall return to this question in Chapter 13.

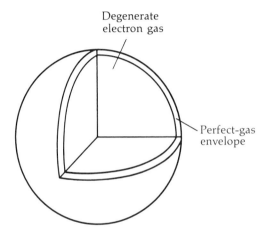

Figure 11.5 A model of a white dwarf.

12

Stellar Evolution

The overwhelming majority of stars, as Chapter 6 emphasized, change their fundamental properties (luminosity and radius) very slowly. At any given time they may be regarded as in a state of equilibrium, a circumstance of which we have made full use in explaining the nature of stellar interiors. But the fact that the changes are gradual does not mean there are none at all. That stars will evolve is altogether inescapable; the whole point is the *pace* of their evolution.

In a general way we can pose the problem of the evolution of a star as follows. Suppose that we have a star with a given mass and radius. Assume further that we know its original chemical composition, which we shall consider to be constant throughout the star. Then a model calculation for the star will yield its luminosity. As the star evolves its chemical composition will unavoidably change: the thermonuclear reactions maintaining its luminosity will diminish the abundance of hydrogen irreversibly with time. Moreover, the chemical composition of the star will cease to be homogeneous. While the percentage hydrogen abundance decreases steadily in the central region, it will remain practically unchanged near the periphery of the star. But this circumstance means that as the star burns its nuclear fuel the stellar model should itself change, and accordingly the structure of the star. Different values for the luminosity, radius, and surface temperature would be expected. Because of these major changes the star will gradually shift its position in the Hertzsprung–Russell diagram. Notice that it will follow a certain trajectory in this diagram—or, to use the customary term, an evolutionary track.

Stellar evolution undoubtedly represents one of the most fundamental problems in astronomy. The question essentially is: how do stars come into being, live, grow old, and die? This is the problem to which our book is devoted. By its very nature it is a complicated problem. It is attacked through purposeful research by representatives of various branches of astronomy, both observationalists and theoreticians. One cannot always tell at the outset which of a group of stars being studied are genetically related. On the whole the problem has proved to be very difficult, and for several decades it did not yield to a solution at all. Furthermore, until comparatively recently the efforts of researchers were often expended in an altogether wrong direction. The mere presence of the main sequence in the Hertzsprung–Russell diagram, for example, led many naïve investigators to think that stars evolve *along* this sequence from the hot blue giants to the red dwarfs. But since there exists a mass–luminosity relation stating that the mass of stars should progressively decrease along the main sequence, these investigators tenaciously argued that the evolution of stars downward along the sequence ought to be accompanied by a continuous and very substantial loss of mass.

All these arguments have turned out to be incorrect. The problem of the evolutionary tracks of stars has steadily become clearer, although certain aspects are still far from fully understood. Particular service in interpreting stellar evolution has been rendered by theoretical astrophysicists specializing in the internal structure of stars, most notably the American expert Martin Schwarzschild and his school.

The early phase in the evolution of stars, associated with their condensation from the interstellar medium, was discussed at the end of Part I of this book. Properly speaking, we were not yet concerned there with stars, but with protostars. As protostars proceed to contract by gravity they become increasingly more compact objects. Their internal temperature continually rises (see Equation 6.4) until it reaches several million degrees in the central regions. At this temperature the first thermonuclear reactions involving light nuclei (deuterium, lithium, beryllium, boron) begin to operate, as the Coulomb barrier is comparatively low in such cases. When these reactions take place the contraction of the protostar slows down. But the light nuclei burn up quite rapidly since their abundance is low, and the contraction of the protostar then resumes at nearly the same rate as before (see Equation 3.6).

The protostar will become stabilized (cease to contract) only after the temperature in its central zone has risen to the point at which the proton–proton chain or the carbon–nitrogen cycle begins to operate. It will take up an equilibrium configuration under the action of its own gravity and gas pressure differential, forces that compensate each other almost exactly (Chapter 6). Indeed, from this time onward the protostar will be a star. The young star will sit in its place somewhere on the main sequence. Its exact place on the main sequence will be determined by the initial mass of the protostar. Massive protostars will sit on the upper part of the sequence; protostars of less mass than that of the sun, on the lower part. Thus protostars will continually arrive on the main sequence over its whole length.

In stellar evolution the protostellar phase is rather fleeting. The most massive stars pass through this phase in just a few hundred thousand years. Thus it is not surprising that the Galaxy contains so few such stars. As a result they cannot be observed very simply, particularly since the sites of the star formation process tend to be buried in dust clouds that absorb light. But as soon as they have become registered at their "permanent address" on the main sequence in the Hertzsprung–Russell diagram, the situation changes abruptly. The stars will remain in that part of the diagram for a very long time, with hardly any change in their properties. This is why most stars are observed to lie on the main sequence.

The structure of a model star that has reached the main sequence comparatively recently may be described by a model calculated under the assumption that its chemical composition is the same throughout (a homogeneous model, as in Figures 11.1 and 11.2). As the hydrogen burns, the state of the star will change very slowly but steadily, and in consequence the point representing the star will follow a certain track in the Hertzsprung–Russell diagram. The manner in which the state of the star changes will depend significantly on whether or not the material in its interior is mixed. If not, then as we have seen for several models discussed in Chapter 11, nuclear reactions will cause the hydrogen abundance to become appreciably lower in the central part of the star than at the periphery. Such a star can be described only by an inhomogeneous model.

But stellar evolution can take another route: mixing may be present throughout the volume of the star, which thereby will maintain a homogeneous chemical composition even though the hydrogen abundance will steadily diminish with time. One cannot tell in advance which of the two alternatives would be realized in nature. Of course in the convection zones of stars there will always be an intensive mixing process, and within these zones the chemical composition ought to be constant. But even in the regions of stars in which radiative energy transfer predominates it is entirely possible that matter may be well mixed. For one should not exclude the possibility that systematic, fairly slow motions of large masses of material may take place, leading to mixing. Motions of this kind can develop because of certain properties of the rotation of a star.

The models computed for any star whose mass remains constant, but whose chemical composition and degree of inhomogeneity vary, will form an evolutionary sequence. By plotting in the Hertzsprung–Russell diagram the points corresponding to successive models along the evolutionary sequence of a star, one can establish its theoretical track in the diagram. It turns out that if the evolution of a star is accompanied by complete mixing of its material, the tracks will be directed leftward from the main sequence. However, the theoretical evolutionary tracks for stars that are not fully mixed (inhomogeneous models) will always move the star rightward away from the main sequence. Which of these two theoretically calculated modes of stellar evolution is correct?

The test of truth is, of course, experience. In astronomy experience means the results gained from observations. Look at the Hertzsprung–Russell diagrams for typical star clusters shown in Figures 1.6, 1.7, and 1.8. We find no stars there located above and to the left of the main sequence. Instead, there are a great many stars to the right of that sequence: red giants and subgiants. Hence we may regard such stars as having departed from the main sequence in the course of their evolution without having experienced complete mixing of the material in their interiors. The explanation of the nature of red giants represents one of the most outstanding achievements in stellar evolution theory.[1] The very existence of red giants implies that, as a rule, the evolution of stars is not accompanied by mixing of material throughout their volume. Calculations reveal that as a star evolves the size and mass of its convective core continually decrease.[2]

Clearly an evolutionary sequence of model stars does not of itself say anything about the rate of stellar evolution. An evolutionary time scale can be obtained by analyzing the change in chemical composition for various members of an evolutionary sequence of models. One can define a certain average abundance of hydrogen in the star, "weighted" so that layers of equal mass make the same contribution to the average. Let X denote this mean hydrogen abundance. Then the time variation of X will evidently determine the luminosity of the star, because the luminosity is proportional to the amount of thermonuclear energy released inside the star each second.

---

[1] In young star clusters a certain proportion of the red giants may be protostars that are still in their contraction phase and are moving toward the main sequence. Nevertheless, in principle they can be distinguished from "regular" red giants, which are fairly old stars (see Chapter 5).

[2] The idea that red giants are formed from main sequence stars after the nuclear fuel in their interior has burned was first put forward by Öpik in 1938.

We may therefore write

$$L = -\alpha \mathfrak{M} \frac{dX}{dt},$$ (12.1)

where α is the amount of energy released through nuclear transmutation of one gram of matter, and the symbol dX/dt represents the change per second in the value of X.

We can define the age of a star as the time interval elapsed since the star first reached the main sequence and hydrogen nuclear reactions commenced in its interior. If the luminosity and the mean hydrogen abundance X are known for various members of an evolutionary sequence, we can readily find from Equation 12.1 the age of some particular model star along its evolutionary track. Anyone who knows the elements of higher mathematics will recognize that Equation 12.1, a simple differential equation, gives the age τ of the star as the integral

$$\tau = \int_X^{X_0} \frac{\alpha \mathfrak{M}}{L} dX,$$ (12.2)

where X_0 denotes the hydrogen abundance in the star at the time when it arrived on the main sequence. For readers not acquainted with higher mathematics we may write a simplified expression for the elapsed time interval between two states of a star with slightly different values of X:

$$\tau_{12} \approx \frac{\alpha \mathfrak{M}(X_1 - X_2)}{L}.$$ (12.3)

By adding up the time intervals τ_{12} we will evidently obtain the elapsed time τ since the star began to evolve, the quantity appearing in Equation 12.2.

Figure 12.1 displays some evolutionary tracks calculated theoretically for comparatively massive stars. These stars begin their evolution near the lower left edge of the main sequence (straight line in the diagram). As they burn their hydrogen the stars move along their tracks, generally *across* the main sequence toward the right, remaining within the band occupied by the sequence. This evolutionary phase, during which the stars stay on the main sequence, is the most prolonged one.

When the hydrogen abundance in the core of such a star has fallen to about 1 percent, the rate of evolution speeds up. In order to maintain the required level of energy release despite the sharply dropping supply of hydrogen fuel, the temperature of the core must rise in compensation. And as in many other cases (Chapter 6) the star here regulates its own structure. The star as a whole contracts, raising its core temperature. For this reason the evolutionary track turns abruptly leftward again; that is, the surface temperature of the star increases.

183

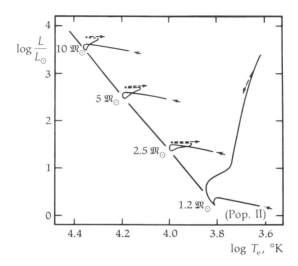

$Figure\ 12.1$ Theoretical evolutionary tracks for massive stars (M. Schwarzschild).

Very soon, however, the star stops contracting, because it has burned all the hydrogen in its core. A new region of nuclear reaction activity now comes into the picture: a thin shell surrounding the core, which is already "dead" (although still very hot). As the star continues to evolve, this shell moves progressively farther out from the center of the star, so that the mass of the dead helium core inside steadily increases. At the same time the core itself proceeds to contract and grow even hotter. As it does so, however, the outer layers of the star begin to swell rapidly and very strongly. The surface temperature therefore declines considerably, even though there is little change in the radiation flux. The evolutionary track accordingly takes another abrupt turn, this time toward the right, and the star shows every sign of becoming a red supergiant. Since stars approach this phase quite swiftly after the close of the secondary contraction process described above, there are scarcely any stars occupying the gap in the Hertzsprung–Russell diagram between the main sequence and the giant and supergiant branch. This gap is very much in evidence in diagrams such as that for a globular cluster shown in Figure 1.8.

The subsequent fate of red supergiants has not yet received adequate study. We shall return to this important subject in the next chapter. Heating of the core may continue up to extremely high temperatures, in the range of 10^8 °K. At such temperatures the triple-alpha process begins to operate (Chapter 8). The energy released in this process halts any further contraction of the core. Afterward the core expands slightly, but the radius of the whole star diminishes. The star becomes hotter and moves leftward once again in the Hertzsprung–Russell diagram.

Evolution takes place rather differently for stars of lower mass, say $\mathfrak{M} \approx 1.1$–1.5 $\mathfrak{M}_\odot$. But for stars whose mass is smaller than the sun's there is little point in considering the evolution at all, because such stars remain on the main sequence for a time span exceeding the age of the Galaxy. This circumstance renders the problem of the evolution of low-mass stars rather uninteresting, or at least not very pressing. We would merely mention that stars of low mass (less than about 0.3 $\mathfrak{M}_\odot$) remain fully convective even while they are on the main sequence. They never develop a radiative core. This tendency is clearly apparent in the case of the evolution of protostars (Chapter 5). A comparatively massive protostar will develop a radiative core even before it arrives on the main sequence. Low-mass objects, however, stay wholly convective in both their protostellar and stellar phases. In such stars the central temperature is not high enough for the proton–proton chain to operate completely. It cuts off with the formation of the He^3 isotope; normal He^4 is not synthesized. After 10^{10} yr, a period comparable to the age of the oldest of these stars, about 1 percent of the hydrogen will be converted into He^3. We would therefore expect the abundance of He^3 relative to H^1 to be abnormally high, about 3 percent.

There is unfortunately no opportunity as yet to verify this prediction of theory by observations. A star with such a low mass is a red dwarf, whose surface temperature is wholly inadequate to excite the helium lines in the optical region. In principle, however, the resonance absorption lines in the far-ultraviolet part of the spectrum might be detectable with instruments carried into space. But the continuous spectra of these stars are so faint that even this dubious possibility is precluded. Still, it is worth noting that many if not most red dwarfs are flare stars of the UV Ceti type (Chapter 1). The phenomenon of rapidly recurring flares in such cool dwarf stars is itself unquestionably associated with convection pervading the whole star. Emission lines are observed during flares. Might the He^3 lines also be detectable in such stars?

If the mass of a protostar, finally, is less than 0.08 $\mathfrak{M}_\odot$, the temperature in its interior would be so low that no thermonuclear reactions at all would be able to arrest the contraction at the main sequence state. Stars of this kind would continue to contract until they have become white dwarfs (or, properly speaking, degenerate red dwarfs). But let us return to the evolution of more massive stars.

Figure 12.2 shows the evolutionary track of a star with a mass of 5 $\mathfrak{M}_\odot$ according to very detailed calculations performed with an electronic computer. The numerals along this track mark significant turning points in the

185

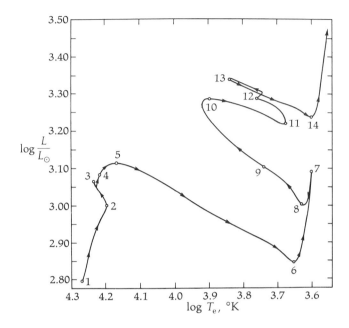

Figure 12.2 Evolutionary track for a star with a mass of 5 $\mathfrak{M}_\odot$. *1–2*, hydrogen burning in convective core, lasting 6.44×10^7 yr; *2–3*, general contraction of star, 2.2×10^6 yr; *3–4*, ignition of hydrogen in shell source, 1.4×10^5 yr; *4–5*, hydrogen burning in thick layer, 1.2×10^6 yr; *5–6*, expansion of convective envelope, 8×10^5 yr; *6–7*, red giant phase, 5×10^5 yr; *7–8*, ignition of helium in core, 6×10^6 yr; *8–9*, disappearance of convective envelope, 10^6 yr; *9–10*, helium burning in core, 9×10^6 yr; *10–11*, secondary expansion of convective envelope, 10^6 yr; *11–12*, contraction of core as helium is depleted; *12–13*, *13–14*, helium shell source; *14–?*, neutrino losses, red supergiant (I. Iben).

evolution of the star. The figure legend gives the duration of each step in the process. We wish to point out here only that the segment *1–2* of the evolutionary track corresponds to the main sequence and the segment *6–7* to the red giant phase. One interesting feature is the decline in luminosity along the segment *5–6* because of the expenditure of energy to "inflate" the star. Analogous tracks calculated theoretically for stars of various masses are shown in Figure 12.3. The numerals marking particular evolutionary phases have meanings similar to those in Figure 12.2.

A simple inspection of the evolutionary tracks illustrated in Figure 12.3 tells us that more or less massive stars move off the main sequence in a rather meandering fashion, forming a giant branch in the Hertzsprung–

Russell diagram. As this transition takes place quite rapidly, there should be only a few stars between the main sequence and the giant branch; we have made this point above. The very fast rise in the luminosity of low-mass stars as they evolve into red giants is a typical feature. Such stars differ in their evolution from more massive ones because the low-mass stars contain a very dense, degenerate core. The pressure of the degenerate gas is so high (Chapter 10) that the core is able to bear the weight of the overlying layers of the star. It will hardly contract at all as the star evolves, and hence will become strongly heated. If the triple-alpha process does begin, it will do so much later. Except for the physical conditions in the zone near the center, the structure of such stars will resemble the structure of more massive ones. Accordingly, after their hydrogen has burned in the central zone their evolution will likewise entail a swelling of their outer envelope, which will bring their tracks into the red giant region of the diagram. But unlike the more massive supergiants, they will have cores consisting of very dense degenerate gas (see the diagram of Figure 11.4).

Figure 12.3   Evolutionary tracks for stars of differing mass. The numerals delimit phases in the evolution like those in Figure 12.2 (I. Iben).

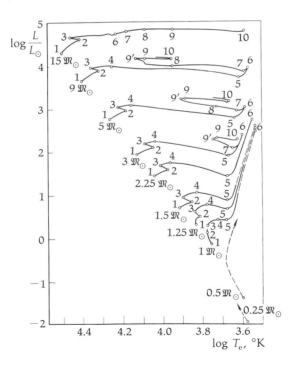

Perhaps the most signal achievement of the stellar evolution theory described in this chapter is its explanation of all the peculiarities in the Hertzsprung–Russell diagram for star clusters. We have already given an account of these diagrams in Chapter 1. As mentioned in that chapter, the age of all the stars in a given cluster should be considered the same. The original chemical composition of the stars in the cluster should also have been the same, for they would all have been formed from an identical (although quite large) gas–dust complex in the interstellar medium. Different star clusters should differ from one another primarily in their age; in addition the primordial chemical composition of globular clusters should have been sharply distinct from the composition of open clusters.

The lines along which the stars belonging to a cluster are arrayed in the Hertzsprung–Russell diagram do not by any means represent evolutionary tracks. These lines are the geometric loci of points in the diagram where stars of differing mass all have the same age. If we want to compare stellar evolution theory with observational results, we should begin by constructing theoretical curves of equal age for stars of differing mass but the same chemical composition.

Figure 12.4   Stellar evolutionary tracks "linked" in time (M. Schwarzschild).

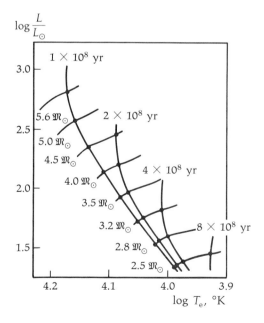

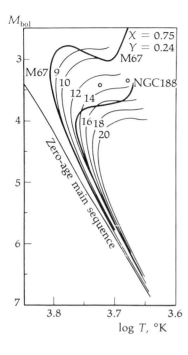

Figure 12.5   Theoretical Hertzsprung–Russell diagram for star clusters.

We can determine the age of a star at various phases of its evolution by applying Equation 12.2. Theoretical tracks of stellar evolution such as those displayed in Figure 12.3 must be used for this purpose. Figure 12.4 shows the results of calculations for eight stars whose masses range from 2.5 to 5.6 $\mathfrak{M}_\odot$. Along the evolutionary tracks of these stars (extending from lower left to upper right), dots mark the positions the stars occupy at various times in their evolution, hundreds of millions of years after their initial state at the lower left edge of the main sequence. The curves that join the dots for various stars are curves of equal age. In the case illustrated here the calculations have been performed for rather massive stars. The computed time intervals of their evolution cover at least 75 percent of their active life—when they are radiating away the thermonuclear energy generated in their interior. For the most massive stars the evolution included in Figure 12.4 extends to the secondary contraction phase, which sets in after the hydrogen in their central zones has been almost completely burned.

If we compare the theoretical curves of equal age with the Hertzsprung–Russell diagram for young star clusters (see Figure 12.5 as well as Figure 1.6), our attention is involuntarily caught by their striking resemblance to the base lines in the cluster diagrams. In full agreement with the chief premise of evolution theory—that the more massive stars leave the main sequence

189

more quickly—the diagram of Figure 12.5 clearly shows that for the stars in a cluster the upper part of this sequence bends to the right. The older the cluster, the lower in the diagram is the point where stars begin to deviate appreciably from the main sequence. With this rule alone we can compare the age of different star clusters directly.

In old clusters the main sequence cuts off somewhere around spectral type A. In young clusters the whole main sequence still remains intact, all the way up to hot massive stars of spectral type B. The diagram for the cluster NGC 2264 (Figure 1.6) exemplifies this situation. Indeed, the curve of equal age computed for this cluster indicates that it has been evolving for only 10 million years. The cluster therefore was born within the memory of the Ramapithecans, ancestors of man. A considerably older star cluster is the Pleiades, whose diagram is shown in Figure 1.4; its average age is about 100 million years, and it still contains stars of spectral type B7. The Hyades cluster (Figure 1.5), though, is quite ancient, about a billion years old, so that its upper main sequence begins only with type A stars.

The theory of stellar evolution explains one other curious property of the Hertzsprung–Russell diagrams for "young" clusters. Dwarf stars of low mass have a very long evolutionary time scale. For example, after 10^7 yr (the period that the cluster NGC 2264 has been evolving) they would not have passed through their gravitational contraction phase, and strictly speaking would not even be stars but still protostars. Such objects, we know, are located to the right of the main sequence (see Figure 5.2, where the evolutionary tracks of the stars begin at an early stage in the gravitational contraction). Thus if the dwarf stars in a young cluster have not yet reached the main sequence, the lower part of that sequence will be shifted to the right in such a cluster, as is actually observed (see Figure 1.6). Even though our sun has been evolving for about 5×10^9 yr and has already depleted a substantial part of its hydrogen fuel reserve, it has not yet emerged from the main sequence band in the Hertzsprung–Russell diagram, as we have said above. Calculations demonstrate that when the sun was young and had newly arrived on the main sequence it was emitting 40 percent less radiation than now, but its radius was only 4 percent smaller than today; its surface temperature then was 5200°K, compared to 5700°K now.

Evolution theory provides a natural interpretation of the Hertzsprung–Russell diagram for globular clusters. Above all these are very old objects. They are only slightly younger than the Galaxy itself. This conclusion follows from the nearly complete absence in their diagrams of stars belonging to the upper part of the main sequence. The lower main sequence, as mentioned in Chapter 1, consists of subdwarfs. Spectroscopic observations indicate that subdwarfs are very deficient in heavy elements, which are tens

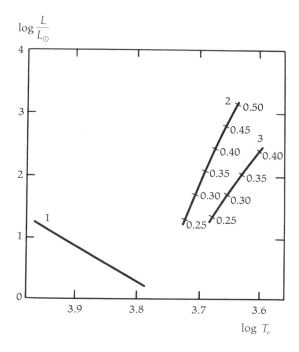

Figure 12.6 Theoretical tracks for stars with $\mathfrak{M} = 1.2\ \mathfrak{M}_\odot$ and a differing heavy element abundance: 1) Main sequence; 2) stellar population II; 3) population I (M. Schwarzschild).

of times less abundant there than in ordinary dwarfs. Hence the primordial chemical composition of globular clusters must have differed greatly from the composition of the material out of which open clusters, richer in heavy elements, have been formed. Figure 12.6 shows theoretical evolutionary tracks for stars with a mass of 1.2 $\mathfrak{M}_\odot$ (approximately the mass of a star that has gone through its evolution in 6×10^9 yr) but with a different initial chemical composition. A core of helium containing one-fourth the mass of the star is steadily growing heavier. Notice that after a star has left the main sequence its luminosity at an equivalent phase in its evolution will be considerably higher if its metal abundance is low (population II). At the same time such stars will have a higher effective surface temperature.

Figure 12.7 plots evolutionary tracks for stars of low mass with a low heavy-element abundance. The points on these three tracks represent the position of each star after 6 billion years of evolution. The three points are joined by a heavy line, which is clearly a curve of equal age. If we compare

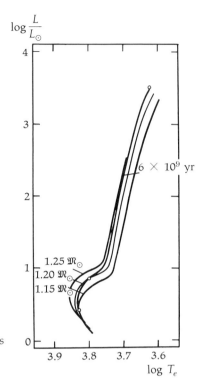

$\log \dfrac{L}{L_\odot}$

4

3

6×10^9 yr

2

1

1.25 $\mathfrak{M}_\odot$
1.20 $\mathfrak{M}_\odot$
1.15 $\mathfrak{M}_\odot$

0

3.9   3.8   3.7   3.6

$\log T_e$

Figure 12.7   Theoretical tracks for low-mass stars deficient in heavy elements (M. Schwarzschild).

this curve with the Hertzsprung–Russell diagram for the globular cluster M3 (Figure 1.8), we immediately observe complete agreement with the curve followed by the stars in this cluster as they leave the main sequence.

The diagram of Figure 1.8 also exhibits a horizontal branch extending leftward from the giant sequence. Evidently this branch represents stars in whose interior the triple-alpha process is taking place (Chapter 8). Thus stellar evolution theory can account for all the properties of the Hertzsprung–Russell diagram for globular clusters in terms of their antiquity and their low heavy-element abundance. One should keep in mind, however, that the abundance of heavy elements varies over a rather wide range in different globular clusters. In fact, even within a single cluster stars may sometimes have different values of Z. Reality is always more intricate than any ground plan.

It is a singular fact that several white dwarfs have been observed in the Hyades cluster but not in the Pleiades. Both clusters are comparatively close to us, so that this interesting difference between them cannot be attributed to different visibility conditions. But we already know that white dwarfs are formed as the terminal phase of red giants with a relatively low mass. Such a giant needs quite a long time, at least a billion years, to evolve fully. This time has passed in the Hyades but it has not yet come in the Pleiades cluster. That is why the Hyades now contains a certain number of white dwarfs, while the Pleiades does not.

Figure 12.8 presents a composite, schematic Hertzsprung–Russell diagram for several clusters, both open and globular. The effect of age differences for different clusters appears very distinctly in this diagram. Thus there is every reason to maintain that the modern theory of stellar structure and the evolution theory based upon it are capable of giving a natural explanation of the basic findings of astronomical observations. Unquestionably this is one of the most notable achievements of twentieth-century astronomy.

Figure 12.8 Composite Hertzsprung–Russell diagram for various star clusters (H. L. Johnson and A. R. Sandage).

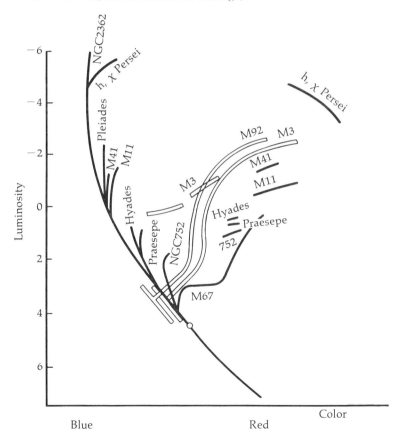

13

Planetary Nebulae, White Dwarfs, Red Giants

During the terminal evolutionary phases of red giants (as well as supergiants), the vast outer atmosphere or envelope begins to shed a great deal of mass. This closing evolutionary step is very difficult to calculate theoretically because the problem is not at all well defined. We simply do not understand exactly how material is ejected from the envelopes of such stars. For the time being we can attempt no more than a qualitative approach.

Throughout the preceding phases of stellar evolution (the gravitational contraction of the protostar, the long pause on the main sequence, and the departure from it after the star has exhausted the nuclear fuel supply in its central regions) we have assumed that no appreciable loss of mass takes place. Admittedly, recent spectroscopic observations carried out in ultraviolet light with rockets and satellites have shown that hot, massive main sequence stars do experience a rather considerable loss of material. But that is another question. In the case of red giants, purely empirical arguments indicate that their existence as stars ends not because they have depleted their nuclear fuel, but merely through the loss of their hydrogen-rich outer envelopes.

Let us mention one simple argument that the author of this book put forward some time ago, in 1956. It concerns a phenomenon long known to astronomers, the objects called planetary nebulae. These are rather dense gaseous structures that surround certain very hot stars of low luminosity. Photographs of two planetary nebulae are shown in Figures 13.1 and 13.2. For several decades astronomers have regarded nebulae of this kind as a natural laboratory in which particular physical processes taking place in the interstellar medium can be studied successfully.

The investigation of planetary nebulae has enriched astronomical spectroscopy with some discoveries of prime importance. Of special note is the study of the interesting process whereby atoms fluoresce through stimulus by "hard" (short wavelength) radiation, as well as the analysis of forbidden atomic transitions represented by lines of ionized oxygen and nitrogen. Planetary nebulae are exceptional in that very accurate determinations have been made of their chemical composition; this is of the greatest significance for all astronomy. However, so utilitarian an approach to these remarkable objects leaves out the main point: Where did they come from? What is their origin? There has been no lack of hypotheses, to be sure, but all have been arbitrary and artificial.

Figure 13.1 The Helix, a planetary nebula (NGC 7293) in the constellation Aquarius.

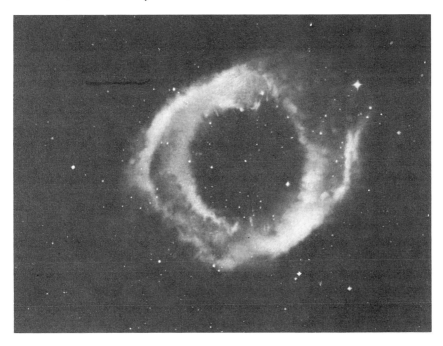

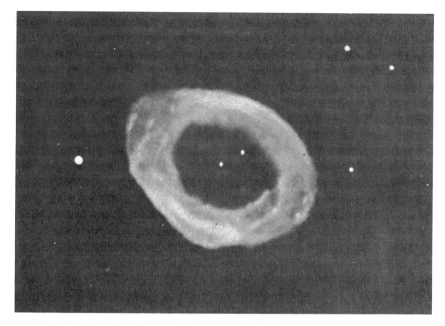

Figure 13.2   The Ring, a planetary nebula (M57) in the constellation Lyra.

When I endeavored to resolve the problem of planetary nebulae I called attention to a circumstance which, from my viewpoint, was fundamental. The gas comprising a nebula is not restrained by a force of attraction. Hence these objects should expand without limit at a relatively low velocity, and quite soon, after only some tens of thousands of years, they should disperse into interstellar space. During this expansion the gas density will drop rapidly. The luminosity of planetary nebulae therefore ought to fall even faster, since their emission per unit volume resulting from electron–ion collisions is proportional to the square of the gas density.

How will these objects look while they are still quite young, with an age of a few thousand years? Analysis has shown that very young nebulae—those which have just become detached somehow from their central stars—will, in the first place, be extremely small, only a few thousand astronomical units across. Second, they will be rather dense; and third, what is especially interesting, their outer layers should represent comparatively cool nonionized gas. At the same time the luminosity of these very young nebulae will be about a thousand times that of the sun. Of course the hot central star (like those appearing in Figures 13.1 and 13.2) could never be visible behind the thick layer of gas.

196

A planetary nebula is a strange object. What does it resemble? One can see without difficulty that in all its basic properties it is similar to the extended cool atmosphere of a red giant. An important argument supporting this fundamental conclusion—that planetary nebulae represent the outer layers of red giant stars which have become disconnected from the hot regions inside, where most of the original mass of the star is concentrated—comes from an anlaysis of the space distribution of these objects. Planetary nebulae are concentrated relatively weakly toward the galactic plane, and they exhibit a strong concentration toward the center of our stellar system. This fact alone implies that the nebulae are the ultimate product of prolonged evolution in very old stars belonging to the galactic disk. Certain red giants of high luminosity have just the same space distribution.

A natural conclusion follows inescapably from this interpretation of planetary nebulae: their very hot nucleus stars represent the bared interiors of red giants. The unsheathing occurred after the outer layers of a red giant became detached from it and, gradually expanding, spread out over a fairly large volume. According to my estimate, which today is generally accepted, the average mass of a planetary nebula is about 0.2 $\mathfrak{M}_{\odot}$. Now imagine how the star, a red giant whose mass is hardly greater than the sun's, would look if it were suddenly deprived of its comparatively cool envelope, its "heavy fur coat." It would be a tiny object with a very high temperature whose outer layers are in a state of violent convection (see the model diagram in Figure 11.4). Calculations of red giant models indicate that at the level above which 0.2 $\mathfrak{M}_{\odot}$ is located the material would have a density of about 10^{-4} g/cm^3, which is a hundred times higher than in the solar photosphere. The temperature here would be about 200,000°K, at a distance out from the center of approximately 10 times the radius of the sun. Evidently the detachment of the outer layers of the red giant will be accompanied by a fairly rapid (but not catastrophic) contraction of its inner regions to a size only a few times larger than the earth. Incidentally, it is entirely possible that red giants of the RV Tauri type, which seem to be the parents of planetary nebulae, have at the closing phase of their evolution a structure different from that described above. For example, their material may be much more strongly concentrated toward the center.

Direct observational proof that planetary nebulae originate from red giants has been obtained very recently by the techniques of radio astronomy. Quite intense radio lines of CO at $\lambda = 2.64$ mm have been discovered in the young planetary nebulae NGC 7087, IC 418, and NGC 6543. The angular size of the corresponding radio sources is considerably larger than that of the planetary nebulae. This means that each radio source comprises an extensive, fairly cool envelope enclosing the H II region observed as the optical nebula, a result which fully agrees with our model. Older, optically thin planetary nebulae should not have such envelopes. Even earlier these nebulae should have produced rather strong infrared radiation emanating from solid dust grains inside the nebula.

Both the molecules and the dust in planetary nebulae are undoubtedly of vestigial origin: they could only have been formed in the comparatively cool atmospheres of red giants. The more durable grains can survive tens of thousands of years in the H II zone without breaking up, whereas the CO molecules will be dissociated there. Hence the dust is distributed throughout the nebula, while the molecules occur only in the outer zone.

We would emphasize that the detachment of the outer envelope from the main body of the star does not take place in an eruptive fashion (as happens, for instance, in the case of supernovae; see Part III), but occurs calmly, almost at zero velocity.

What causes the envelope to be shed? This problem is still very far from being solved. Some possibilities will be discussed below.

The very hot object remaining after the outer envelope has been detached should be in an unstable intermediate state. It will evolve rapidly, entering a stable state of some kind. What will this state be? There can be no question but that the stable object into which the nucleus of a planetary nebula evolves should be a white dwarf. For individual nuclei we may draw such a conclusion at once. To take an example, the very faint nucleus of the planetary nebula NGC 7293 shown in Figure 13.1 (this is the closest such object to us) has an absolute magnitude of 13.5 and a temperature above 100,000°K. Accordingly, its linear size is only slightly larger than that of the earth, which for a mass of about 1 $\mathfrak{M}_\odot$ implies a mean density of several hundred thousand grams per cubic centimeter. This is the typical density of a white dwarf!

Moreover, a curious tendency is observed: the older planetary nebulae are (and their age can be estimated quite easily), the more their nuclei resemble white dwarfs. It seems that during the comparatively short lifetime of planetary nebulae their nuclei tend to have difficulty settling down and becoming more or less normal white dwarfs.

Important evidence supporting our claim that planetary nebulae, red giants, and white dwarfs are genetically related comes from an analysis of statistical data. At any one time our Galaxy contains tens of thousands of planetary nebulae, although only a few of them are directly accessible to observation. But their average lifetime is just a few tens of thousands of years. About once a year, then, a planetary nebula is formed somewhere. And as a byproduct just as many white dwarfs, the final stage in the evolution of the nuclei of these nebulae, originate every year. This mechanism is very efficient: during the time span our stellar system has been evolving, it has produced several billion white dwarfs. That is just the order of magnitude of the whole white dwarf population in the Galaxy! On the other hand, the statistics of red giants of the RV Tauri type tells us that they number about a million in the Galaxy. Thus if we regard them as the parents of planetary nebulae, the lifetime in this phase should be about a million years—a perfectly acceptable value.

In Chapter 11 we explained that the material in the innermost zone of red giants has properties (degeneracy) identical to the material in white dwarfs. Now we see that this circumstance is no coincidence. Like an egg in a hen, a white dwarf gradually matures at the center of a star in order that it may hatch when the time is right. The newborn chick, the white dwarf, will be surrounded by a suitable kind of shell and other appurtenances of its birth. We call it the nucleus of a planetary nebula. After tens or hundreds of thousands of years we will have a normal white dwarf; the planetary nebula formed along with it will long since have dispersed in interstellar space.

The qualitative picture we have just outlined, linking the closing phase in the evolution of red giants to the simultaneous formation of planetary nebulae and their nuclei—tiny, hot, dense stars that evolve rapidly into white dwarfs—has been developed extensively over the past few years in a number of investigations resting on advances in stellar evolution theory. Many details of this process, so important for stellar cosmogony, have now become understood.

To begin with, let us consider more fully the process whereby the nucleus of a planetary nebula evolves into a white dwarf. In 1956 the author of this book pointed out that as the nuclei go through their rapid evolution the temperature of their surface layers will rise at first. Since the luminosity will not change very much in the meantime, we may conclude that the nuclei will rapidly contract. More precise theoretical calculations based on planetary nebulae observed in the Magellanic Clouds[1] yield an empirical relation between the luminosity of the nuclei of planetary nebulae and the effective temperature T_e of their surface layers. This relation is illustrated schematically in the upper left corner of Figure 13.3. The dashed line at the lower left shows the analogous relation for white dwarfs that have cooled. Also included are the luminosity–temperature relations for main sequence stars and red giants, and the horizontal branch of the Hertzsprung–Russell diagram for globular clusters. The drop in the luminosity of the nuclei of planetary nebulae after it reaches a maximum with the rising temperature signifies a further rapid contraction. In the region of the diagram between the very hot nuclei and the white dwarfs, other faint stars are observed. Emis-

---

[1]These are the closest galaxies to us. Planetary nebulae in them are all at practically the same distance, so their luminosity can be determined quite easily from their apparent magnitude.

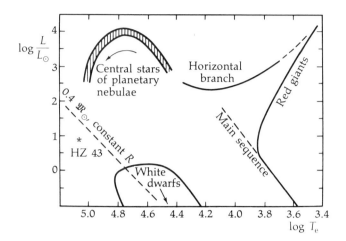

Figure 13.3 Empirical relation between the luminosity of the nuclei of planetary nebulae (as well as other objects) and their surface temperature (E. E. Salpeter).

sion and absorption lines are absent from their spectra, and the violet part is much enhanced. Almost surely these are highly evolved nuclei of planetary nebulae, for which the nebulae themselves have expanded and dispersed. Thus the luminosity–temperature diagram demonstrates strikingly (and purely empirically!) the genetic relationship between the nuclei of planetary nebulae and white dwarfs.

One can grasp the theoretical principles of this type of evolution from the following considerations. Imagine a star of uniform chemical composition which, having exhausted its nuclear energy sources, contracts in a characteristic time determined by the Kelvin scale (Chapter 3). Then the density of the matter at its center will rise according to an R^{-3} law. By calculating theoretically the evolution of this idealized star one can find its bolometric luminosity, central temperature, and surface temperature, all as functions of its central density. Moreover, one can obtain a theoretical (L, T_e) relation. Figure 13.4 displays the corresponding curves. Notice that the (L, T_e) relation for this model provides a good fit to the empirical relation shown in Figure 13.3. Detailed calculations have also been performed for more complicated model stars devoid of nuclear sources (at very high temperatures, for example, allowance should be made for the generation of a great many neutrinos that freely remove energy from the stellar interior). The lower curve in Figure 13.5 shows the (L, T_e) relation calculated for a model star with a mass of 1.02 $\mathfrak{M}_\odot$ consisting wholly of a homogeneous mixture of

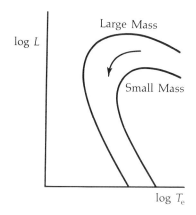

$\log L$

Large Mass

Small Mass

$\log T_e$

Figure 13.4   Theoretical relation between the luminosity of the nuclei of planetary nebulae and their surface temperature.

carbon and oxygen. A model in which 5 percent of the material of the star forms an outer helium envelope yields quite a different curve (Figure 13.5). Nevertheless, over a wide range of variation in the model parameters the dependence of bolometric luminosity on surface temperature changes little in general character and remains consistent with the empirical behavior shown in Figure 13.3.

Under what conditions, though, are we entitled to assume that a star can evolve without nuclear energy sources? The amount of energy released

Figure 13.5   Theoretical (L, T_e) relation for the nucleus (1.02 $\mathfrak{M}_\odot$) of an evolving carbon–oxygen planetary nebula (E. E. Salpeter).

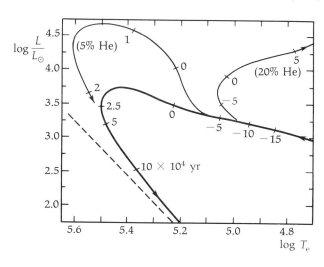

in nuclear reactions depends primarily and very strongly on the temperature. Hence a condition for nuclear reactions to be inoperative is that the central temperature be comparatively low. But what would impose such an upper limit on the temperature of stellar interiors? Mainly their chemical composition. If, for example, the central temperature is 5×10^6 °K and the star consists only of helium, then, of course, no nuclear reactions will take place there (see Chapter 8); but if the star consists of hydrogen, the proton–proton chain will begin to operate at that temperature. In turn, as we saw in Chapter 6, the central temperature is determined by the mass of the star (Equation 6.4). Thus if the chemical composition of the star is specified, then in order for its evolution to be described by the tracks depicted in Figure 13.4 its mass must remain below some critical value. For example, a star consisting of pure hydrogen should have a mass less than 0.08 $\mathfrak{M}_\odot$; a helium star, less than 0.35 $\mathfrak{M}_\odot$; a carbon star, less than 1.04 $\mathfrak{M}_\odot$. The corresponding central temperatures are 4×10^6, 1.2×10^8, and 6×10^8 °K.

In view of this situation we should regard the nuclei of planetary nebulae as objects in which nearly all the helium has been transmuted by nuclear reactions into carbon, oxygen, or neon. Otherwise their mass would have to be less than 0.35 $\mathfrak{M}_\odot$, and this limit would conflict with the fact that planetary nebulae are observed to have a comparatively low mass (0.2 $\mathfrak{M}_\odot$). For the mass of the red giant from which the nucleus and the planetary nebula itself were formed should be somewhat greater than the solar mass. Furthermore, the observations (although rather scanty) indicate that for the most part white dwarfs have a mass in the range 0.5–1.0 $\mathfrak{M}_\odot$. Probably the nuclei of planetary nebulae are coated with a thin crust that has not been able to burn helium, or possibly even hydrogen. If one allows for this crust in theoretical calculations one finds a lower value for the temperature of the nucleus.

Thus the modern theory of internal stellar structure brings us to the conclusion that stars deprived of their nuclear energy sources (as is undoubtedly the case for the nuclei of planetary nebulae) should evolve into normal white dwarfs. A second, no less important question remains to be considered quantitatively: Just how are such stars formed? What was the preceding phase of their evolution?

The fact that the nuclei of planetary nebulae consist primarily of a mixture of carbon, oxygen, and heavier elements at once implies that we are dealing with thoroughly evolved objects. Beyond doubt, then, the ancestors of planetary nebulae cannot have been among the main sequence stars. That is, they would have moved far away from the sequence. But only red giants belonging to the horizontal branch of the Hertzsprung–Russell diagram for sufficiently old clusters (see Chapter 12) could be such objects. At the present epoch the stars coming off the sequence in old clusters have a mass of about 1.1 $\mathfrak{M}_\odot$ (for population I stars, which are comparatively rich in heavy elements) or 0.85 $\mathfrak{M}_\odot$ (for population II stars, in which the heavy element abundance is low, as in globular clusters). For young population I clusters these masses are higher, but they generally do not exceed 1.5 $\mathfrak{M}_\odot$. We may therefore conclude that the ancestors of planetary nebulae, members primarily of population I (the disk population), should have masses in the range 1.1–1.5 $\mathfrak{M}_\odot$. With such a mass, a star should develop a degenerate core.

As we saw in Chapter 12, the red giant phase of evolution closes with a helium flash, when the triple-alpha process for converting helium into carbon is ignited throughout the helium core. Calculations show that by this time the mass of the helium core will be in the range 0.4–0.5 $\mathfrak{M}_\odot$, almost independently of the total mass of the evolving star. After the helium flash begins the star will evolve along the horizontal branch in the Hertzsprung–Russell diagram. Some time after much of the helium in the core has been transformed into carbon and heavier elements, the helium nuclear reactions will be concentrated in a thin layer surrounding a core that has burned out once again. Furthermore, such stars have another layer outside in which hydrogen nuclear reactions take place. Stars of this kind, then, have a double shell source of nuclear energy; Figure 13.6 schematically illustrates the structure of such a star. These stars will greatly increase their luminosity, which will rise to thousands of times the solar luminosity, whereas the luminosity on the horizontal branch is only a hundred times that of the sun.

Figure 13.6 Diagram of a star with a double shell source of nuclear energy.

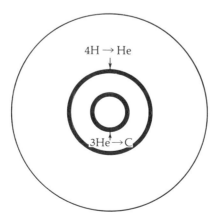

The structure of such stars, in which no more nuclear reactions can operate in the burned-out central zone, should resemble the structure of red giants. As a result the star will again become swollen and its radius will extend to a distance comparable to the earth's orbit, whereas the radius of the hydrogen-burning zone will be far smaller, only about 3×10^9 cm. By the end of this evolutionary phase the star will once more have become a red giant. It may have tens of thousands of times the solar luminosity, while the mass of the thin helium shell in its interior will be no more than a few hundredths the solar mass. The hydrogen-burning zone will ascend far outward, so that the region inside this envelope will now contain about 70 percent of the mass of the star. This whole phase of evolution, with two layers of nuclear energy production, will last roughly a million years.

During the very last steps of this phase the star will be fully prepared to shed its hydrogen-rich outer envelope and thereby give rise to a planetary nebula plus a nucleus. Indeed, the luminosity of young planetary-nebula nuclei is 10^4 times the solar luminosity, practically the same as the luminosity of double-shell giant stars. The radius, structure, and mass of the nuclei are nearly identical to the same parameters for this type of star beneath the hydrogen-burning shell (that is, the "inert" central part, consisting of heavy elements and covered with a thin helium crust). Thus the future nucleus of a planetary nebula is welded into the interior of a highly evolved star. Now it merely remains to understand the mechanism responsible for detaching the extended, hydrogen-rich outer envelope.

First of all, planetary nebulae are observed to have a low expansion velocity, averaging about 30 km/s. It is natural to infer that the nebular material has been torn from the star at a point with a parabolic velocity of the same order. If we take the mass of the inner part of the star to be 0.8 $\mathfrak{M}_\odot$, we find that the detachment would have occurred at a distance of about one astronomical unit (1.5×10^{13} cm) from the center of the star— just the radius of the giant star that was the ancestor of the nebula. Although a quantitative theory for the shedding of an envelope by a distended, highly evolved star has not yet been worked out (the problem is very difficult), at least three possible causes of the phenomenon may be mentioned:

1. Because of the special character of instability in the extended envelope, strong oscillatory processes should develop, accompanied by changes in the thermal regime of the star. Such oscillations would have a period of roughly 10^4 years.

2. Ionization of hydrogen in a certain zone of the star beneath its photosphere could produce strong convective instability. A similar phenomenon occurring on the sun represents the initial cause of solar activity.

In cool giant stars convective motions may be incomparably stronger than on the sun.

3. So great is the luminosity of the stellar ancestor of a planetary nebula that the *pressure* of its radiation flux on the outer layers may serve to expel them. Calculations too complex to give here demonstrate that envelope material will be thrown off by the action of radiation pressure, and the rate of flow may reach 10^{21}–10^{22} g/s. This fact means that practically the whole stellar envelope can be shed in a few thousand years, resulting in the formation of a planetary nebula. Most likely all three of these envelope detachment mechanisms are effective, as though they were helping one another.

Persuasive support for the idea that planetary nebulae develop from the outer layers of red giants comes from the discovery that practically all planetaries emit surplus infrared radiation. Dust grains, mixed more or less uniformly with hot gas in the nebulae, are responsible for this radiation. The physical conditions in a planetary nebula—especially the high temperature of the plasma there—preclude the formation of grains from the gaseous medium. The grains found in a nebula must accordingly be left over from the time when the nebula itself was formed, having survived such gradual destructive processes as collisions with protons and with photons of hard radiation. On the other hand, the cool, extended atmospheres of red giants, where much of the gas can persist in a molecular state, provide all the conditions favoring grain formation. A large proportion of cosmic dust presumably enters the interstellar medium in this very way: it is dispersed by planetary nebulae.

Emission in the CO radio line at 2.64-mm wavelength (see Chapter 5) has recently been detected in the infrared objects CRL 2688 and CRL 618. Analysis of the CO line indicates that we are observing fairly dense gaseous envelopes expanding at a velocity of 20 km/s. Evidently the quite hot stars observed in the central parts of these objects managed to peep through the envelopes around them only a few thousand years ago. At that time the objects would simply have been red giants with carbon-rich atmospheres. But we can then infer that CRL 2688 and CRL 618 are proto–planetary nebulae.

The point sources of soft x rays lately discovered with the tools of space astronomy appear to be related to the problem of planetary nebulae. These sources, of which four are known, have proved to be unusually hot white dwarfs (their surface temperature is of order 10^5 °K). Probably they are nuclei of old planetary nebulae whose expanding envelopes have faded to a negligible surface brightness. It would be interesting, on the one hand, to detect optical vestiges of planetary nebulae near sources such as these, and on the other, to attempt measurements of the soft x-ray flux from the closest planetary nebulae, including the Helix Nebula illustrated in Figure 13.1. One of these stars, the very hot white dwarf HZ 43, is marked by an asterisk in Figure 13.3. This object evidently occupies an intermediate position between hot white dwarfs and the nuclei of planetary nebulae.

Thus from the standpoint of modern stellar evolution theory, the formation of planetary nebulae and their nuclei constitutes a legitimate process in the evolution of red giants.

Is this the only possible route leading to white dwarfs? We can merely say that such a path is very often followed. Yet it could hardly have been responsible for every white dwarf. We may, for example, conceive of a gradual outflow of material from the outer layers of some red giants, unlike the discrete shedding of an envelope that results in the formation of a planetary nebula. Finally, the classical white dwarf, discovered before all the others—the celebrated companion of Sirius—belongs to a binary system. And the circumstances of stellar evolution in binary systems are highly specialized. We turn to this subject in the next chapter.

14

Evolution in Close Binary Systems

Stellar evolution has been discussed at some length in the preceding chapters. An important reservation is necessary, however: we were speaking of the evolution of individual, isolated stars. How will evolution proceed for stars comprising a double system, or, more generally, a multiple system? Will one of the stars disturb the normal evolution of its companion? This question is of fundamental significance because multiplicity is an extremely common occurrence among stars. About half of all main sequence stars belong to multiple systems. For the upper part of the main sequence, including hot and massive stars of spectral types O and B, the proportion of stars belonging to multiple systems is at least 70 percent. In the case of population II stars (see Chapter 1), though, multiplicity is a fairly rare if not unparalleled phenomenon.

A further stimulus to research on stellar evolution in binary systems has been the fact that some remarkably curious stars are observed only in binary systems. Outstanding among them are the novae, whose outbursts have long attracted astronomers' closest attention. Today x-ray stars are of special interest, as we shall see in Chapter 23. They too seem to occur only in close binary systems. No less striking is the discovery that certain categories of cosmic objects pointedly avoid binary systems. Evidently something there "annoys" them. The famous pulsars, which we shall meet in Part IV of this book, are examples of such objects.

The chief property of a star, the one determining the whole course of its evolution, is its mass. If a star has a large mass it will evolve faster, burning up the hydrogen in its interior more rapidly and entering the red giant and supergiant phases sooner. However, in 1951 the Soviet astronomers Pavel P. Parenago and Alla G. Masevich pointed out that in close binary systems the more luminous component generally has the *smaller* mass. At that time the theory of stellar evolution described in Chapter 12 had not been developed at all. Nevertheless, the situation in close binary systems appeared to be peculiar: the more massive component in a system would ordinarily be a main sequence star, while the less massive one would have an excess luminosity; that is, it would be almost a giant, or a "subgiant." As stellar evolution theory progressed it became clear that subgiants are stars so far along in their evolution that they have left the main sequence. But a perfectly legitimate question then arises: how could stars with a demonstrably lower mass have advanced further in their evolution than more massive stars? This notable result in stellar astronomy has been called the Algol paradox after the famed eclipsing double star in which the paradox is conspicuously apparent.

In 1955 this paradox was explained in a very plausible way. The more luminous star in a binary was originally the more massive. But after exhausting a substantial part of its nuclear fuel it began to swell. As it did so, much of its mass flowed off to the other component, whose mass thereby grew larger than the mass of the more rapidly evolving star. Thus the most important process governing the evolution of stars in a binary system is mass exchange between them. We cannot regard stars in binary systems as evolving with constant mass.

How does the exchange of matter between the components of a binary system operate? Consider a binary system whose components have masses $\mathfrak{M}_1$ and $\mathfrak{M}_2$ and travel in a circular orbit of radius a. Then according to simple gravitational theory, for each of the components there will be a surface in space outside of which particles of matter can no longer be retained by the gravitational attraction of the corresponding star. This surface is determined both by the attraction of the other star on the particles and by the centrifugal force caused by the rotation of the system as a whole. If the particles are located on the surface itself, they need only be given a tiny velocity directed outward and they will leave the sphere of attraction of the star within the surface. But if the particles are located in the neighborhood of the point L_1 (Figure 14.1), then upon leaving the first star they will be captured by the attraction of its companion. A surface possessing these properties is called the zero-velocity surface or the critical Roche surface,

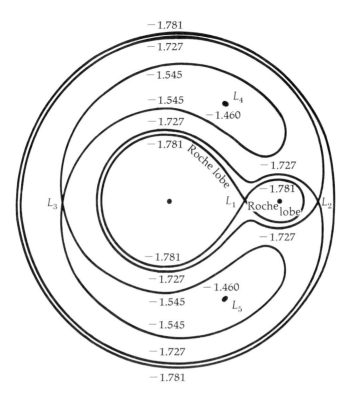

Figure 14.1 Diagram of a Roche surface.

and the point L_1 through which matter can flow from one star across to the other is called the inner Lagrangian point.

The Roche surface consists of two closed cavities or lobes which surround the two stars and have the point L_1 in common. The radius of each lobe may be represented by the approximate expression

$$\frac{r}{a} = 0.38 + 0.2 \log \frac{\mathfrak{M}_1}{\mathfrak{M}_2}. \tag{14.1}$$

Equation 14.1 is quite accurate enough for systems with mass ratios in the range $0.3 < \mathfrak{M}_1/\mathfrak{M}_2 < 20$.

Now let us take the following model for the evolution of stars in a close binary system. So long as the two components of the system remain on the main sequence, their radii will be smaller than the radii of the corresponding Roche lobes, as given by Equation 14.1. When a substantial part of the hydrogen fuel has been exhausted in the central zone of the more rapidly evolving and more massive star (the primary component), the radius of that star will begin to increase, while the radius of the secondary component (the other star) holds the same. Thus the more massive component will become more and more swollen, until its outer layers fill its Roche lobe (see Figure 14.1). When that happens the expansion of the primary component will practically cease, for its excess mass will escape from the Roche lobe and begin to pour into the second component, whose mass will now begin to grow.

The rate at which the evolving star loses mass will rise very rapidly with any further increase in the radius R of that star after the radius R_1 of the Roche lobe is attained. Calculations show that the mass loss q per unit time conforms to the law

$$q \propto \left(\frac{R - R_1}{R_1} \right)^{n+3/2}, \tag{14.2}$$

where the quantity n, called the polytropic index, is a parameter describing the structure of the star. We may adopt $n = 3$; then in order for the mass exchange between the components of the binary system to proceed at a more or less reasonable pace the ratio $(R - R_1)/R_1$ must be less than 0.03. In other words, during the evolutionary phase when mass is flowing from one component to the other, the radius of the evolving star must continually stay very close to the radius of the Roche lobe.

To a first approximation we may assume that the gas shed by the evolving star will not leave the confines of the binary system in the course of the evolution; that is, the total mass $\mathfrak{M} = \mathfrak{M}_1 + \mathfrak{M}_2$ will be conserved. It is then natural to suppose that the distance separating the components will vary with the evolution according to the relation

$$a = \frac{\text{constant}}{\mathfrak{M}_1^2 (\mathfrak{M} - \mathfrak{M}_1)^2}. \tag{14.3}$$

One can show that the components will be closest to each other when enough mass has been transferred to the unevolving component for the masses of the two stars to be comparable.

How will evolution proceed in such a system? For definiteness, let us consider the case in which the mass of the evolving component is 5 $\mathfrak{M}_\odot$ and the components have a mass ratio of 2. The theoretical time dependence of the radius of such a star (if it were single) is illustrated in Figure 14.2. There are three steps in the swelling of the star as it evolves:

    A. An initial phase, in which hydrogen is burned at the center of the star and the radius gradually increases (see Figure 12.2) until the star has left the main sequence.

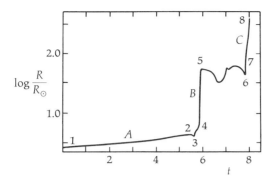

Figure 14.2   Theoretical dependence of the radius of a star on time.

B. Rapid expansion of the stellar envelope, accompanied by contraction of the core after the hydrogen there has all burned. This phase continues until the temperature of the contracting core has risen to the point at which the triple-alpha process discussed in Chapter 8 begins to operate.

C. A phase that sets in after the helium has burned. The core starts to contract again and becomes further heated until the nuclear reactions involving carbon begin.

The evolution described above refers to an isolated star, but in a binary system it will be violated, the more so if the components are close together. For example, if in our case the period of the binary system is about one day, the star will already fill its Roche lobe in phase A and will begin to discharge mass toward the secondary component. If the period measures some weeks, the mass exchange will occur in phase B. And finally, if the period is longer than three months it will happen in phase C. This last phase, which certainly not every star in the single state reaches, has been very inadequately investigated, and we shall not be concerned with it here.

In recent years quite a few numerical calculations of evolution with mass exchange in binary systems have been performed. These calculations show that two steps should be distinguished in the mass exchange. At first the rate of mass transfer from the evolving component to the secondary is very high. The evolving component loses a large part of its mass in a time comparable to the Kelvin–Helmholtz scale for contraction (see Chapter 3):

$$t_K = \frac{G\mathfrak{M}^2}{R_1 L} = 3 \times 10^7 \, \frac{\mathfrak{M}^{*2}}{R^* L^*} \, \text{yr}, \tag{14.4}$$

211

where an asterisk means that the quantity is expressed in solar units. The average rate of mass exchange will therefore be

$$q = \frac{\mathfrak{M}^*}{t_K} = 3 \times 10^{-8} \frac{R^* L^*}{\mathfrak{M}^*} \, \mathfrak{M}_\odot/\text{yr}. \tag{14.5}$$

Such binary stars as the remarkable β Lyrae system are evidently passing through this rapid mass exchange state.

By the close of this stormy period in the evolution of a binary system, the two components will have an opposite mass ratio. If at the outset the evolving component was twice as massive as its companion, by the end of the period it will have become only half as massive. Thenceforth evolution will proceed considerably more gradually in the system, and the rate of mass transfer will fall off sharply. Little further change will take place in the luminosity of the now less massive evolving component. It will now be a subgiant, and it will continue to evolve for a period approximately as long as the time required for the originally more massive star to evolve while it was calmly resting on the main sequence. But the subgiant will have 10 times the luminosity of a main sequence star with a mass equal to that after the exchange process.

Thus far we have described the evolution of a binary system during phase A. Phase B proceeds differently for stars of relatively high and low mass. This distinction is present because, as we have seen in Chapter 12, an exceedingly dense, degenerate core is formed during the evolution of less massive stars. The period of rapid mass exchange will be similar for stars of any mass if the exchange begins during phase B. But afterward there will be differences. In the more massive stars the subsequent evolution will take place considerably faster. If the original mass of the evolving component exceeds 3 $\mathfrak{M}_\odot$, then once the triple-alpha reaction process has started in the core, the star will stop expanding and the mass loss from its surface will slow down greatly or cease altogether. Such a star, according to the Polish astronomer Bohdan Paczyński, who has done much work in this field, will resemble a Wolf–Rayet star, a very hot object whose spectrum exhibits broad emission lines.

On the other hand, if the original mass of the primary star is comparatively low, the rapid expansion of its envelope in the red giant phase will stop for another reason: degeneracy will set in within the core of the star. This development will also sharply inhibit the mass outflow. The star will radiate because of hydrogen reactions inside a thin shell surrounding the core (see Chapter 12). The luminosity of the evolving star will be quite high, 100 times the luminosity of a main sequence star of the same mass. It is interesting to recognize that by the close of this phase the mass of the evolved component will have diminished greatly: it may become five or even

212

ten times smaller than the mass of the secondary component, which will then have absorbed into itself the bulk of the original mass of its companion. But even though it has swallowed material from the first star, the secondary component will remain on the main sequence; because of its greatly increased mass, however, it may even become more luminous than the evolved component. Just such a situation is observed in close binary systems of the Algol type. Ultimately, when only a little mass is left in the evolving star, its radius will begin to shrink and it will evidently turn into a white dwarf.

The picture sketched above for the evolution of binary systems in phases A and B is supported by a great many observations. In particular, this interpretation explains in a natural way the long-known empirical fact that the smaller the mass ratio $\mathfrak{M}_1 / \mathfrak{M}_2$, the greater the excess luminosity of the evolving component. Interesting problems arise when the subsequent evolution of the secondary component is analyzed. Matters are complicated by the circumstance that the gas transferred to the secondary component carries a large angular momentum along with it because of the orbital motion of the evolving star that is shedding mass. For this reason a rapidly rotating gaseous disk may form around the secondary star and, so to speak, store much of the excess angular momentum within itself (Figure 14.3). In the limiting case in which most of the mass and orbital angular momentum of the evolving component are communicated to the secondary component and the gaseous disk spinning around it, the distance between the components will become smaller by a factor of three to four. It has been estimated that more than half the mass of the binary system and at least half

Figure 14.3   Schematic configuration of the gas streams and disk in the β Lyrae system.

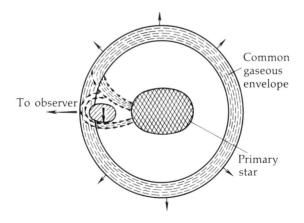

its angular momentum would be concentrated in the gaseous disk. Indeed, the observations prove that dense gaseous disks and rings are present in binary systems.

We have outlined the basic trends of stellar evolution in close binary systems. The term *close* as used here does not at all signify a geometric proximity of the components. We call a system close if its evolving component fills the corresponding Roche lobe at any phase. But we have already seen that in phase C (Figure 14.2) the lobe can be filled even though the components are as far apart as one astronomical unit and their orbital period is measured in years. From the standpoint of stellar evolution, then, most binary systems are close. It is important to understand that the evolution of these systems as described here is gradual, quiescent, by no means catastrophic in character.

However, astronomers have long been aware of a class of highly unstable stars that always belong to binary systems and are never observed to be solitary. These are the novae and novalike stars, as well as flare stars of the U Geminorum type. The binary systems in which these strange objects are found differ in several interesting respects. First of all, they generally possess very short periods, typically just a few hours. Both components are dwarf stars, and they are situated very close together. Such objects often have a rather late spectral type, K or M, on which are superimposed broad emission lines as well as the continuous spectrum of a very hot component. Everything suggests that the red component in a system of this kind (a dwarf!) has filled its Roche lobe and is shedding mass through the inner Lagrangian point toward the hot component of the binary system. As in the case of a binary system experiencing calm evolution, a rapidly spinning, fairly massive gaseous disk forms around the hot component and produces the emission lines observed. The active component in such a system is the compact hot star surrounded by the disk into which matter streams from the star inside. It seems that the hot component has already passed through its evolution and, at some epoch in the past, transferred much of its material to its companion star. But now the companion is returning the favor by restoring to the evolved star the material "borrowed" many millions of years ago.

One distinctive characteristic of the outbursts in novae and novalike stars is their repeatability, or recurrence. The interval between flares in novalike stars is about 100 years. Presumably, in the more violently eruptive novae these intervals stretch out to thousands of years. That nova outbursts are actually recurrent follows simply from the fact that several dozen novae appear every year in the Galaxy. Were it not for recurrence, then after a couple of billion years *all* the stars in the Galaxy would flare up as novae, which is patently absurd. Thus in a certain class of stars outbursts can occur repeatedly.

There can be little doubt that the well-evolved, compact, hot star represents an object allied to a white dwarf and very poor in hydrogen (see Chapter 12). Yet the red component filling its Roche lobe is continually sending hydrogen-rich gas to the evolved star.[1] After this gas has collected in the surface layer of the hot star for hundreds or thousands of years, it may become responsible for a thermal explosion of a "local" character, that is, not embracing the whole structure of the star. A sizable amount of mass, perhaps 10^{-5}–10^{-4} $\mathfrak{M}_{\odot}$, is ejected in such an outburst, as may be inferred from spectroscopic observations of novae. This is approximately the mass that is transferred to the compact hot star from its companion during the interim between successive outbursts.

Much still remains unclear in this purely qualitative picture of nova outbursts. Above all, what kind of nuclear reactions are stoked by the hydrogen that has accumulated in the surface layers of the evolved star? How many outbursts can the resources of the binary system sustain? What will such a system be like when the eruptive phase has ended? All these interesting questions are still awaiting their answer.

U Geminorum stars have flares with a considerably higher recurrence frequency and a smaller amplitude than novae (Figure 14.4). Like novae between outbursts, stars of this type are very compact, hot objects of low luminosity. It is remarkable, however, that no traces of ejected gas are observed in the flares of U Geminorum stars. Nevertheless, spectra of these stars recorded during quiescent periods between flares show emission lines, just as novae do, indicating that a gaseous disk is present. Apparently the mechanism responsible for the quick sequence of flares in U Geminorum stars is completely different from the nova mechanism. These stars too await their investigators, both observationalists and theoreticians.

Thus evolution in close binary systems can result in the creation of "Siamese twins," certain freaks that we observe as novae, U Geminorum stars (is Gemini the constellation of Siamese twins?), and the like. Some even more astonishing binary systems will be discussed in Part IV of this book. Enough has been said to conclude that the doubling of a star decisively controls its evolution.

---

[1]Recent polarimetric observations have shown that a former nova (the one which appeared in the constellation Hercules in 1934) is rotating about its axis with a period of 142 seconds, and that its rotation speed is gradually increasing. This fast spin is easily explained by gas flowing out of the companion star, a red dwarf, and carrying a large angular momentum along with it.

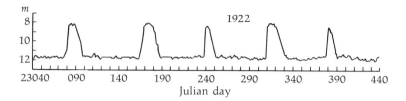

Figure 14.4 A light curve for SS Cygni, a star of the U Geminorum type.

To close this chapter we consider the interesting phenomenon of runaway stars—hot, massive stars of spectral type O whose velocities with respect to neighboring stars are very high, around 100 km/s, whereas normal O stars have velocities closer to 10 km/s. Another peculiarity of runaway stars is that they are always single, even though multiplicity is especially common among O stars.

Some years ago the Dutch astronomer Adriaan Blaauw suggested that the runaway stars may have resulted from the breakup of wide double systems caused by an explosion of the more massive component; its loss of material would sharply weaken the gravity that had held the rapidly orbiting pair together. Various evidence now indicates, however, that the mass exchange process we have described will postpone the explosion until it can no longer disrupt the binary; rather, the explosion will impart a strong thrust to the system as a whole.

The exploded component in the fast-moving system will probably have collapsed to a neutron star, emitting x rays for a while. (X-ray sources indeed seem to be high-velocity objects.) But the neutron star would persist as a radio pulsar (Chapter 19) for a much longer time, little inhibited by the "stellar wind" flowing from its optical companion. This interpretation would receive decisive support if pulsars should be discovered at the sites of runaway stars, and we would gain a new technique for probing stellar winds.

Part III

Stars Explode

On the twenty-second day of the seventh moon of the first year
of the period Chih-ho, Yang Wei-tê said: "Prostrating
myself, I have observed the appearance of a guest-star
in the constellation T'ien Kuan; on the star
there was a slightly iridescent yellow color.
Respectfully, according to the dispositions for
Emperors, I have prognosticated, and the result said:
The guest-star does not infringe upon Aldebaran;
this shows that a Plentiful One is Lord, and that
the country has a Great Worthy. I request that
this prognostication be given to the Bureau
of Historiography to be preserved."

—*Sung Hui Yao*

*(In this chronicle the Chief Calendrical Computer
of Imperial China, Yang Wei-tê, reports his
observations of a supernova in the summer
of 1054. History has indeed preserved
vestiges of this explosion. Today
we recognize it as the Crab Nebula,
perhaps the most interesting
object in all the Galaxy.)*

15

Supernovae:
A General Survey

From time immemorial sky watchers have known that once in a while stars appear where none had been noticed before. Whenever a star flares up and becomes bright enough, it destroys the familiar configuration of the constellation in which it has emerged and attracts the involuntary attention of people who, while no specialists in astronomy, nevertheless know the stars in the sky. But flares as bright as these are rare events. Historical chronicles bring us reports of remarkable phenomena of this kind that happened many centuries ago (see the epigraph to this part of the book). In most cases, though, the strange stars are too faint to be seen with the unaided eye. These extraordinary stars have long been known as *novae.*

That novae are galactic objects was established quite some time ago. At maximum light their absolute magnitude reaches -7 or even brighter. Accordingly, they become tens or hundreds of thousands of times as luminous as the sun. After a few months their brightness drops sharply, and eventually they become stabilized as hot little dwarf stars of very low luminosity. Some years ago it was shown that the great majority of novae (if not all) are close binary systems. Several dozen novae flare up annually in our Galaxy, but few of these are accessible to astronomical observation, since most of them are very far away and are hidden from us by the light-absorbing interstellar dust medium.

In Chapter 14 we spoke briefly of these stars in connection with the problem of evolution in binary stellar systems. Individual novae, we recall, flare up repeatedly at fairly protracted time intervals, amounting to hundreds or thousands of years. It is important to understand, however, that for all its grandeur such an outburst does not entail any fundamental structural change in the star, let alone its disruption. After each successive outburst the star reverts to approximately the same state as before the event. On occasion, however, astronomers observe an incomparably more magnificent phenomenon: the explosion of a star, accompanied by radical changes in its structure. But astronomers did not by any means come to this interpretation all at once.

The story begins on August 31, 1885, at practically the oldest observatory in the Soviet Union of today—the one in Tartu, Estonia. There the German astronomer Ernst Hartwig discovered a nova located fairly close to the nucleus of the Andromeda Nebula. This star had a brightness of about 6.5 magnitudes, so people with keen vision could see it without any optical instruments. The apparent magnitude of the whole Andromeda Nebula is about $4^m.5$; thus the radiation flux from the brilliant nova was only 6.3 times smaller than from the entire nebula. Since there could be no doubt that the star had blazed up within the nebula itself, its intrinsic luminosity must have been only 6.3 times smaller than the luminosity of the Andromeda Nebula.

Neither Hartwig nor his contemporaries yet recognized that the Andromeda Nebula is not simply a patch of luminous matter located relatively close to the sun, but a gigantic stellar island numbering several hundred billion stars. At that time the word *galaxy* was not even in the language. To be sure, ever since the eighteenth century the German scientist Johann Heinrich Lambert's concept of "island universes" had been widespread in astronomy. Spiral nebulae, which then were already familiar, were thought to represent enormous collections of stars embedded in a highly rarefied medium of gas and dust. According to this concept our Galaxy, observed as the Milky Way belt in the sky, was considered to be an island universe of the same kind as spiral nebulae very remote from us. However, Lambert's deep idea was purely speculative in character. No procedure for determining the distances of the spiral nebulae had yet been substantiated physically. It was not until the early 1920s that the concept of island universes was proved, to become a lasting conquest of science.

We therefore ought to be astonished above all at the grandeur of the phenomenon observed by the astronomer at Tartu. Just imagine! Here is the Andromeda Nebula, which we now know is 650,000 parsecs away from us, or more than 2 million light years. And yet at this awesome distance a star flares up that can be seen by the unaided eye! For comparison, note that even the brightest regular stars discovered in this nebula (which, by the way, has furnished the main proof that the island universe concept is correct) have an extremely faint apparent magnitude, about 20^m. There are hundreds of billions of stars in that giant galaxy (larger than our own Galaxy, which is itself a giant object); yet altogether they were emitting only 6.3 times as much light as the single brilliant star. What a fireworks!

From Hartwig's observations we can reconstruct the light curve of this star—the dependence of its magnitude on time. For example, two weeks before maximum it had a brightness of 9^m, whereas a year earlier nothing could be detected at the position of the star, so it was then fainter than 15^m. Beginning in March of the following year, 1886, there was no trace of the star even in the largest telescopes.

During recent decades ordinary novae have systematically been observed in the Andromeda Nebula. At maximum light they reach 17–18^m. Every year several dozen novae are observed there, approximately as many as appear in our own Galaxy. The nova of 1885 must therefore have represented a truly uncommon event, as it was about 12^m brighter than ordinary novae. In other words, its luminosity was tens of thousands of times that of ordinary novae at maximum light.

Between 1885 and 1920 several outbursts of bright novae were observed in the extragalactic nebulae, or galaxies, closest to us. An outburst of exceptional interest was the star that appeared in July 1895 in the nebula NGC 5253. This star, known as Z Centauri, had a brightness of $7^m.2$ at maximum light. Remarkably, the galaxy NGC 5253 itself was 5^m (fully 100 times) fainter. That galaxy admittedly is a dwarf, no match for the Andromeda Nebula or our Galaxy, but even so it contains several billion stars. Thus for a short time a single star emitted 100 times as much radiation as all the billions of stars in the whole galaxy! Something to wonder at. And history repeats itself: in the same galaxy, NGC 5253, another star flared up in 1972, reaching a brightness of 8^m. It has been contributing much to the growth of our understanding about the nature of such objects. After all, the techniques of astronomical observation are at an immeasurably higher level today than in 1895. The second nova in NGC 5253 has been furiously assaulted by a whole army of astronomers, who have carefully investigated its radiation in many different parts of the spectrum. There has already been substantial progress in the interpretation of these objects now that analyses have begun to be published.

Altogether about 10 such outbursts were recorded in various galaxies during the 1885–1920 period. Events of this kind were observed in galaxies of every shape—elliptical, spiral, irregular. From this incomplete series of observations alone we may draw the very important conclusion that outbursts of such phenomenal power occur extremely rarely. A rough estimate shows that in any given galaxy there will be one outburst every few centuries, on the average.

In 1919, on the basis of the observational evidence just described, the eminent Swedish astronomer Knut Lundmark put forward the hypothesis that along with ordinary novae, whose outbursts are fairly common, on rare occasions stars flare up that are tens of thousands of times as bright at maximum. The Swiss- and German-American astronomers Fritz Zwicky and Walter Baade proposed in 1934 that such stars be called *supernovae.* Although this term is, in my opinion, rather absurd, it came into very wide favor and is now universally used to designate the mighty event of a stellar explosion.

Supernova outbursts should also occur from time to time in our Galaxy, a giant stellar system only a little inferior to the Andromeda Nebula. But, as we have said, the phenomenon is very rare. Still, if a supernova outburst can just be seen with the unaided eye even from so immensely far away as the Andromeda Nebula, what could we expect of a supernova flaring up nearby, within our Galaxy? A point of some importance should be made here. Because of absorption by interstellar dust grains, the light of objects distant from us in our Galaxy will be very strongly attenuated. The uninitiated reader might of course ask, "Why doesn't this happen when supernovae arise in foreign galaxies far away?" The answer is that our Galaxy, like other spiral galaxies, is a markedly flattened system. A good model might be a very thin disk shaped like a phonograph record. The light-absorbing dust particles are concentrated in a very thin layer. Hence if a supernova were to flare up at the far end of the disk, its light would have to travel many thousands of parsecs through the dust layer and could be completely absorbed. But other galaxies are generally so oriented that our line of sight forms a large angle with the planes of their disks. Light from supernovae in other galaxies, therefore, will travel a comparatively short distance, perhaps a few hundred parsecs, through an absorbing medium.

Nevertheless, the outburst of a supernova in our Galaxy should as a rule be accompanied by very striking optical effects. Lundmark made a special study of ancient historical chronicles in an effort to find accounts of sudden appearances of stars, which in some cases might have been supernovae. This captivating type of research had also been carried on prior to Lundmark (by Edouard Biot and Alexander von Humboldt, for example, in the mid–nineteenth century). But only the Swedish astronomer knew what to seek: he was looking for old supernova outbursts. His quest yielded outstanding results.

Studies by Lundmark and his successors have established that over the past 1000 years at least six supernovae have been observed in our Galaxy: in 1006, 1054, 1181, 1572, 1604, and 1667. The supernova of 1054 has played a special role in the history of astronomy, for the famed Crab Nebula (see Chapters 17 and 19) stands in its place. But clearly, of course, ancient chronicles are entirely inadequate for investigating this unique phenomenon. Special observing patrols must be organized to watch for supernovae appearing right before our eyes in other stellar systems. The idea behind a search of this kind is very simple: if a supernova outburst occurs only once every few centuries in any given galaxy, then by systematically monitoring many hundreds of galaxies we might hope to observe one or two supernovae annually, on the average (unless we miss some). The search faces the difficulty, however, that we cannot possibly tell in advance in which galaxy an outburst will happen.

The first supernova patrol was instituted in 1933 by Zwicky, using a very modest 10-inch telescope. He conducted systematic searches in 175 areas of the sky containing most of the galaxies comparatively close to us. In this way he carefully monitored 3000 galaxies brighter than 15^m, of which 700 were brighter than 13^m. Results were not long in coming from this regular program. During the period 1936–1939 he observed 12 supernovae in different galaxies. Allowing for unavoidable imperfections in the patrol system, Zwicky reached the important conclusion from his observations that a galaxy will experience on the average one outburst every 360 years.

Zwicky's work was carried on in closest cooperation with other astronomers. The supernovae he discovered were investigated as thoroughly as possible, both photometrically and spectroscopically. Light curves (Figure 15.1) as well as spectra were obtained for these supernovae. After a break due to World War II the surveys were resumed in the 1950s with better instrumentation. Thus, while only 54 supernovae had been discovered from 1885 to 1956, in the next seven years 82 were found, and by 1977 more than 450 supernovae were on record.

Figure 15.1 Light curve of the supernova that appeared in the galaxy NGC 1003 in 1937.

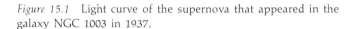
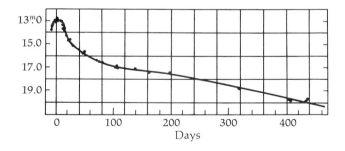

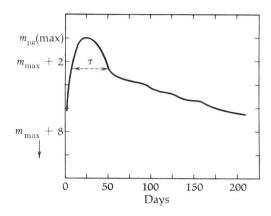

Figure 15.2   Schematic light curve of a type I supernova (F. Zwicky).

These studies have shown that supernovae by no means represent a homogeneous group of objects. First of all, their light curves are very diverse. To a certain approximation supernovae may be divided into two types according to their observed properties. Figure 15.2 schematically illustrates the light curve of a type I supernova. Following a swift rise, the brightness remains nearly constant for some time. Then the supernova begins to decline in brightness fairly rapidly, after which its numerical apparent magnitude proceeds to increase almost linearly, corresponding to an exponential decay in luminosity. The close resemblance among the light curves of different supernovae after maximum is noteworthy. Type II supernovae display light curves of a quite distinct sort (Figure 15.3). They occur in a wide variety, but as a rule their maxima are narrower (of shorter duration) than for type I supernovae. During the terminal phase type II objects have considerably steeper light curves. Secondary maxima or other features are sometimes observed. Supernovae of this type are very likely not a homogeneous group.

If we know the distances of the galaxies in which supernovae have appeared, we can find their absolute magnitudes at maximum light. These are close to -19^m, which corresponds to a luminosity as high as 3×10^{43} erg/s, or nearly 10 billion times the luminosity of the sun! From the light curves one can estimate that during the course of its outburst such a star will radiate up to 10^{50} erg of energy. The sun needs a billion years to radiate this much energy, but here it is liberated in just a few months.

An interesting and unquestionably important circumstance is the dependence of the supernova type on the properties of the galaxy in which the outburst has occurred. Type II supernovae appear only in the arms of spiral galaxies, whereas in elliptical and irregular galaxies we find only type I supernovae. But in spiral galaxies such as our own, both type I and type II supernovae occur. The fact that only type I supernovae flare up in elliptical galaxies is itself highly significant. According to current views based on stellar evolution theory and observational evidence (Chapter 12), the star population of such galaxies includes no stars whose mass exceeds a certain limit close to the mass of the sun. Elliptical galaxies contain very little interstellar matter, so that the star formation process there ended long ago. Hence the population of these galaxies comprises very old stars of low mass (not much more than the solar mass). About 10 billion years ago, when star formation was actively taking place in elliptical galaxies, massive stars too were produced there. But such stars have a comparatively short evolutionary time scale, as we have seen in Chapter 12, and they would long since have passed through the red giant phase and turned into white dwarfs and other "dead" objects, which we shall describe in Part IV of this book. An important conclusion follows: before their outburst type I supernovae were very old stars whose mass was at most only slightly (say 10 to 20 percent) greater than the mass of the sun. Since the light curves and spectra (see

Figure 15.3   Light curves of several type II supernovae (W. Baade).

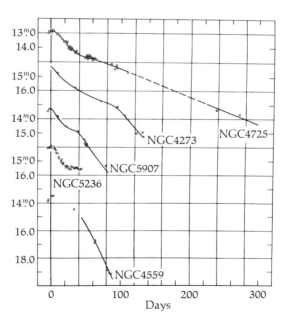

below) of all supernovae of this type are remarkably similar, we may state that even in spiral galaxies such as our own, stars exploding as type I supernovae are very old objects of comparatively low mass.

As for the stars that become type II supernovae, it is logical to infer that they are young objects. This conclusion follows from the simple fact that they are located in spiral arms, where stars are formed out of a gas–dust medium. They must be located fairly close to the site of their formation, simply because they could not have traveled very far during their lifetime. Since the random velocity of stars and gas clouds in spiral arms is about 10 km/s and the arms are several hundred parsecs thick, we find that the stars representing future type II supernovae are no more than a few tens of millions of years old.

But on the other hand, even without knowing anything about the specific mechanism responsible for the outburst (actually an explosion) of a star, we can say that it cannot go through such an unpleasant event until it has left the main sequence and begun the very complicated closing phase of its evolution (Chapter 12). What type of stars, then, stay no longer than a few tens of millions of years on the main sequence? Evidently they can only be stars with a mass of at least 10 $\mathfrak{M}_\odot$ (Chapter 12). Thus from very simple considerations we may conclude that the stars becoming type II supernovae are massive young objects. When they were located on the main sequence they were stars of spectral types O and B, or hot blue giants. However, the fact that only type I supernovae appear in irregular galaxies such as the Magellanic Clouds would seem inconsistent with the picture we have outlined, for these galaxies contain a great many hot, massive stars. Why is it that type II supernovae are not observed there?

We have emphasized above that in the case of type I supernovae very old stars are exploding, with a mass only slightly greater than that of the sun. The regular outcome of the evolution of such a star is its conversion into a white dwarf accompanied by the formation of a planetary nebula (Chapter 13). Several planetary nebulae are produced every year in our Galaxy, and just the same number of stars having a mass scarcely greater than the sun's end their lives by turning into white dwarfs. Yet a type I supernova outburst occurs only about once a century (or perhaps a little more often; see Chapter 16), and the exploding star should have the same mass as the ancestors of the planetary nebulae. But this situation implies that only one out of a hundred stars with the same comparatively low mass will become a type I supernova as its evolution comes to an end. Why? What stellar "pathology" controls this altogether singular dénouement in the life of a star, so dramatically different from the fate of the great majority? Alas, we must say forthwith that modern theoretical astrophysics has not yet given a persuasive answer to this very clear question. There are, to be sure,

226

quite a few rather elegant hypotheses in this regard. And it is understandable that so magnificent a phenomenon of nature would attract the interest of theoreticians. We shall explain some of these hypotheses in Chapter 18.

As with the light curves, the spectra of type I and II supernovae also differ from each other. Type II supernovae have spectra of a more orthodox form. Against a very bright background continuous spectrum they exhibit broad emission and absorption bands. The energy distribution in the continuous spectrum near maximum light shows that the gases are radiating at a very high temperature, above 40,000°K. In this regard the spectra of type II supernovae resemble those of hot stars of spectral type B. The principal emission and absorption bands can be reliably identified with lines of hydrogen, helium, and several other elements. These bands are exceptionally wide because of the Doppler effect due to the enormous velocities of the gases moving in different directions and producing the emission and absorption. The velocities may reach 10,000 km/s. Type II supernovae have spectra similar to ordinary nova spectra, which have been rather fully studied and are attributable to the ejection of a substantial amount of gas during the outburst. There is a distinction, however, in that type II supernovae have far wider emission and absorption bands.

By applying dependable astrophysical techniques for analyzing such spectra, we can estimate the mass of the gas ejected during an outburst. For type II supernovae it exceeds 1 $\mathfrak{M}_\odot$, whereas in ordinary novae only 10^{-4}–10^{-5} $\mathfrak{M}_\odot$ is ejected in each outburst. The sharp rise in the brightness of a supernova after the flare starts actually results from the continuous growth of the radiating surface of the opaque gaseous envelope as it expands at great speed. After maximum light the ejected envelope becomes transparent, and further expansion merely serves to decrease its luminosity, since the density of the envelope drops rapidly. The gaseous material expelled during the outburst forever breaks its connection with the exploding star and travels out into interstellar space, interacting with the interstellar medium. Huge envelopes are formed after supernova explosions; they persist as nebulae for tens of thousands of years and constitute very important objects of astronomical research. We shall discuss these objects in some detail in Chapters 16 and 17, but for the present we merely wish to point out the large amount of mass ejected in a type II supernova outburst, clearly demonstrating that the stars were quite massive before their explosion—the same result obtained above by indirect arguments.

The spectra of type I supernovae have a rather different appearance. They are very similar to one another, and curiously enough the time dependence is the same for different supernovae. One can even determine from the form of the spectrum the time elapsed since the outburst. These spectra typically have very broad, partially overlapping emission bands. As time passes individual bands gradually disappear, with new bands taking their place; their relative intensity also varies. For a long while astronomers could not comprehend the spectra of type I supernovae, although of course there was no lack of hypotheses. It was not until 15 years ago that the late American astronomer Dean B. McLaughlin came across the right trail. Previous investigators had been "hypnotized" by the broad emission bands in the spectra of these supernovae; attempts to identify them were in vain. The American astronomer called attention to the dips between the bands, interpreting them as absorption bands eating away at the continuous spectrum. He obtained several new spectrograms of type I supernovae in which the presence of absorption bands is so distinct as to leave no room for doubt. McLaughlin identified these absorption bands with lines of helium, calcium, and certain other elements; the radiating gases move in a disordered fashion at velocities of about 10,000 km/s.

Subsequently, McLaughlin's method was successfully applied by the Soviet astronomers Eval'd R. Mustel' and Yurii P. Pskovskii to decipher type I supernova spectra. In his identification of the absorption lines in these spectra, Pskovskii started by assuming that the vast photospheres of supernovae are essentially similar to the photospheres of supergiants of spectral types A and B. The only difference was considered to be that the gas densities are many orders of magnitude higher in the case of supernovae. This similarity implies in particular that lines sensitive to the luminosity of a star (such as strong lines of singly ionized metals that are fairly abundant) should be enhanced. Only lines of this kind will not be washed out by the Doppler effect resulting from the enormous expansion velocities of supernova envelopes.

In the red part of the spectrum the lines forming the $\lambda\lambda$ 6347, 6371 Å doublet of ionized silicon are examples. At expansion velocities of around 10,000 km/s this doublet can account for the rather deep minimum observed at 6150 Å in the spectra of nearly all supernovae of this type. Another ionized-silicon doublet is responsible for the minimum in the blue part of the spectrum at 4000 Å. Pskovskii has also identified the fairly deep minimum at 3800 Å in the near-ultraviolet part of the spectrum with the prominent H and K resonance lines of ionized calcium. The identification suggested by Pskovskii has been fully vindicated with the discovery by the American astronomers Robert P. Kirshner and John B. Oke and coworkers that a very deep minimum at 8600 Å in the near-infrared region is undoubtedly due to the $\lambda\lambda$ 8498, 8542, 8662 Å triplet of ionized calcium.

Today it appears that the spectra of type I and type II supernovae do not differ drastically from each other. Both types of spectra exhibit broad, diffuse absorption lines framed on the redward side by a wide, more or less intense emission feature (see Figure 15.4). Lines of this character are formed

228

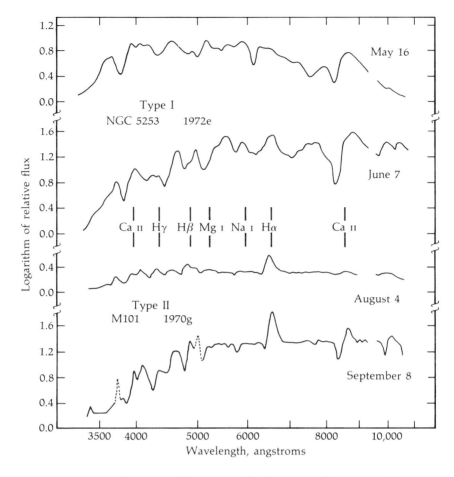

Figure 15.4 Tracings of the energy distribution in the spectra of representative type I and type II supernovae at comparable phases of development (Hale Observatories).

in extended, moving stellar envelopes. Evidently the main distinction between the spectra of type I and II supernovae is that in the latter case the hydrogen lines are strongest, while in type I supernovae the hydrogen lines are either very faint or entirely absent. This important disparity is most naturally explained by differences in the chemical composition of the corresponding envelopes.

By analyzing the spectroscopic data we can arrive at a rough estimate for the mass of the ejected envelope. It is of the order of one solar mass.

A similar estimate has been obtained by the author from analysis of the oxygen emission lines observed late in the spectral development of the bright type I supernova that appeared in 1937. We shall see presently that the masses of the envelopes expelled in outbursts can also be determined by other methods.

16

Supernova Remnants as X-Ray and Radio Sources

When a star explodes in a supernova event, a nebula is formed which expands at an immense rate, generally of the order of 10,000 km/s. This high expansion velocity is the chief sign that distinguishes the remnants of supernova outbursts from other nebulosity, such as planetary nebulae. Planetaries expand at a moderate speed—a few tens of kilometers per second, or about the velocity to be expected for hot gas expanding into a vacuum (see Chapter 13). But supernova remnants are a different story: everything here points to an explosion of enormous power, blowing the outer layers of the star off in all directions and imparting great velocities to the separate pieces of the expelled envelope.

Many hundreds or thousands of years later, the gas clouds ejected in the explosion will begin to be slowed down by the ambient medium with which they are interacting; their velocities will start to fall, sinking to just hundreds or even tens of kilometers per second. Well before that happens, all traces of the exploded star observable in the optical range will have vanished. But through tens of millennia a distinctive nebula will persist, the product of a gigantic cosmic catastrophe—the explosion of a star. After some hundreds of millennia, however, this catastrophic residue in the interstellar medium will also have been completely erased: the supernova remnants will have become fully dissolved in that medium. Only the pulsars, those still puzzling objects into which most exploding stars are transformed (see Part IV), will continue to emit radio waves for many millions of years more.

We may regard the outburst of a supernova as a very strong local disturbance of the interstellar medium surrounding it. In this sense we need not understand what has caused the star to explode or what the specific properties of the explosion process are. We need only know the total amount of energy released in the explosion as kinetic energy of the ejected gaseous envelope, and the density of the surrounding interstellar medium. A comparable problem pertaining to strong explosions in the earth's atmosphere (which fortunately are now forbidden by most civilized countries) has been solved by the Soviet hydrodynamicist Academician Leonid I. Sedov. In 1960 the author of this book applied Sedov's solution to the problem of supernova outbursts.

Let us suppose that the surrounding interstellar medium is homogeneous with a constant gas density of n_1 atoms per cubic centimeter. Sedov's theory assumes that the explosion is adiabatic; that is, no energy escapes from the site of the explosion by means of radiation. A supernova outburst may be treated as an instantaneous release of thermal energy E at a point that we shall take to be the origin of coordinates at time $t = 0$. At some future time t the disturbance from the explosion will permeate all the interstellar medium within a sphere of radius R_2. Inside this sphere the interstellar gas through which the shock wave generated by the explosion is propagated will have a very high temperature. Beyond the sphere, the temperature will drop abruptly to its normal undisturbed value. Right at the boundary of the sphere, where $R = R_2$, the interstellar gas will be four times as dense as in the undisturbed region outside.

Now if we wish to apply Sedov's theory to our problem we must regard the interstellar medium as a continuous compressible fluid. This assumption is perfectly legitimate, because even though the interstellar medium is exceedingly rarefied, the mean free path of the atoms and ions there nevertheless is far shorter than the distance R_2. According to Sedov's theory the following basic equations will be satisfied:

$$R_2 = \left(\frac{2.2E}{\rho_1}\right)^{1/5} t^{2/5} = 10^{15}\left(\frac{E}{E_0 n_1}\right)^{1/5} t^{2/5} \text{ cm}, \tag{16.1}$$

$$T_2 = \frac{3}{25}\frac{2.2E}{kn_1}R_2^{-3} = 1.45 \times 10^{21}\left(\frac{E}{E_0 n_1}\right)^{2/5} t^{-6/5} \text{ °K}, \tag{16.2}$$

$$\rho_2 = 4\rho_1, \tag{16.3}$$

where $E_0 = 7.5 \times 10^{50}$ erg, k is the Boltzmann constant, t is expressed in seconds, and the density of the interstellar medium is given by $\rho_1 = m_H n_1$.

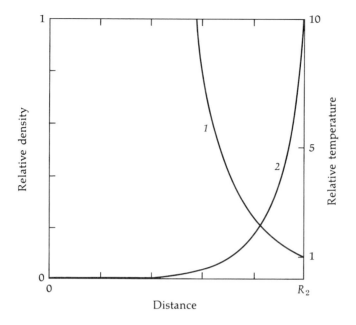

Figure 16.1 Schematic distribution of (1) the temperature and (2) the density in a shock wave (C. Heiles).

Equation 16.2 has a simple meaning: all the energy E released in the explosion will be disseminated to the gas particles inside the sphere of radius R_2, heating the gas to the temperature T_2. More detailed calculations yield profiles for the density and temperature distribution within the sphere that encompasses the disturbance from the exploding star. These profiles are illustrated in Figure 16.1, which indicates that the gas density is very low in the central part of the sphere. In effect the gas forms a layer about 0.1 R_2 thick. The temperature of this gas increases toward the center of the sphere.

From Equations 16.1 and 16.2 we can obtain the rate at which R_2 increases, that is, the expansion velocity of the shock front:

$$v = \frac{2}{5}\left(\frac{2.2E}{\rho_1}\right)^{1/5} t^{-3/5}. \tag{16.4}$$

We thus have the simple relation

$$R_2 = \frac{5}{2} vt. \tag{16.5}$$

233

The practical significance of Equation 16.5 is very great; we can measure the expansion velocity of a supernova remnant (see below) and, knowing R_2, find the age of the remnant, the time elapsed since the explosion.

We should emphasize that Sedov's theory does not apply to the comparatively early stage of the disturbance of the interstellar medium by the explosion. During the later phases, which the theory describes very satisfactorily, all traces of the gas clouds expelled at enormous velocity during the explosion will have disappeared. They will have dissolved into the ambient interstellar gas, giving up their energy to it. The mass of gas contained inside a sphere of radius R_2 is tens or hundreds of times the mass of the gas ejected in the explosion. It consists mainly of the interstellar matter that the explosion has disturbed. On the other hand, the energy radiated by the hot gas behind the shock front is considerably smaller than the original energy E of the explosion.

Still later in the expansion of the nebula the explosion can no longer be regarded as adiabatic, and Sedov's theory once again becomes inapplicable. The gas behind the shock front is able to cool comparatively rapidly. Under such conditions it is not the energy of the moving gas that will be conserved (as in the case of an adiabatic explosion), but its momentum: $\frac{4}{3}\pi R_2^3 \rho_1 v =$ constant. The radius will depend very weakly on time: $R_2 \propto t^{1/4}$. Most radio nebulae representing supernova remnants are either in their adiabatic expansion phase or in a transition phase in which radiative processes are beginning to play a role. To a first approximation, then, Sedov's theory may be applied to the remnants of supernova outbursts.

As indicated above, we have here treated the problem of the disturbance of the interstellar medium by a supernova explosion in an idealized fashion. For example, we have neglected the magnetic field in the interstellar medium as well as the pressure of the relativistic particles confined within the expanding nebula (see below). One can show, however, that during the adiabatic expansion phase these factors are not of controlling significance. Far more important is the circumstance that, unlike the case in our idealized model, the interstellar medium is not homogeneous. As a result, condensations present in the medium will be squeezed by the shock wave traveling outward from the explosion. Dense clumps of gas will form, often with an elongated, filamentary shape. Because of their high gas density, these filaments will cool rapidly to a temperature of some tens of thousands of degrees, and accordingly they will become observable by means of optical astronomy.

Thus the site of the explosion will be bordered by a system of fine filamentary nebulae. These nebulae will be distributed very irregularly around the site, reflecting the original distribution of clouds in the interstellar medium surrounding the exploded star. Several decades ago optical

234

astronomers identified a system of filamentary nebulosity in the constellation Cygnus—the first evidence for the existence of great disturbances in the interstellar medium due to stellar explosions. This interpretation of the filamentary nebulae was first proposed by the eminent Dutch astronomer Jan H. Oort, who pointed out the absence of hot stars capable of exciting these nebulae to glow in the normal way (through the action of ultraviolet radiation).

Figures 16.2 and 16.3 show two of the best-studied filamentary nebulae. The system of filaments in the constellation Cygnus (Figure 16.2) has the

Figure 16.2   The Cygnus Loop, also called the Veil or Network Nebula, a system of delicate filaments representing the remnant of a supernova outburst.

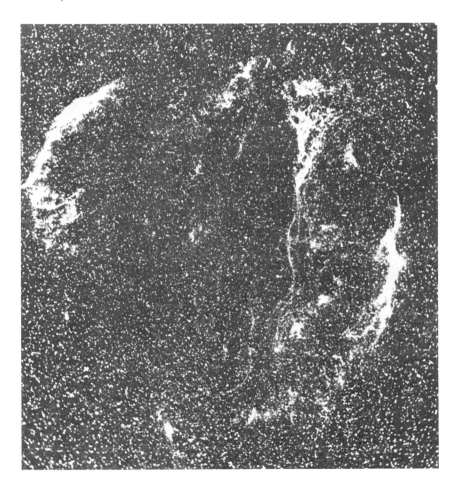

Figure 16.3   The nebula IC 443 in the constellation Gemini, the remnant of a supernova outburst. At the right is the bright star η Geminorum.

huge angular diameter of about 3°. Since the distance of these nebulae is known to be about 800 pc, the linear diameter $2R_2$ of the system is about 40 pc, a very large value. Indeed, a sphere 20 pc in radius encloses several thousand stars! From this example we can grasp how powerful must be the explosion associated with a supernova. The spectrum of the filaments comprises series of emission lines of hydrogen, ionized oxygen, nitrogen, sulfur, and other elements. By analyzing the shifts in the wavelengths of these lines caused by the Doppler effect, one finds that the whole system of filaments shown in Figure 16.2 is expanding at a velocity v of about 115 km/s. Identifying this value with the velocity of the shock front, we can find the age of the filamentary system from Equation 16.5; it is about 70,000 yr. For the other nebula, shown in Figure 16.3, the expansion velocity is about 100 km/s and the age is 65,000 yr. Recently, however, analyses of the soft x rays emitted by these nebulae (see below) have yielded considerably higher velocities for the shock fronts, so the objects are now believed to be several times younger.

The temperature of the gas at the periphery of the system of filamentary nebulae in Cygnus should be about 3×10^6 °K, according to Equation 16.2. Imagine a vast envelope 20 pc in radius in which the interstellar gas is heated to this high a temperature, but with relatively cool, dense, threadlike filaments included inside, as illustrated in Figure 16.2. The bulk of the gas within the envelope of radius $R_2 = 20$ pc has a high temperature; the cool threads are only small impregnations. Other supernova remnants have a similar structure. Thus the great majority of the gas contained in supernova remnants was unobservable until recently, since highly rarefied, very hot gas emits very little optical radiation.

The advent of x-ray astronomy has fundamentally changed this situation. In 1970 a source of soft x rays was discovered at the position of the filamentary nebulosity in Cygnus. The source had an angular size comparable to that of the system of nebulae. The shape of the x-ray spectrum implies that the radiating gas has a temperature of several million degrees. It is interesting to recognize that plasma at such a temperature with a chemical composition similar to the composition of the interstellar medium should radiate strong emission lines in its spectrum—mainly lines of highly ionized oxygen atoms that retain only one or two of their inner electrons. These lines fall in the soft x-ray region of the spectrum and have a wavelength of about 20 Å. Apparently they have actually been detected in the x-ray spectrum of the filamentary nebulae in Cygnus. In the near future x-ray spectroscopy of such objects should provide us with exceptionally valuable information on the physical conditions in supernova remnants.

Although the cosmic x-ray detectors presently available still have a very low resolving power (see the Introduction), the angular size of the filamentary nebulae in Cygnus is so large that we can obtain a perfectly genuine if crude x-ray picture of this source. Figure 16.4 shows two images of this kind. Notice above all the distinct shell structure of the emission region, which is fully consistent with the theory we have described. Although its distribution is highly irregular, the radiating material is located at the periphery of a vast quasi-spherical region. This circumstance arises, as we have said, from the nonuniform density distribution in the interstellar medium surrounding the exploded star. A rough correspondence between the distribution of x rays and optical radiation can also be discerned.

Soft x rays have lately been detected in several other comparatively old supernova remnants, most notably the object Vela X. In fact, in the soft x-ray range (with photon energies of a few hundred electron volts) these are the brightest objects in the sky (other than the sun, of course). Much work

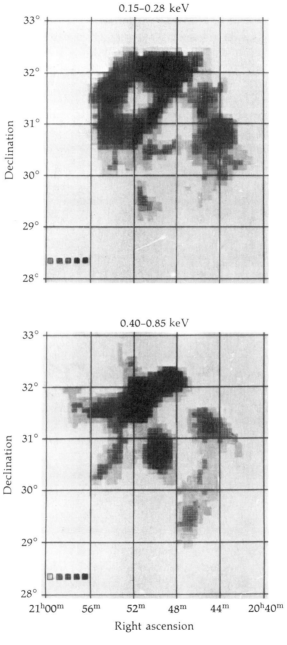

Figure 16.4 X-ray images of the filamentary nebulosity in the constellation Cygnus, in two parts of the spectrum.

238

is still ahead of us, however, before we can hope to gain thorough mastery of this important part of the spectrum.

Thus far we have spoken of the optical and x-ray emission of the nebulae that are generated by supernova outbursts. Both these types of radiation result simply from the high temperature in the plasma formed behind the front of the shock wave traveling out from the site of the explosion into the interstellar medium. But at the very dawn of radio astronomy it was found that supernova remnants are powerful radio sources of a very special kind. The discovery of radio waves coming from supernova remnants unquestionably represents a vital step forward in our study of these objects. As we shall see presently, the analysis of radio emission is a most effective tool for investigating the physical conditions in the expanding envelopes ejected when stars explode. And this knowledge in turn brings us a better understanding of the stellar explosion process itself. A further circumstance of particular interest is that radio astronomy offers us an opportunity to measure the distances of the sources—a matter of very great significance for an interpretation of their nature. Let us turn now to some of the main results gathered from observations of the radio waves emitted by supernova remnants.

In 1948 the British radio astronomers Sir Martin Ryle and F. Graham Smith detected in the constellation Cassiopeia an unusually bright radio source that they called Cassiopeia A. At that time radio astronomy was going through the opening, "heroic" phase of its development. Noteworthy discoveries by these former officers of the radar corps followed one after another. Two years before Cassiopeia A was discovered, another group of British radio astronomers found the first discrete source of radio emission in the sky, the celebrated Cygnus A, which turned out five years later to be a remote galaxy. It was the first of the radio galaxies.

At meter wavelengths the flux density of the radio emission from Cassiopeia A is almost twice that from Cygnus A and not too different from the radio flux density of the quiet sun—when no sunspots, flares, or other manifestations of solar activity are present. The fact that an extremely distant cosmic object sends us nearly as much radio-wave radiation as the sun right next to us is remarkable in itself. It indicates that extraordinary cosmic processes, fundamentally distinct from the optical phenomena, are taking place in the radio range. Today, more than a quarter century after Cassiopeia A was discovered, radio astronomy has come a long way. At the limit of its capabilities it can record radio waves millions of times weaker than those from Cassiopeia A. The great majority of weak sources are extragalactic objects. Only a small fraction (about 100) of the comparatively bright sources known have been identified with supernova remnants. But let us return to Cassiopeia A.

Upon discovery of this brightest of radio sources, astronomers were struck by the fact that in optical light nothing at all peculiar could be seen at that position. They had the impression that this very strong flux of radio waves was coming to us from empty space. Three years later, though, in 1951, Smith obtained a much better measurement of the position of Cassiopeia A, and as a result Baade and another German-American astronomer, Rudolph Minkowski, detected at that place a very faint, altogether singular nebula, undoubtedly associated with the radio source. Further inspection showed that the radio source, while small, has a perfectly definite angular size: about 4' across. Wisps and patches of faint optical nebulosity fill up the whole area occupied by the radio emission.

Cassiopeia A has a very distinctive radio spectrum. It may be represented closely by the power law

$$F_\nu \propto \nu^{-\alpha}, \tag{16.6}$$

where ν is the frequency of the radio waves and $\alpha \approx 0.8$ at all wavelengths from the meter to the centimeter range. The quantity α is called the spectral index, while the spectral flux density F_ν is defined as the amount of energy crossing unit surface area per unit time within a unit frequency interval. A power law spectrum is typical of most cosmic radio sources. Although sources differ in the value of their spectral index α, it generally remains within a limited range. This behavior of the spectrum is intimately related to the mechanism of radio emission, which we shall discuss below.

Soon after 1948 several radio sources associated with supernova remnants were found in our Galaxy. The very next year radio astronomers in Australia discovered radio emission from the Crab Nebula, the remnant of a supernova explosion in 1054. Three years later radio waves were detected from the remnants of the supernovae of 1572 (Tycho's star) and 1604 (Kepler's star). Then an extended radio source (angular size about 3°) was discovered at the site of the system of filamentary nebulae in the constellation Cygnus. At almost the same time another extended radio source was found in the constellation Gemini where the filamentary nebula IC 443 is located. This discovery provided evidence that IC 443 is also a supernova remnant. A substantial number of such objects were recorded in the years following. They are all situated near the galactic equator, indicating that they are very strongly concentrated toward the galactic plane.

Among the fairly large radio sources of low surface brightness that represent supernova remnants, Cassiopeia A stands out sharply. It is a compact object with an enormous surface brightness (in radio emission), while the optical nebula associated with it differs strikingly from the filamentary nebulae observed in "old" supernova remnants. This nebula has such an unusual shape that when it was first discovered investigators persistently

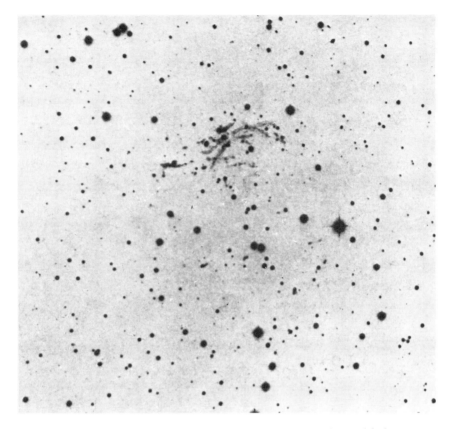

Figure 16.5 The Cassiopeia A nebula, photographed in red light by W. Baade with the 200-inch telescope.

refused to regard it as a supernova remnant. Indeed, the appearance and spectrum of the nebula have nothing in common either with the Crab Nebula or with the systems of filamentary nebulae in Cygnus and Gemini illustrated in Figures 16.2 and 16.3.

A photograph of the nebula corresponding to Cassiopeia A, taken in red light, is shown in Figure 16.5. There is a rather elongated filament extending for about 3' at a distance of 2' from the center of the nebula, together with many starlike spots covering the whole area occupied by the radio source. However, these little spots are not stars at all but fairly dense gaseous condensations. In addition to the spots there are also tiny (up to 20") elongated filaments. Some of them are rather bright, while others are scarcely perceptible. All these patches of nebulosity are confined to a circle slightly more than 6' in diameter.

241

The spectra of the individual filamentary condensations are especially interesting. The emission lines of the diffuse filaments exhibit huge radial velocities, reaching nearly 8000 km/s. On the other hand, the starlike spots show scarcely any significant radial velocities.

The whole pattern observed for the optical Cassiopeia A nebula may be explained as follows. The diffuse nebulae represent gas clouds ejected at the time of the supernova explosion, which are moving at great speed through the surrounding interstellar medium. We should emphasize that the rapidly moving filaments and the interstellar medium have a very different chemical composition. Such elements as oxygen, sulfur, and argon are tens of times more abundant (relative to hydrogen) in the filaments than in the interstellar medium. This circumstance implies that the material expelled in the outburst has undergone a complicated chemical transformation involving nuclear reactions. Observations made over the past 20 years have shown that the moving filaments are very unstable: they seem to form in empty space, last for a decade, and disappear. Large filaments sometimes break up into small ones, and separate parts of the large filaments may have very high relative velocities. Many aspects of the physical processes taking place in the filaments of Cassiopeia A are not yet understood.

From the observed expansion velocity of the Cassiopeia A filamentary system we can estimate the age of the object. By this method the star responsible for producing Cassiopeia A is found to have exploded around 1667 (say, during the period 1659 to 1675). It would seem surprising that European astronomers, who had successfully observed Tycho's and Kepler's stars nearly a century earlier, failed to notice anything unusual at that time in the constellation Cassiopeia. How could this have happened? Why was the outburst of this supernova missed in an era when there were already observatories in Europe? Of course the apparent brightness of a star depends not only on the power of its radiation but also on its distance. What, then, is the distance of Cassiopeia A?

The first reliable estimate of the distance of this source was obtained by radio astronomy. The method entails studying the absorption line in the radio spectrum of the source at 21-cm wavelength. This line is formed when radio waves are absorbed by interstellar hydrogen atoms. Since these atoms are concentrated primarily in the spiral arms of the Galaxy, which have different velocities relative to one another, the "profile" of the line splits into several components corresponding to the absorption of hydrogen in various arms. Now there are three spiral arms in the direction of Cassiopeia A, but the absorption line profile comprises only two clearly defined dips in intensity. We may at once conclude that the radio source is located somewhere between the second and third arms of the spiral structure (Figure 16.6), so its distance should be about 3000 parsecs (or 10,000 light years).

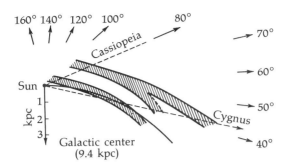

Figure 16.6 The radio astronomy method for determining the distance of the Cassiopeia A nebula.

The same value for the distance is obtained by comparing the rate at which the filaments of the nebula are observed to creep across the celestial sphere (as measured in angular units, such as seconds of arc per year) with the velocity of the filaments along the line of sight determined from the Doppler shifts of their spectral lines.

Thus Cassiopeia A is about 3000 pc away. With no interstellar absorption, the apparent magnitude of the exploding supernova (whose absolute magnitude should have been about -19^m; see Chapter 15) would have been -6^m, perhaps even brighter than the supernova of 1054 that so astonished the Chinese, the Japanese, and even the aboriginal peoples of North America. The phenomenon occurred in a region of the sky that never sinks below the horizon. In order for so remarkable an event to pass unnoticed, we would have to assume an absorption amounting to at least $7-8^m$ (a factor of 1000 or more), and further suppose that a stretch of several weeks of inclement weather prevailed throughout Europe just when the supernova happened to be at maximum light. In principle, of course, this could have been the case. But although there is substantial absorption in the direction of Cassiopeia A, it is not so very great: about $4\overset{m}{.}3$. The author of this book believes that another factor might have rendered Cassiopeia A invisible.

For it is conceivable that an enormous amount of dust might have been formed in the expanding envelope of this supernova, absorbing all the light and converting it into invisible infrared radiation. Such a process might very well occur as the gas rapidly expanding from the explosion cools off. Under these conditions simple molecules might first be formed there, followed by intricate molecular complexes and then grains of dust. A similar phenomenon is observed in the outbursts of certain novae, which emit a great amount of infrared radiation after maximum light. This radiation is undoubtedly attributable to dust. In some type II supernovae (Cassiopeia A

243

evidently belonged to this class) the quantity of dust needed for the absorption could develop even before maximum light. As the envelope of the exploding star subsequently expands, the dust located there will be dispersed, and in the case of Cassiopeia A the optical radiation of the filaments might very well be observable by now.

If this explanation of the invisibility of Cassiopeia A at the time of the outburst is correct, we may logically draw the following conclusion. Many, perhaps most, type II supernova outbursts are unseen because of the dust grains formed there during the explosion. To some extent this interpretation might account for the remarkable fact that all the supernovae recorded in old chronicles belong to type I. One should recognize, however, that the radiation of type II supernovae is absorbed particularly strongly by *interstellar* dust, because unlike type I supernovae they are located very close to the galactic plane (see Chapter 15).

We stated above that, in addition to the rapidly moving filaments, almost stationary condensations are observed in Cassiopeia A. These condensations probably represent interstellar gas compressed by the shock wave. It appears, however, that the condensations have a rather extraordinary chemical composition: nitrogen is anomalously abundant there relative to oxygen. If this is the case, we would have to infer that the shock wave from the explosion was propagated not through the interstellar medium but through the envelope expelled from the star, which then exploded as a supernova. Actually it is still premature to make any categorical judgments on this point. At any rate, all the peculiarities of the very unusual supernova remnant Cassiopeia A can be ascribed to the youth of the object.

In 1966 x rays were discovered coming from Cassiopeia A. Unlike the x rays from other, considerably older supernova remnants, the x rays from Cassiopeia A are quite hard. Both the power and the spectrum of the Cassiopeia A x rays find a natural explanation in the theory outlined above. We note in this connection that the interstellar gas in the neighborhood of Cassiopeia A is overdense ($N_e \approx$ 10–20 cm^{-3}), accounting for the requisite x-ray power (proportional to $N_e^2 R^3$, where R is the radius of the nebula). The hardness of the thermal x rays from Cassiopeia A is due to the extreme temperature (about 3×10^7 °K) of the plasma behind the shock front, which in turn results from the high expansion velocity of the nebula or, in the final analysis, from its youth.

We come now to the fundamental question of the nature of the radio waves emitted by supernova remnants. Radio emission can presently be observed in practically all ionized gaseous nebulae, both diffuse and planetary. But this radiation is, if we may say so, trivial in character. It is purely thermal, and its intensity and spectrum are given by the standard Kirchhoff law:

$$I_\nu = B_\nu(1 - e^{-\kappa_\nu l}), \tag{16.7}$$

where I_ν is the observed intensity, $B_\nu(T) = 2kT/\lambda^2$ is the radiant intensity of an ideal black body, κ_ν is the absorption coefficient at frequency ν, and l is the distance the source extends along the line of sight. The quantity $\kappa_\nu l$ is called the optical depth, τ_ν.

If the optical depth is large enough, then

$$I_\nu = B_\nu(T) = \frac{2kT\nu^2}{c^2}; \tag{16.8}$$

thus whatever the physical properties of the source the intensity will have a perfectly definite value depending only on the frequency and on the temperature of the radiating ionized gas. But if the optical depth $\tau_\nu \ll 1$, Equation 16.7 will take the form

$$I_\nu = B_\nu \tau_\nu \propto N_e^2 T^{-1/2} l, \tag{16.9}$$

since $\kappa_\nu \propto N_e^2 T^{-1/2}$. A characteristic feature is that the radiant intensity is independent of frequency in this case. An elementary emission event itself entails collisions between electrons and ions moving at thermal velocities. The thermal radio emission from H II regions in the interstellar medium (Chapter 2) is explained by this very process.

A mere glance at such nebular remnants of supernova outbursts as Cassiopeia A shows that their radiation has nothing in common with thermal radiation. Within a restricted range of ν the latter radiation may be represented by a power law of the type $F_\nu \propto \nu^{-\alpha}$, where α is between 0 and -2; but for supernova remnants the spectral index is positive ($\alpha \approx 0.5$–1.0) over a wide frequency band (see Figure 16.7). Furthermore, the radio intensity itself reaches enormous values, especially at low frequencies. According to Equation 16.8 we can always find a certain temperature, called the brightness temperature (see Chapter 4), corresponding to any intensity. At meter wavelengths the intensity of Cassiopeia A corresponds to a brightness temperature of hundreds of millions of degrees. But as Equation 16.8 indicates, in the case of thermal radiation the brightness temperature would simply be the temperature of the gas, which as a rule is of the order of 10,000°K. Nor can we regard the observed radio emission as thermal radiation coming from very hot gas behind the front of the shock wave propagated through supernova remnants (see above). The radio intensity calculated from the observed x rays (which are thermal) turns out to be negligible. And finally, one should not forget the complete disagreement between the observed radio spectra of supernova remnants and the spectra of thermal radio sources.

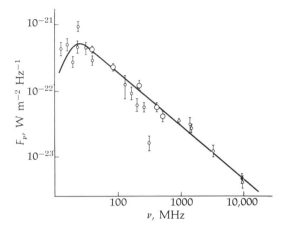

Figure 16.7 Radio spectrum of the Cassiopeia A nebula.

The correct idea explaining the radio emission of supernova remnants (as well as most other sources of cosmic radio waves) was proposed in 1950 by the Swedish physicists Hannes Alfvén and Nicolai Herlofson, and independently by the German astrophysicist Karl Otto Kiepenheuer. During the years following, this idea was worked out in full detail, primarily in the USSR, and raised to the level of a very refined theory. Its application to specific astronomical objects, particularly to supernova remnants, has proved most fruitful. The new theory has afforded an interpretation of many astronomical observations and has predicted a variety of new effects that have been confirmed completely by special observing programs. What, then, is this theory?

From physics it has long been known that an electron moving in an external magnetic field H will emit radiation at a characteristic frequency $\nu_H = eH/2\pi m_e c$, where e is the charge of the electron and m_e is its mass. This is the frequency at which the electron gyrates about magnetic lines of force perpendicular to its direction of travel. If the energy E of the electron is very high, exceeding the rest energy $m_e c^2$ of the electron (such an electron is said to be relativistic), then the character of the radiation will undergo qualitative changes. First of all, such an electron will emit not just one definite frequency but a continuous spectrum—radiation at a great many closely spaced frequencies. This radiation will reach its maximum intensity at the frequency

$$\nu_m = \frac{eH}{4\pi m_e c}\left(\frac{E}{m_e c^2}\right)^2. \tag{16.10}$$

246

On the low-frequency side, for $\nu \ll \nu_m$, the intensity will rise gradually with frequency as $\nu^{1/3}$, while for $\nu \gg \nu_m$ it will fall off steeply. Another important property of radiation from relativistic electrons is its directivity. Nearly all the radiation will be concentrated within a cone whose axis coincides with the direction of the electron's instantaneous velocity and whose vertex angle $\theta = m_e c^2 / E$. Radiation of this type has long been familiar to physicists working with accelerators. It has aptly been named synchrotron radiation.

In order to give some feeling for the order of magnitude of the quantities appearing in the synchrotron radiation equations, we note that the rest energy of an electron is $m_e c^2 = 5 \times 10^5$ eV and $\nu_H = eH/2\pi m_e c = 2.8 \times 10^6$ H. Suppose that an electron with an energy $E = 10^9$ eV, the energy of soft cosmic rays, is moving through a magnetic field. Then it will radiate a continuous spectrum whose maximum intensity will occur at the frequency $\nu_m = 1.4 \times 10^6 (E/m_e c^2)^2 H = 5.4 \times 10^{12} H$. If we take a magnetic field $H = 2 \times 10^{-5}$ gauss—a strength of the same order as that of interstellar fields—then $\nu_m \approx 100$ MHz, corresponding to a wavelength of 3 m. This is a typical value for galactic radio waves. If the electron were nonrelativistic, it would emit only the single frequency $\nu_H \approx 50$ Hz, or a wavelength of 6000 km. Such radiation cannot be detected with ground-based radio telescopes; the ionosphere admits only radio waves shorter than about 30 m. Indeed, even radio telescopes which presumably will soon be carried into space by specialized satellites could hardly be capable of recording such ultralong radio waves from nonrelativistic electrons. Thus the capacity of relativistic electrons in very weak magnetic fields to emit comparatively high frequencies is useful because it makes them accessible to observation.

We have considered the synchrotron radiation of just one relativistic electron with a given energy. Actually, research on primary cosmic rays, which also consist of relativistic particles, has shown that the particles are distributed with respect to energy according to a certain power law. This law has the form

$$N(E > E_0) = \frac{K}{\gamma - 1} E_0^{-(\gamma+1)}, \qquad (16.11)$$

where $N(E > E_0)$ designates the number of particles per unit volume with an energy above E_0. The quantity γ characterizes the energy spectrum of the relativistic particles.

Assuming that the velocities of the relativistic electrons are oriented at random relative to the magnetic field, one can derive a theoretical equation for the intensity and spectrum of the synchrotron radiation from a group of relativistic particles whose energy spectrum is given by Equation 16.11. This important equation is

$$I_\nu = 1.3 \times 10^{-22}(2.8 \times 10^8)^{(\gamma-1)/2}u(\gamma)KH^{(\gamma+1)/2}l\lambda^{(\gamma-1)/2}, \qquad (16.12)$$

where $u(\gamma)$ is a numerical factor of order unity, l as before denotes the length along the line of sight of the region in which the relativistic electrons are moving, and λ is the wavelength of the radiation. If we replace λ by the frequency $\nu = c/\lambda$, we at once see that Equation 16.12 yields a power spectrum for the synchrotron radiation in full accord with the observations. The spectral index α (see the relation 16.6) is thereby

$$\alpha = \frac{\gamma - 1}{2}. \qquad (16.13)$$

Equation 16.13, first obtained by the Soviet radio astronomer A. A. Korchak, relates the exponent in the power law energy spectrum of relativistic electrons to the spectral index of their synchrotron radiation. For example, in the case of Cassiopeia A the spectral index $\alpha = 0.8$, so $\gamma = 2.6$, a typical value for cosmic rays. One may say that the power law radio-emission spectrum of supernova remnants reflects the power law energy spectrum of the relativistic particles responsible for the radio waves we observe.

We have arrived, then, at a most important conclusion: the expanding nebulae that constitute the remnants of supernova outbursts contain a huge number of relativistic particles—or, in other words, cosmic rays! An opportunity has at last opened up to observe primary cosmic rays (properly speaking, their electron component) not just near the earth but in the depths of the Galaxy and even throughout the universe, for the radio emission of galaxies and of the quasars discovered in 1963 is synchrotron in nature. In making this opportunity available and inaugurating a new era in the study of cosmic rays, radio astronomy has achieved one of its greatest successes.

By applying the theory of synchrotron radiation to such real sources as supernova remnants one can find the total number of relativistic electrons located there and their energy, as well as the magnetic field strength. The following procedure is used. First of all, one must recognize that radio sources should contain not only electrons but other relativistic particles, especially protons. The heavy relativistic particles emit very little radiation, however; their mass is too great. Hence a peculiar situation prevails in radio sources: of all the relativistic particles there, only electrons can be observed by their synchrotron radiation.

Nevertheless, using another technique developed recently, observations can also be made of the basic proton–nucleon component of cosmic rays in supernova-remnant envelopes. We are indebted to progress in gamma-ray astronomy for this major achievement. The specialized American satellite SAS-2 has recorded a flux of hard γ rays from one of the closest supernova remnants, Vela X, 450 pc away. The γ rays of energy above 50 MeV emitted by this source have a flux density of roughly 10^{-5} cm^{-2} s^{-1}. These photons are formed when relativistic protons (the principal component of cosmic rays) collide with the nuclei of interstellar gas atoms in the source. Under such conditions π^0 mesons will be produced, each decaying rapidly into two γ-ray photons.

In any event, let W_r denote the energy of all the relativistic particles contained in unit volume (that is, the energy density of the particles). Then the energy density of the relativistic electrons may be represented as $W_e = kW_r$, where the factor k is less than unity. The difficulty with the problem is that the value of k is generally very uncertain. Analyses of the composition of the primary cosmic rays observed in the immediate vicinity of the earth indicate that $k \approx 0.01$. It is worth noting in this connection that searches for relativistic electrons in cosmic rays have in fact been undertaken just because of the successful application of synchrotron theory to radio astronomy. But we certainly cannot infer that in all sources, including supernova remnants, the value of k should be the same as near the earth. Moreover, in Chapter 17 we shall argue that in the Crab Nebula the bulk of the relativistic particles ought to be electrons. For lack of better alternatives, two assumptions are usually made in calculations: $k = 0.01$ and $k = 1$. Fortunately this uncertainty in the value of k does not have too strong an effect on the fundamental conclusions regarding the nature of supernova remnants. Incidentally, for Vela X the data now available from γ-ray astronomy are reasonably consistent with the assumption that $k = 0.01$.

An important circumstance is that the energy density W_r of the relativistic particles cannot exceed the energy density W_m of the magnetic field, which is equal to $H^2/8\pi$. Otherwise the magnetic field would not be able to retain the relativistic particles for long, and they would fly off in all directions at nearly the speed of light. On the other hand, we may perfectly well have a situation in which $W_m > W_r$. The condition that the relativistic particles be retained by the magnetic field may be written mathematically as

$$W_m = \frac{H^2}{8\pi} \geq W_r = \frac{W_e}{k}. \tag{16.14}$$

One can show that if $W_m \approx W_r$, the total energy of the relativistic particles as well as the magnetic field energy will be at a minimum. Accordingly, the auxiliary condition 16.14 is usually adopted in calculations, setting a lower limit on the energies of the relativistic particles and the field in the source. For most radio sources the condition 16.14 appears to be satisfied.

Supernova remnants form extended radio sources with a rather complicated intensity distribution. Figure 16.8 displays a radio picture of the Cassiopeia A nebula, obtained with a radio interferometer having good resolution, about 20″. Although the structure of the radio image shown in Figure 16.8 is quite intricate and abounds in detailed features—including at least ten small condensations—on the whole it has a definite "shell" character. The radio emission is concentrated at the periphery of a spheroidal region, and the thickness of the emission layer amounts to a few tenths of its radius. We can easily estimate the radius from the measured angular radius (about 2.5) and the distance of Cassiopeia A, which, as we have explained, is about 3000 pc. The radius of the spheroidal shell filled with relativistic particles at the periphery is slightly more than 2 pc.

In Figure 16.9 we illustrate some cruder radio images of "old" supernova remnants, obtained with lower resolving power. Even in these cases the shell structure of the radio sources is distinctly apparent. We should emphasize, however, that all these envelopes have a most irregular structure.

Figure 16.8 A radio map of the Cassiopeia A nebula (Cavendish Laboratory, Cambridge University).

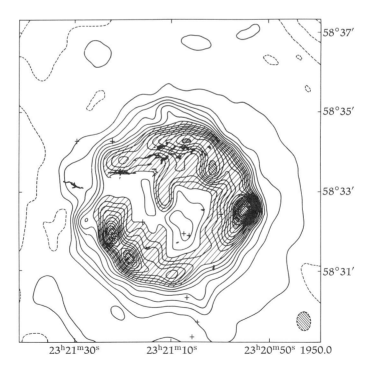

250

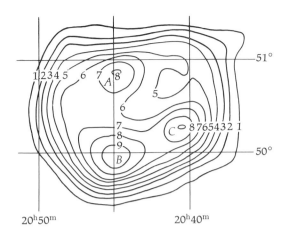

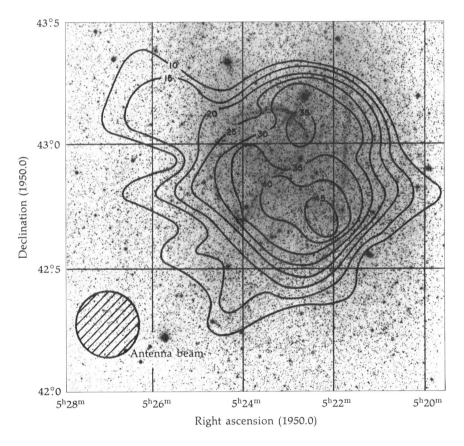

Figure 16.9 Radio maps of two supernova remnants, in Cygnus (Cavendish Laboratory) and Auriga (University of Illinois).

251

The linear size of the envelopes can be determined from their measured angular size if we know their distance. In the case of the system of filamentary nebulae in Cygnus (see Figure 16.2), the distance has been reliably determined from the measured radial velocities of the filaments and the rate at which they spread out over the sky, expressed in seconds of arc per year (see Chapter 15). One finds that the radius of this source is about 20 pc, or 10 times that of Cassiopeia A. For the sources whose radio images are shown in Figure 16.9 the radius has been obtained from the distance, as derived by a radio-astronomy method that we shall discuss presently. Most sources of this kind have a radius close to 10 pc.

We now have all the information necessary to determine the energy of the relativistic particles in supernova remnants and the strength of the magnetic field there. The radio emission intensity required for the calculations can be obtained from the measured flux density and angular size. If the intensity (or brightness) remains constant within a given range in angle, we will have approximately

$$I_\nu = F_\nu / \Omega, \tag{16.15}$$

where Ω is the solid angle subtended by some feature in the source and F_ν is the flux density from that feature. Having measured F_ν for all the features, we can find I_ν for each and then sum to obtain the total energy of the relativistic particles in the source. At the same time we find the distribution of the magnetic field over the source. For rough calculations we can even ignore the source image structure altogether, regarding it as an object of constant intensity, which in this case can also be determined from Equation 16.15. The quantity F_ν would now denote the flux density measured for the whole source, and Ω the solid angle it subtends.

In the case of Cassiopeia A, calculations for $k = 1$ yield the value $W_e = 2 \times 10^{48}$ erg. By convention this quantity is equal to the energy W_H of the magnetic field in the source, whence $H = 2.5 \times 10^{-4}$ gauss. But if—as in the neighborhood of the earth or in the source Vela X—relativistic electrons comprise only 1 percent of the total energy of the relativistic particles ($k = 0.001$), then the value of W_r will increase by a factor of 7 according to the calculations, and the field will become 2.7 times as strong. In this case, then, $W_r \approx 10^{49}$ erg and $H \approx 7 \times 10^{-4}$ gauss. The relativistic particles confined in the region occupied by a supernova remnant have a prodigious amount of energy—equal to nearly all the energy radiated during the outburst of the star. The magnetic field strength in a supernova envelope is also quite high, a hundred times the strength of the interstellar magnetic field.

Analogous calculations performed for older supernova remnants give similar values for W_r, of order 10^{48} erg, although the field strength is appreciably lower. For example, in the case of the source associated with the filamentary nebulae in Cygnus, $W_e = W_r = 2.5 \times 10^{48}$ erg and $H = 2 \times 10^{-5}$ gauss if $k = 1$, while $W_e = 1.7 \times 10^{49}$ erg and $H = 5 \times 10^{-5}$ gauss if $k = 0.01$. Thus the relativistic particles and magnetic fields represent an important attribute of supernova remnants, in many respects controlling the evolution of these objects. This circumstance arises in particular from the major role that the pressure of the magnetic field and relativistic particles plays in the dynamics of remnants. The interaction of the relativistic particles with the plasma present in supernova remnants takes place by means of the magnetic field: the particles exert pressure on the magnetic lines of force, which in turn guide the motion of the plasma.

We see that the physical conditions in supernova envelopes are determined by a complex linkage: an interaction among cosmic rays, magnetic fields, very hot plasma formed behind the shock front, and dense, comparatively cool gaseous filaments embedded in the plasma. Thus the synchrotron theory has fully explained all the properties of the radio emission of supernova remnants. In particular, we can understand the linear polarization of the radio emission observed in Cassiopeia A, the filamentary nebulae in Cygnus, and kindred supernova remnants. This polarization reaches substantial values, up to 10 percent in some cases. In relatively weak magnetic fields, thermal radio emission would not produce any perceptible polarization. But evidently synchrotron radiation should nearly always be polarized, because the radiating object always has a physically preferred direction associated with the magnetic field. Only in sources for which the magnetic fields are distributed at random and the scale of field irregularities is much smaller than the size of the source will the synchrotron radiation flux be nearly unpolarized. The interpretation of the polarization observed in cosmic radio sources is a signal achievement of the synchrotron theory.

However, no theory can be definitely accepted until it can be used to predict some altogether new phenomenon that later is indeed observed. A major role in the history of astrophysics and radio astronomy has been played by the prediction that the optical radiation of the Crab Nebula should be polarized, an idea brilliantly confirmed by the observations. This prediction was made by interpreting the very familiar optical emission of the nebula as synchrotron radiation, a matter that we shall discuss in detail in the next chapter. For the present let us consider another consequence of the synchrotron theory, one which enables us to make another important prediction that has been confirmed experimentally.

Supernova remnants represent objects that expand without limit, ultimately dispersing into the interstellar medium. The Cassiopeia A nebula, which has been described in some detail above, is a young object. The gas clouds thrown out in this supernova explosion have scarcely begun to be decelerated by the interstellar medium. They have retained nearly all the velocity they acquired at the time of the outburst. On the other hand, such objects as the filamentary nebulae in Cygnus, IC 443, and similar ones represent quite old supernova remnants. Their linear size is five to ten times that of Cassiopeia A. Their velocity of expansion has dropped sharply. Finally—and perhaps this is the most interesting point—their radio emissions are considerably weaker than those of Cassiopeia A.

The power of a source is proportional to the square of its distance and to the flux density we observe. Since the filamentary nebulae in Cygnus are almost four times closer than Cassiopeia A while their radio flux density is almost a hundred times lower, Cassiopeia A is emitting radio waves at a power some 1500 times as great as an old object such as the filamentary nebulosity in Cygnus! Thus we have reached the purely empirical conclusion that as a supernova remnant expands its radio power decreases strongly. There is an even greater decline in the surface brightness of old remnants. For example, the surface brightness of the radio source associated with the filamentary nebulae is 100,000 times lower than that of Cassiopeia A.

Yet the number of relativistic electrons responsible for the synchrotron radiation of these objects hardly diminishes at all as the objects evolve. For the relativistic particles are, so to speak, trapped in the tangled magnetic field confining them, which in turn is firmly coupled to the expanding gaseous filaments. But why is the evolution of supernova remnants accompanied by such a marked drop in the power and intensity of the radio emission?

This question was investigated by the author in 1960. It was found that the strength of the magnetic field should decline as the radio nebula expands. One would expect the magnetic flux, or the product of the square of the radius R of the nebula by the average intensity of the magnetic field, to be conserved during the expansion process. If this is the case, the magnetic field of such an object should in the first approximation vary as R^{-2} while the radio nebula expands. Two other factors serve to decrease the power of the radio emission. The first is that as the region containing the relativistic electrons expands, the energy of each electron declines in inverse proportion to the radius of the region. This behavior is fully analogous to the cooling of an expanding gas. Furthermore, the angle between the direction of travel of the relativistic electrons and the magnetic field will continually decrease. With synchrotron radiation only the field component perpendicular to the direction of travel of the electron is important[1] (a

---

[1]Hence in Equation 16.12 for the synchrotron radiation intensity the H should be replaced by the field component $H_\perp$ perpendicular to the velocity of the relativistic electron. Usually $H_\perp$ is nearly equal to H, so this correction has no appreciable significance.

relativistic electron moving strictly along the magnetic field would not radiate), so this decrease in angle is tantamount to a further drop in the magnetic field, causing an additional decrease in the power of the synchrotron radiation of supernova remnants.

There are several quite definite reasons, then, for the steady decline in the synchrotron radiation power of radio nebulae as they expand. Taking all these factors into account, we can write the following simple relation for the power of an evolving synchrotron radiation source as a function of the radius of the radio nebula during its adiabatic expansion phase:

$$L \propto R^{-2(2\alpha+1)},$$ (16.16)

where α as before denotes the spectral index of the source. The mean surface brightness of an expanding radio source should decline according to a steeper law:

$$\bar{I} \propto R^{-2(2\alpha+2)}.$$ (16.17)

To a first approximation, these relations describe very well the decay in the flux density and brightness of the radio emission from expanding supernova remnants. For instance, we may infer that when Cassiopeia A has expanded to a linear size equal to the present size of the system of filamentary nebulae in Cygnus, the power of its synchrotron radiation will be many thousands of times lower. On the other hand, when the supernova remnant in Cygnus was as small as Cassiopeia A is today, its radio power should have been several times that of the present Cassiopeia A (since its spectral index is $\alpha = 0.47$).

Thus theory predicts that the radio flux from supernova remnants should continually decrease. Now suppose that a source is expanding at a constant rate, as is true of young sources such as Cassiopeia A that have not yet been decelerated. In this case the radius of the source will be proportional to its age, and we may write the relation 16.16 in the form

$$F_\nu \propto t^{-2(2\alpha+1)},$$ (16.18)

because the flux density F_ν is proportional to the power L_ν and the distance r of the source remains unchanged while it evolves.

Let ΔF_ν represent the annual decrease in the flux density of such a source. Then we readily obtain the following expression for the relative drop in the flux each year:

$$\frac{\Delta F_\nu}{F_\nu} = -\frac{2(2\alpha+1)}{T},$$ (16.19)

where T is the age of the source in years. For Cassiopeia A we have $\alpha = 0.8$ and $T \approx 300$ yr, whence $\Delta F_\nu/F_\nu \approx 1.7$ percent per year: the flux density of

Cassiopeia A would be expected to show a steady ("secular") decrease of rather more than 1.5 percent annually! This is a very large amount, but the main thing is that it can easily be measured. A systematic application of synchrotron radiation theory to Cassiopeia A therefore allows us to predict that this radio source, the brightest in the sky, should melt away before our eyes!

As soon as this prediction was published in 1960, the British radio astronomers repeated their observations of Cassiopeia A with the same telescope with which they had discovered this remarkable source in 1948. By using an identical instrument and techniques to make observations at different times, the unavoidable errors of measurement can be reduced. The results of these observations could not help but catch the imagination: from 1948 to 1960 the flux from this brightest of sources had dropped by nearly 15 percent! The annual decrease in the flux was found to be 1.1 ± 0.14 percent. Since then measurements of this kind have often been repeated. Various authors quote values of $\Delta F_\nu / F_\nu$ ranging from 1.1 percent to 1.7 percent per year. Probably the value is close to 1.2 percent per year, which is 30 percent smaller than the theoretical value obtained from Equation 16.19. One should remember, however, that the equation has been derived under the simplifying assumption that $R \propto t$, that is, that the nebula is not decelerated at all. But there is definite observational evidence that deceleration of the Cassiopeia A filaments by the interstellar medium has already begun. For example, the positive radial velocities of the filaments of this nebula are systematically larger than the negative velocities, which indicates that the side of the envelope facing us has begun to slow down. If we correct for this circumstance, the quantity $\Delta F_\nu / F_\nu$ drops to the observed value of 1.2 percent per year. In 30 years the radio flux of Cassiopeia A decreases by a factor of almost 1.5, while toward the beginning of the past century it was nearly 10 times as great as now. Too bad there was no radio astronomy then!

The detection of the theoretically predicted rapid decline in the radio flux of Cassiopeia A affords a direct proof of the synchrotron theory and all its implications. In older radio sources that are supernova remnants it is not possible to observe the secular decrease in the flux, because their radius changes too slowly.

A few years ago American radio astronomers confirmed observations by the Soviet radio astronomers K. S. Stankevich and N. M. Tseitlin at Gor'kii, who established that the expansion of Cassiopeia A is accompanied by changes not only in its flux but also in its spectrum. The spectrum becomes progressively flatter. At this rate the spectral index of Cassiopeia A ought to be only half as great after about 1000 years, approaching the spectral index of old remnants. Although the cause of this interesting effect

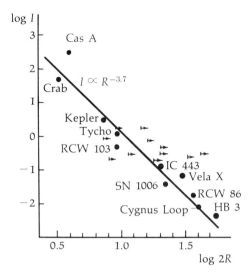

Figure 16.10   Relation between radio surface brightness and radius of supernova remnants (L. Woltjer).

is not entirely understood, it is probably related to the source's youth, as acceleration of relativistic particles is still taking place there (see below).

Figure 16.10 shows how the surface brightness of supernova remnants depends on their radius for all objects with well-estimated distances. The relationship clearly conforms to a power law, although it is less steep than for $\alpha = 0.8$. One finds empirically that

$$I_\nu \propto R^{-4}. \tag{16.20}$$

We should not be troubled that the empirical law for the decline in radio luminosity with increasing radius is more gentle than that given by the theoretical relation 16.17. In the first place, not all the nebulae are in the adiabatic expansion phase; second, most nebulae have comparatively small spectral indices.

We shall use the empirical relation 16.20 to develop a new method for determining the distances of supernova remnants that, by other methods, can be estimated only with much difficulty if at all. Our method, which is statistical in character, is in principle fully analogous to the procedure for determining the distances of planetary nebulae that we suggested in 1956— a technique that soon came into general acceptance. In both cases the luminosity of the objects decreases fairly rapidly as they grow in linear size.[2] Hence the difference in the size of various objects is not so very great, and in the first approximation the distance will be inversely proportional to the angular size φ.

Just as in the case of planetary nebulae, the difference in the linear sizes of radio nebulae is described by a factor that depends on their measured intensity (or surface brightness). The final equation also contains such properties of the supernova explosion as the total energy E released and the original magnetic field H_0 in the envelope. These properties are wholly similar to the mass of the envelope expelled in planetary nebulae. The expression for the distances of radio nebulae ultimately takes the form

$$r \propto E^{0.20} H_0^{-0.15} I_{\nu}^{-0.20} \varphi^{-1}, \tag{16.21}$$

where the exponent β in the relation $L_{\nu} \propto R^{-\beta}$ is taken equal to 3. At present this relation or some modification of it is considered the basic one for determining the distances of supernova-remnant radio nebulae, which now number more than a hundred known objects.

From the distances of supernova remnants we can estimate the total number of such nebulae in the Galaxy. Their population turns out to be about 500, with over 100 directly observed and catalogued. Since the average age of the remnants is around 20,000 years (see the beginning of this chapter), we can estimate the average frequency with which supernovae explode in the Galaxy: about one outburst every 40 years.

How are the relativistic particles in expanding supernova envelopes actually formed? The most natural explanation might seem to be that they are somehow accelerated there very early in the evolution of the envelope, perhaps immediately after the supernova explosion. Although we now have unquestioned evidence that the stellar remnants of an outburst are powerful generators of cosmic rays (see Chapter 17 as well as Chapters 20 and 21), one can show that no such process could account for the amount of cosmic rays observed to occur in supernova remnants. In fact, because of the expansion of the remnants the relativistic particles trapped there will lose their energy very quickly, for the same reason that gas expanding into a vacuum will cool.

Quite recently the young British astronomer Stephen F. Gull demonstrated convincingly that the relativistic particles arise fairly late in the expanding envelope formed after the supernova outburst, at a stage when the envelope is about 100 years old and its radius is about 1 pc. Gull showed

---

[2]For planetary nebulae the radiation per unit volume is proportional to N_e^2, and for the luminosity of a whole expanding nebula we have $L \propto N_e^2 R^3$. Since $N_e \propto \mathfrak{M} R^{-3}$ (where $\mathfrak{M}$ is the mass of the nebula), $L \propto \mathfrak{M}^2 R^{-3}$ and the intensity $I \propto \mathfrak{M}^2 R^{-5}$. Accordingly, the distance r of the nebula is proportional to $\mathfrak{M}^{2/5} I^{-1/5} \varphi^{-1}$, because $r = R/\varphi$.

that a convective layer will develop in such an envelope by that time, and in this layer charged particles will be accelerated to relativistic energies. According to the theory, about 1 percent of the initial kinetic energy E of the envelope will be transmitted to relativistic particles, in excellent accord with the observations. Indeed, data acquired from x-ray astronomy indicate that $E \approx 10^{51}$ erg, while $W_r \approx 10^{49}$ erg (for $k = 0.01$). Very likely this acceleration process has not yet concluded in Cassiopeia A, which would evidently explain the observed secular change in its spectrum.

Among all the supernova remnants one remarkable object stands out, although its nature has yet to be definitely established. We are speaking of the renowned galactic spur. Even the first radio observations of the Galaxy, carried out at the beginning of radio astronomy, revealed one very prominent feature in the distribution of radio brightness over the sky. The intensity of cosmic radio emission clearly shows a substantial concentration toward the galactic equator and center. But 30° away from the center a rather bright, comparatively narrow band of radio emission extends out almost perpendicular from the equatorial region, stretches a vast distance nearly to the north galactic pole, and, describing a giant loop, returns to the galactic equator. Altogether the galactic spur forms a small circle on the celestial sphere about 110° in diameter. Its spectrum indicates that the spur is emitting synchrotron radiation. No extended optical objects, not even very faint ones, have been detected in the neighborhood of the spur.

At various times hypotheses have been put forward seeking to explain this extraordinary feature in the radio emission of the Galaxy. The most intriguing suggestion is that the spur might be the remnant of a supernova outburst which occurred some tens of thousands of years ago. Since the spur radio emission has approximately the same surface brightness as the radio nebula associated with the filamentary nebulae in Cygnus, on this interpretation the linear diameter of the spur would be about 35–40 pc. Hence the supernova would have exploded very close to the sun, only about 25 pc away. This event might have happened perhaps 20,000 years ago, within the memory of Cro-Magnon man.

A strong argument for the supernova nature of the spur has been the fairly recent discovery of soft x rays along the whole belt. We already know that such x rays are an important attribute of old supernova remnants. The absence of optical filamentary nebulae from the spur region should not be regarded as a serious impediment to the supernova interpretation. Optical supernova remnants are very diverse. Practically speaking, we still have only a poor idea how the remarkably delicate gaseous filaments are formed in the shock wave that is propagated through the interstellar medium after a supernova outburst. There is, however, one important argument against such an explanation: no rise in the intensity of hard γ rays has been found in the vicinity of the spur. This circumstance implies that few relativistic protons are present there.

Nonetheless, we cannot exclude the possibility that a supernova exploded in the solar neighborhood about 20,000 years ago. This is a very improbable occurrence, because on the average such an event should happen only once every few million years. How would the phenomenon have looked to an observer in Cro-Magnon time? Above all the observer would have seen an extraordinarily bright star with an apparent magnitude of about -17. It would have shed roughly a thousand times as much light on the earth as the full moon does, although it would still have been a thousand times less bright than the sun. Since the color temperature of a type II supernova at maximum light is about 40,000°K, the bulk of the star's radiation would have been concentrated in the ultraviolet part of the spectrum. The flux density of the ultraviolet radiation, which is capable of ionizing the upper layers of the earth's atmosphere, would have been about 1000 erg cm^{-2} s^{-1}, a thousand times the ultraviolet flux received from the sun. As a result the upper atmospheric layers would have become tens of times more strongly ionized. But this radiation, disastrous to all life, would not have reached the ground. The atmospheric armor of our planet is simply too thick.

The most important effect from a supernova outburst close to the sun would be the rise in the level of primary cosmic rays. This would happen when the expanding radio nebula reaches the solar system, which would then become surrounded by the nebula. The geometry of the spur indicates that no such event has yet taken place. Our solar system should be only 5–10 pc, a very small distance, away from the closest point of the radio nebula. A few tens of thousands of years would still have to pass before we are inside the radio nebula. And what would happen then? Nothing special: the intensity of the soft component of the primary cosmic rays would rise by several times, that's all.

There would be considerably more serious consequences for the earth if a supernova were to explode less than 10 pc away from us. In this event the cosmic rays might become tens of times as intense. But such phenomena have occurred very rarely in the history of our sun, about once every 100 million years. The rise in the cosmic ray intensity in the solar neighborhood would last around 20,000 years, after which the envelope of the radio nebula would withdraw until, hundreds of thousands of years later, the cosmic ray background would revert to its original undisturbed value.

What impact would such a strong rise in the density of primary cosmic rays near the sun have on the earth? Above all it might have (or may have actually had) serious genetic consequences for many species of animals and plants that populate (or once populated) our planet. The evolution of species is, of course, controlled by natural selection, the motive force behind biological evolution. But natural selection is governed by the conditions in the environment. Mutations unavoidably taking place will be preserved for posterity if they are favorable for the survival of a species. The presence of radioactivity in the air layer near the ground is one cause of spontaneous mutations. A substantial increase in the frequency of mutations because of high radioactivity in the air could have the most severe consequences for many species. Different species, however, react in different ways to hard irradiation. For species with a short reproduction cycle, the radiation dose would have to rise by a factor of 1000 in order to double the mutation frequency, whereas an increase in exposure of three to ten times would suffice for long-lived species.

In collaboration with Valerian I. Krasovskii, we proposed some time ago that the notorious extinction of the dinosaurs at the end of the Cretaceous period might have resulted from a supernova outburst near the sun. At the present time, however, the available paleontological evidence is inadequate to confirm—or refute—this hypothesis. Further, one should recognize that a rise in the level of hard radiation could have enhanced the evolution of certain species of animals and plants. Might such an event have produced the luxuriant vegetation of the Carboniferous period? Finally, the origin of life on the primordial earth could itself have been stimulated by a high level of radiation.

17

The Crab Nebula

It would hardly be an exaggeration to say that no object in the night sky has given astronomy such valuable, fundamentally new information as the Crab Nebula. In particular, the Crab Nebula was the first galactic object to be identified as a radio source (this happened in 1949). And in 1963 it was the first galactic object to be identified as a source of x rays. A completely new type of optical radiation hitherto unknown to astronomers was first discovered in the Crab Nebula (see below). Before that time the only kind of optical radiation recognized by astronomers was the thermal emission of stars and nebulae.

Together with analyses of the radio emission, x rays, and gamma rays, a new interpretation of the optical light revealed for the first time the vital role that charged particles of ultrahigh energy and the magnetic field play in the dynamics and evolution of many cosmic objects, especially extragalactic ones. The insight gained into the nature of this remarkable nebula has furthered the development of modern astrophysical concepts, according to which on the scale of galaxies, clusters of galaxies, and the universe, cosmic rays represent at least as important a factor as the classical components of matter—stars and the interstellar medium.

Quite recently, in early 1969, the Crab Nebula bestowed on astronomers the next surprise: the most extraordinary of all known pulsars is located at its center. We will have much to say about this object in Part IV of this book.

262

For now we only wish to point out that the Crab Nebula was the first celestial object found to be genetically related to pulsars, that entirely new type of population in the Galaxy. The author of this book awaits new surprises from this nebula, which has already done so much work for astronomy. In particular, he would not be astonished if the pulsar in the Crab Nebula should prove to be the first object in which gravitational radiation is detected (see Chapter 24).

The Crab Nebula is generally regarded as the remnant of a type I supernova outburst. The claim is supported, for example, by the light curve of the supernova of 1054, as reconstructed from ancient Chinese chronicles. Further, as we shall see below, the mass of the gas ejected during the explosion of this star could scarcely exceed a few tenths of a solar mass. On the other hand, the Crab Nebula is situated close to the galactic plane, and we cannot exclude the possibility that a star several times as massive as the sun erupted. Finally, it is also noteworthy that the Crab Nebula differs greatly from the remnants of the 1572 and 1604 supernovae, which were undoubtedly formed during type I outbursts. One should recognize, however, that the Kepler and Tycho supernova remnants differ from each other as well. The same applies to the remnant of the type I supernova that appeared in 1006. Thus the radio nebulae produced by type I supernova outbursts are very diverse. In a situation of this kind there is not much point in disputing whether or not the Crab Nebula stands out among other nebulae formed through explosions of type I supernovae. One important property of this nebula is perhaps worth emphasizing, however: its filaments are expanding at the abnormally low velocity of about 1500 km/s.

Back in the nineteenth century the Crab Nebula was an object of study by a number of eminent astronomers. But most of these observations were visual. The famed Irish observer William Parsons, the third Earl of Rosse, may have been the first to record the filamentary structure of the Crab Nebula. He was indeed the "godfather" of this nebula, naming it the Crab after its distinctive shape. Figure 17.1 shows a drawing of the nebula made by Lord Rosse in 1844; it does look rather like a crab. For all its lack of sophistication this drawing is fully consistent with modern photographs (Figure 17.2). In particular, the prominent gulf in the nebula, clearly visible in the photograph, outlines the two "claws" of the crab.

The first photograph of the Crab Nebula was taken in 1892. Its spectrum began to be investigated in 1913–1915 by the noted American astronomer V. M. Slipher. He was the first to draw attention to the doubling of the emission lines in the spectrum (Figure 17.3), but he wrongly attributed the phenomenon, of all things, to the Stark effect, which had been discovered in the laboratory not long before his observations. Of course such an interpretation would make us smile today. We do not, however, want to be too hard on this outstanding astronomer, to whom we owe quite

263

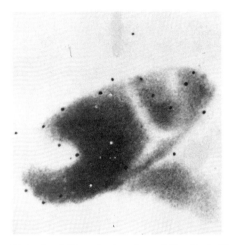

Figure 17.1   Lord Rosse's drawing of the Crab Nebula.

Figure 17.2   The Crab Nebula, photographed by W. Baade at the Mount Wilson Observatory.

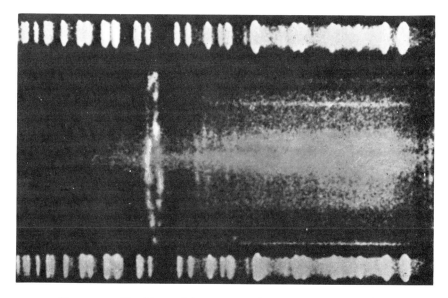

Figure 17.3 Doubling of the emission lines in a spectrum of the Crab Nebula obtained by N. U. Mayall.

a few important discoveries,[1] for astrophysics was in an embryonic state at that time. Slipher was also the first to point out the bright continuous spectrum of the Crab Nebula, on which the emission lines are superimposed. Spectroscopic studies of this nebula were later conducted by several eminent astronomers.

We shall now briefly describe the spectrum. To a first approximation it resembles the spectra of planetary nebulae. The brightest emission lines are the characteristic lines of ionized oxygen, nitrogen, and sulfur. Fainter lines of hydrogen are also observed. However, unlike all known gaseous nebulae, including planetaries, the Crab Nebula exhibits a very bright continuous spectrum. Of course gaseous nebulae do show such a continuum— a comparatively faint one, formed particularly through the simultaneous emission of two photons (called the two-photon process, on which we cannot dwell here). But in the Crab only a few percent of the total radiation is concentrated in the lines, while in planetary nebulae the behavior is just the reverse.

By the late 1930s Walter Baade, the renowned observational astronomer, had taken some exceptionally interesting photographs of the Crab through filters. Figure 17.4 shows a photograph made with the 100-inch telescope

---

[1]In particular, he was the first to discover the red shift in the spectral lines of remote galaxies.

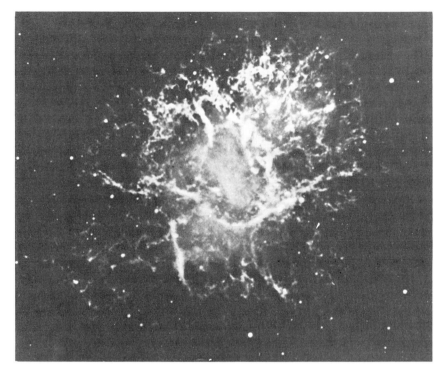

Figure 17.4   The Crab Nebula, photographed by W. Baade through a filter transmitting the Hα line.

of the Mount Wilson Observatory through a filter transmitting the Hα line (the prominent red line of hydrogen) and lines of ionized nitrogen nearby. This photograph differs significantly from the photograph in blue light shown in Figure 17.2. Delicate filaments now form a network of wondrous beauty, a tracery covering the whole nebula along its periphery. The Hα photograph demonstrates that the spectral lines are emitted not by the entire volume of the nebula but only by the filamentary network, whereas the continuum is radiated by the nebula as a whole.

The structure of regions radiating only in the continuous spectrum was revealed by a photograph taken through a special filter whose passband contains no strong emission lines (Figure 17.5). This photograph differs strikingly from the one of Figure 17.4. It might seem to the nonspecialist that we are looking at two entirely different objects! In this picture the nebula appears to have a far more diffuse or amorphous structure than in Figure 17.4. The bright features are also distributed in a different manner. These photographs demonstrate, then, that the Crab Nebula comprises two separate parts: an open network of thin gaseous filaments serving as an envelope at the periphery of the nebula, and an amorphous substance occupying practically the whole volume and emitting the continuous spectrum—a component of the nebula that remained a puzzle for many years.

The nebula is shaped like a fairly regular ellipse whose angular size is about 4′ × 6′. A noteworthy circumstance is that two inconspicuous stars of the 16th magnitude are located near the center of this ellipse, about 5″ away from each other. The southerly star (the lower one in the photograph) has played an extraordinary role in the history of astronomy (Chapter 19). The nebula itself has an apparent magnitude of about $8\overset{m}{.}5$; thus it is emitting a thousand times as much radiation as each of these little stars.

In 1921 the American astronomer Carl O. Lampland compared photographs of the nebula taken eight years apart and found that changes were occurring there. Bright individual details in the amorphous mass had shifted noticeably, and the brightness distribution did not remain constant but seemed instead to "breathe." For more than 30 years these remarkable changes could not be explained. This variability in structural detail is a unique property of the Crab Nebula. Neither in planetary nor in diffuse

Figure 17.5 The Crab Nebula, photographed by W. Baade at the Palomar Mountain Observatory through a filter that transmits no bright emission lines.

nebulae is anything of the kind observed. From the known distance of the Crab (see below) and the angular shift of the features after eight years one could infer that certain parts of the nebula were moving at a velocity close to 0.1 of the speed of light, a fantastically fast motion! As this behavior was thought to be completely irrational, astronomers simply ignored it. The nature of these variations did not become clear until the mechanism responsible for the optical radiation of the Crab was finally understood. We shall return to this point presently.

The position of the gaseous filaments of the nebula changes considerably more slowly. Observations separated by 30 years have established that the whole filamentary system is expanding. It is, so to speak, spreading out over the sky at an angular rate of about 0″.23 annually. But the angular radius of the nebula is about 180″. We may at once conclude that the age of the nebula, assuming that it has continued to expand at the same velocity, is almost exactly 800 years. This value is of course close to the age measured from the time of the supernova outburst (about 920 years), but nonetheless it is somewhat smaller. An important conclusion follows: the motion of the filaments in the Crab Nebula is *accelerating.* Only by ascertaining the nature of the radiation from the amorphous mass of the Crab has it been possible to explain this acceleration of its filaments (see below).

We turn now to an analysis of spectroscopic observations of the Crab. The doubling of the emission lines originally found by Slipher has been confirmed by subsequent research. In the spectrum of Figure 17.3 the spectrograph slit is oriented along the major axis of the nebula. It is clear that the bright emission lines are double at the center of the nebula, while at the edges of the nebula there is no doubling of the lines. The lines look patchy—an effect that results from the filamentary structure of the regions of the nebula emitting the lines. The peculiar arched shape of the doubled lines graphically proves that the doubling is caused by an expansion of the system of gaseous filaments in which these lines are emitted. Because of their expansion the filaments in the central part of the disk of the nebula are moving toward us and away from us. Due to the Doppler effect, the former will radiate lines shifted toward the violet side of the spectrum, and the latter will give lines shifted toward the red. At the edge of the nebula, on the other hand, the filaments are moving nearly perpendicular to the line of sight, so that no appreciable shift in the wavelengths of the lines will be observed.

By measuring the separation between the components of the doubled lines one can obtain the expansion velocity of the filamentary system; it is about 1200 km/s, on the average. Knowing the angular rate at which the filaments spread out in a plane perpendicular to the line of sight (0″.23 per year) and the linear expansion velocity, one can determine the distance of the Crab Nebula, which (allowing for its ellipsoidal shape) is about 1700 pc. This value accords with the estimated apparent magnitude of the "guest star" of 1054. After correcting for the absorption of light one finds that the absolute magnitude of the star at its maximum brightness was about -18^m, which is comparable to the absolute magnitude of supernovae.

From the measured radial velocities of the filaments and their "proper motions" across the celestial sphere, a rough three-dimensional model of the Crab Nebula has been developed, as illustrated in Figure 17.6. The oddly interlacing network of gaseous filaments of this nebula seems to border the amorphous mass at its periphery. Analysis of the width and relative intensities of the spectral emission lines enables us to understand the physical conditions prevailing in the gaseous filaments. For an individual filament the electron density is found to range from several hundred to several thousand electrons per cubic centimeter, while the temperature of the gas is close to 20,000°K. The chemical composition of the filaments, so far as can be judged from their spectra, is not so very different from the composition of other gaseous nebulae. There is, to be sure, a suspicion that

Figure 17.6   A spatial representation of the filaments in the Crab Nebula, derived from velocity measurements by N. U. Mayall at the Lick Observatory.

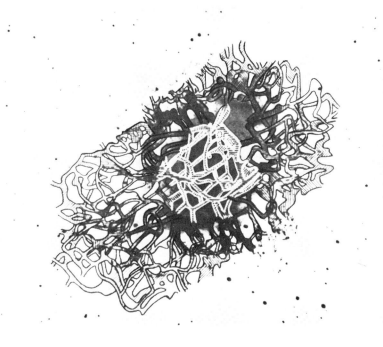

helium has a relatively high abundance in the Crab, but uncertainty as to the temperature of the filaments makes this conclusion very unsafe. The total mass of the gas confined in the filaments of the Crab Nebula is hardly more than a few tenths of a solar mass.

Before proceeding to an analysis of the main component of the optical radiation of the Crab Nebula with its continuous spectrum, we should consider the radio emission. As mentioned above, radio waves from the Crab were discovered by John G. Bolton in Australia in 1949. This discovery was unquestionably a landmark on the road toward an understanding of the nature of the Crab Nebula. Even the earliest observations revealed major differences between the radio spectra of the Crab and of the other radio sources known at the time. The Crab was found to have a far more gentle spectrum than the others. An abnormally small spectral index α was found for the radio emission of the Crab Nebula, only 0.28. In other words, the radio flux decreases much more slowly toward higher frequencies than is true for other sources. Thus, for example, while at 3-m wavelength Cassiopeia A has 10 times the flux density of the Crab, in the centimeter wavelength range the situation is reversed.

The Crab Nebula has practically the same angular size in the radio as in the optical range. And another important circumstance is worth emphasizing. As we saw in Chapter 16, the radio sources associated with old type II supernova remnants have a shell structure, with their radio-emitting regions located at their periphery. Nothing of the sort is observed in the distribution of radio emission in the Crab Nebula. In this case the radio sources fill the whole volume of the nebula, and are in fact concentrated toward the center.

Radio observations when the Crab is occulted by the moon are of great interest. Since it is located near the ecliptic, the Crab is from time to time covered by the moon traveling around the celestial sphere. These eclipses afford rich opportunities for studying the brightness distribution of the radio emission—a great advantage because until recently the resolving power of radio telescopes was inadequate. As the edge of the moon crosses the nebula it seems to turn off the bright features one after another, so that their coordinates can be measured with great precision. A lunar occultation of the Crab Nebula takes place about once every 10 years.

During the radio observations of the 1964 occultation of the Crab by the moon a very important discovery was made. Right at the center of the nebula, not far from the southerly little star mentioned above, a radio source was found with a very small angular size, no more than 0″.1. This source is especially intense at low frequencies. For instance, at a frequency of about 25 MHz (a wavelength of about 12 m) it supplies from 30 to 50 percent of the entire radio flux of the Crab Nebula. The existence of such a radio source had been suggested even earlier by observations of the scintillation of the radio emission from the Crab during passage of the radio waves through the plasma of the outer solar corona—an event that occurs every year about June 15, when the sun's annual motion along the ecliptic brings it very close to the Crab. Not until five years after these observations did it become clear that the tiny radio source is actually the remarkable pulsar discovered at the center of the Crab in 1968.

At wavelengths in the centimeter range the radio emission of the Crab exhibits considerable linear polarization. This fact by itself is strong evidence that the radio emission of the Crab is synchrotron in nature. In 1953, even before the discovery that the radio emission is polarized, the author of this book proposed the theory of synchrotron radiation to account for the radio emission of the Crab Nebula. A calculation of the type described in Chapter 16 indicates that the observed radio emission would compel the Crab to have a magnetic field with a strength of about 10^{-3} gauss. The radio waves are emitted by relativistic electrons in the nebula with energies of the order of several hundred million electron volts. The total energy of the magnetic field permeating the Crab is very large, about 3×10^{48} erg. It is informative to compare this energy with the kinetic energy of the gaseous filaments of the Crab, which, as we have seen, are dispersing at a velocity of about 1500 km/s. Taking the mass of the filaments as 0.1 $\mathfrak{M}_\odot$ or 2×10^{32} g, we find that their kinetic energy is about 1.5×10^{48} erg, almost the same as the energy of the magnetic field. We have the important result that the magnetic field of the Crab should play a substantial part in the dynamics of the expansion of the filaments, a consideration to which we shall return below.

In the same year, 1953, the author of this book interpreted the amorphous optical radiation of the Crab Nebula as resulting from the same synchrotron mechanism that is responsible for the radio emission. Previous attempts to explain this long-familiar radiation had encountered essentially insuperable difficulties. On the classical interpretation of the optical continuum of the Crab, involving the only mechanism known at that time—the thermal radiation of hot, ionized gas—one would have to assume that the nebula contains an enormous amount of that gas, say 20–30 $\mathfrak{M}_\odot$. And the temperature of the gas would have to reach hundreds of thousands of degrees, a most unusual value for gaseous nebulae, whose temperature is tens of times lower. But the mass was itself too large. In fact, if the 1054 supernova was of type I, as indicated by its light curve, then the mass of the exploding star should have been only slightly greater than the mass of the sun. Finally, a much lower temperature is observed in the gaseous filaments, around 10,000–20,000°K. It is very hard to image how such comparatively cool filaments could survive surrounded on all sides by a substantially hotter plasma of the same density. Just the pressure of the hot gas outside would cause them to contract indefinitely!

The idea that the optical continuum radiation in the Crab, like its radio emission, is produced by the synchrotron mechanism removes all these difficulties and irreconcilable contradictions in a radical way. For if the nebula contains relativistic electrons with energies in the 10^8–10^9 eV range, then there should also be some relativistic electrons (although considerably fewer) with energies as high as, say, 10^{11}–10^{12} eV. Since the frequencies at which relativistic electrons radiate by the synchrotron mechanism are proportional to the square of their energy (see Equation 16.10), if electrons of energy 10^8–10^9 eV in a magnetic field $H \approx 10^{-3}$ gauss radiate at decimeter and centimeter wavelengths then electrons of energy $\approx 10^{11}$ eV will actually radiate in the optical range, at frequencies hundreds of thousands of times higher. The concept is quite simple.

The spectral flux density in the continuous optical spectrum of the Crab is nearly 400 times lower than in the radio frequency range. On the other hand, the radio spectral flux *decreases*, although slowly, toward higher frequencies. It would be logical to conclude that the synchrotron radiation of the Crab might extend beyond, toward millimeter and infrared wavelengths. And finally, why might it not stretch still farther, to optical and even higher frequencies? In other words, we would expect this nebula to have a single synchrotron spectrum that ought to extend from radio to optical frequencies and beyond.

Calculations show that the Crab Nebula indeed has a single energy spectrum for the relativistic electrons, incorporating both the far more numerous electrons with energies of 10^8–10^9 eV that are responsible for the radio emission and the much more energetic electrons in the 10^{11}–10^{12} eV range, tens of thousands of times less abundant, that radiate the optical and ultraviolet photons. The numerical density of these high-energy particles is altogether insignificant: just one electron per 100 cubic meters of space! Yet the total mass of all the relativistic particles in the Crab, taking one relativistic proton for each relativistic electron, is about 10^{27} g, of the same order as the mass of the earth! If we were to try quantitatively to explain the optical radiation of the Crab by the thermal mechanism, we would have to suppose that it has about 10^{35} g of hot gas, or 100 million times more matter than in the case of the relativistic particles. We see how efficient the synchrotron radiation mechanism is: a very few relativistic particles are enough for the nebula to radiate powerfully over a rather long time.

Thus the new concept turned the amorphous mass of the Crab Nebula into a bubble containing only a negligible (on a cosmic scale, of course) amount of matter in the form of relativistic particles. In addition to solving the many difficulties that had confronted the old thermal theory, the synchrotron theory resolved a further puzzle known since the time of Lampland. It directly explained the rapid variations this astronomer had observed in the brightness distribution of the Crab (see above). Separate clouds of relativistic electrons can indeed travel inside the nebula at velocities reaching an appreciable fraction of the speed of light!

Circumventing the problems with the old theory is, of course, very important in itself. No less attractive is the logical order and elegance of the new theory. But all these features do not prove that the new theory is correct. In order to establish the truth of a theoretical construct firmly, one must on the basis of the new ideas predict further effects which either have been completely unknown or have seemed absolutely incomprehensible. We may regard the changes observed in the Crab as having been of the latter type. As for completely new effects, these too were not wanting.

On the basis of the new theory, the Soviet physicists Isaak M. Gordon and Vitalii L. Ginzburg predicted that if the optical radiation of the Crab is synchrotron in nature it should be linearly polarized. This effect implied by the new theory was first detected in 1954 by the Soviet astronomers Mikhail A. Vashakidze and Viktor A. Dombrovskii, who made observations independently of each other and by different methods. They used telescopes of very modest size and could obtain only the average polarization of the nebula. They were unable to investigate the polarization of individual features in the Crab. Nevertheless, their results were very impressive. The mean polarization was found to be about 10 percent, a very large amount. For comparison, note that the light from distant stars becomes polarized as it passes through dust clouds, but the polarization rarely exceeds 1 to 2 percent and the measurements are very unreliable.

Two years after the discovery by the Soviet scientists, polarimetric observations of the Crab Nebula were made in California with the 200-inch telescope, the largest in the world at the time. The American astronomer Walter Baade, whom we have mentioned on several occasions, took a series of excellent photographs of the amorphous mass of the nebula, using a special filter for that purpose together with polaroids in various orientations. These photographs, reproduced in Figure 17.7, cannot but capture the imagination. Certain bright details appear in one picture, then disappear in another taken through a polaroid oriented at right angles. This behavior indicates that the light of such features is almost 100 percent linearly polarized. When the polaroid is turned by 90° some regions of the nebula change until they are unrecognizable. It is as though we were looking at different nebulae. By measuring the original photographic plates it has been possible to determine the amplitude and direction of the polarization at every point of the nebula (Figure 17.8).

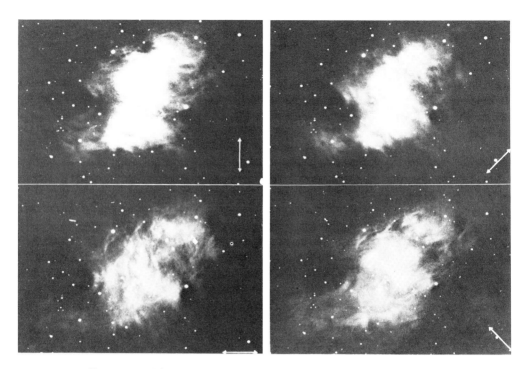

Figure 17.7 The Crab Nebula, photographed by W. Baade at the Palomar Mountain Observatory through polaroids oriented in four different ways.

According to synchrotron radiation theory, the polarization should be directed perpendicular to the magnetic field in the emission region. Since the polarization of the light of the Crab is comparatively regular in character, we can draw a system of lines of force representing the magnetic field there (Figure 17.9). Notice that the direction in which individual features of the amorphous mass are elongated conforms to the direction of the magnetic field. We have graphic evidence, then, that separate clouds of relativistic particles tend to flow along the lines of force. The magnetic field lines serve as conduits for such motion, preventing the particles from spreading in a direction perpendicular to the field. A behavior of this kind can occur only if the energy density of the magnetic field is comparable to or greater than the energy density of the relativistic particles. Thus the character of the magnetic field itself and its relationship to the amorphous mass strikingly demonstrate the synchrotron nature of the optical radiation of the Crab Nebula.

The structure of the centermost part of the Crab (Figure 17.10) merits special attention. This region had been carefully examined in 1942–43 by Baade, who discovered remarkable changes taking place there. Tiny, fairly bright condensations, generally elongated in shape, appear from time to time at the very center of the nebula and move swiftly from their place of origin

274

in a direction away from the center of the nebula. Figure 17.10 schematically illustrates the structure of this region in the middle of the Crab. The two circles mark the central stars. The dashed areas *b* and *a* symbolize bright patches in the nebula. Although feature *b* shows hardly any variations, the same cannot be said of the wisp *a*, which continually changes its position (relative to the central stars), shape, and brightness. It appears quite suddenly in the space between the lower of the two stars (the one which has proved to be a pulsar; see Chapter 18) and the patch *b*. Then it moves toward *b* and sometimes merges with it. The whole cycle described above takes three to four months. If we know the angular displacement of the wisp *a* and the time consumed by that shift, we can easily compute the velocity of *a*, or properly speaking its projection on a plane perpendicular to the line of sight. This velocity is more than 40,000 km/s!

It looks very much as though a compression wave of magnetic field lines is traveling at great speed through the central magnetized region of the

Figure 17.8 Polarization in the Crab Nebula, according to L. Woltjer. Photograph by W. Baade.

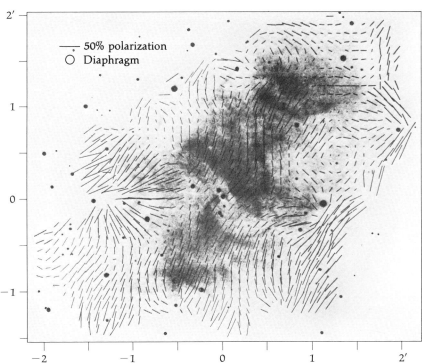

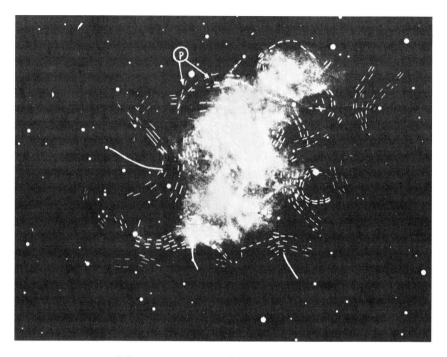

Figure 17.9   Schematic structure of the magnetic field in the Crab Nebula, according to J. H. Oort and T. Walraven.

Crab. This result follows from an analysis of the polarization of the wisp *a*, from which we can establish that the magnetic field is aligned with the direction of elongation of the wisp. Hence the wisp *a* is moving perpendicular to the direction of the magnetic field. All these extraordinary phenomena indicate that activity has continued in the central part of the Crab Nebula up to the present time. Historically this was the first evidence that the remnant of the 1054 supernova is not dead—that physical processes of tremendous power are in operation there, causing the very rapid variability observed in the central part of the nebula.

In 1963 a new page was turned in the history of research on the Crab. In the spring of that year a group of investigators from the U.S. Naval Research Laboratory under the direction of the noted American scientist Herbert Friedman, a founder of space astronomy, discovered that the Crab is emitting x rays. The experiment was performed with a small rocket of the Aerobee class. The x rays were detected by a battery of proportional photon counters whose total area was only 65 cm^2. This detector recorded photons in the 1.5–8 Å wavelength range. Quite a strong flux density was measured,

276

1.5 × 10⁻⁸ erg cm⁻² s⁻¹, only an order of magnitude lower than the flux density of the brightest x-ray source, discovered shortly before in the constellation Scorpius. Today, more than a dozen years later, the techniques of x-ray astronomy can record a flux density hundreds of times weaker than that of the Crab Nebula. Altogether more than 200 x-ray sources have now been found in the sky. But the source identified with the Crab remains one of the brightest.

As soon as this x-ray source was discovered, a problem arose: what does it represent? One might have thought that the x rays were coming from the Crab Nebula itself, which, although small, has a perfectly definite angular size of about 5'. On the other hand, one could not exclude the possibility that the x-ray source might be a star that had once exploded as a supernova. In this event the angular size of the x-ray source would be negligible; it would seem to be a point source.

At that time, as today, x-ray astronomy had a very low resolving power (the capability of distinguishing a point source from a small but extended object). But a fortunate circumstance enabled the problem to be cleared up quickly. We have already explained that a very effective method of studying the radio brightness distribution of the Crab is to analyze its radio emission when it is occulted by the moon. A similar method may well serve to analyze the x-ray emission of the nebula. At an appropriate time when the nebula is being eclipsed one would have to launch a rocket carrying a detector pointed toward the Crab. The detector should make a continuous record of the level of x rays coming from the nebula. If the x rays are emitted by a point source, then as soon as the source is covered by the edge of the moon the flux should drop sharply to zero. If the source is extended, the x-ray flux would fall off gradually as the edge of the moon advanced across the source.

At our suggestion an experiment of this kind was carried out by Friedman on July 7, 1964. The moon occulted the nebula at a rate of about 0'.5

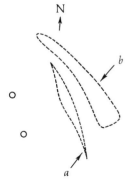

Figure 17.10 Wisps at the center of the Crab Nebula (schematic).

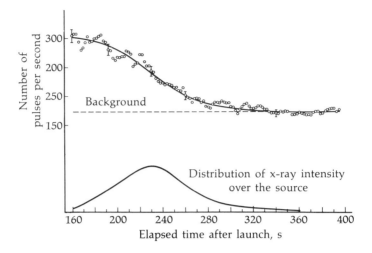

Figure 17.11   X-ray measurements during an occultation of the
Crab Nebula by the moon. Top, time dependence of the counting
rate of x-ray photons. U.S. Naval Research Laboratory.

per minute. The maximum angular size of the Crab Nebula is 6', so the
eclipse lasted about 12 minutes in all. But the rocket could spend only about
5 minutes at heights above 100 km, in the zone where cosmic x rays can be
recorded. For this reason the observing program was planned so that only
an area about 2' across at the center of the Crab would be monitored. The
results of these observations are presented in Figure 17.11, which shows the
time dependence of the detector readings as the nebula was occulted by the
moon. From the upper curve in this figure we see at once that the x-ray
source is extended, since the detector readings fall off gradually as the moon
crosses the nebula. The lower curve reconstructs the brightness distribution
of the x-ray source in the Crab Nebula, determined from the measurements
by the detector. The source has an angular size of about 1', considerably
smaller than the size of the nebula in optical light.

Subsequently, many observations of the x rays of the Crab were carried
out in different parts of the spectrum. With equipment borne into the upper
atmosphere by balloons it has been possible to follow the hard spectrum of
the Crab as far as the gamma-ray range, to photon energies of hundreds
of millions of electron volts. We will recall that in the optical range the
photon energy is 2–3 eV, while photons at radio wavelengths have an energy
hundreds of thousands or millions of times lower still. Thus the photons at
the far ends of the entire band of electromagnetic waves received from the

Crab Nebula have energies in the ratio of 10^{14}, or 100 trillion. No other cosmic object (except the sun, which is extremely close to us and therefore very bright at all wavelengths) has been investigated over so wide a range of the spectrum.

This nebula exemplifies the full-range scope of modern astronomy. There still remain two "white patches" of unexplored territory in the spectrum of the Crab: the far infrared and the ultraviolet. These parts of the spectrum are extremely difficult to observe. Ultraviolet radiation, for example, is strongly absorbed by the interstellar medium. Nevertheless, the electromagnetic spectrum of the nebula can today be outlined with full assurance. It is plotted in Figure 17.12, where the abscissa is the frequency of the radiation (proportional to the photon energy) and the ordinate is the spectral flux density.

First of all, this spectrum is completely unlike the spectrum of a thermal radiation source at any temperature, in particular a high temperature. Thermal radiation is described by the standard Planck formula; such a spectrum is illustrated schematically in Figure 17.13. But the spectrum of the Crab Nebula closely matches a power law of the form $F_\nu \propto \nu^{-\alpha}$. From the radio to the near-infrared range the spectral index α remains very small, about 0.3. But somewhere near frequencies of 10^{13}–10^{14} Hz (at wavelengths of perhaps 10–30 microns) there is a marked change or break in the spectrum. Beginning at a frequency of around 10^{14} Hz and extending all the way to the exceedingly high γ-ray frequencies of some 10^{22} Hz, the spectral index has a constant value close to 1.

Figure 17.12   The synchrotron spectrum of the Crab Nebula.

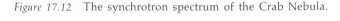

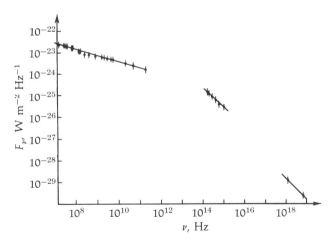

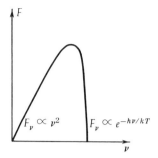

Figure 17.13   Behavior of the Planck spectrum.

A spectrum of this kind uniquely demonstrates the synchrotron nature of all the electromagnetic radiation of the Crab. Direct support for this claim is provided by recent measurements of polarization in the x rays arriving from the nebula. If relativistic electrons with energies of hundreds of millions of electron volts are responsible for the meter radio waves of the Crab and electrons of about 10^{11} eV energy for its optical radiation, then the γ rays should be produced by the motion of electrons with energies up to 10^{15} eV. All these relativistic electrons are traveling in the fairly regular magnetic field of the nebula, whose structure is outlined in Figure 17.9; the mean strength of the field is nearly 10^{-3} gauss.

The power of the synchrotron radiation emitted by the relativistic electrons is proportional to the square of their energy. But the radiation is generated at the expense of the energy of the electrons. Accordingly, the lifetime of the relativistic electrons (the time they require to lose a substantial part of their original energy by emitting synchrotron radiation) is inversely proportional to their energy:

$$t_1 = \frac{0.00835}{H^2 E} \text{ yr,} \tag{17.1}$$

where H is the magnetic field strength and E is the energy in billions of electron volts. A very important inference may be drawn from this equation: the lifetime of electrons with energies above about 10^{11} eV is shorter than 1000 years, the age of the Crab Nebula. Thus unless the supply of high-energy relativistic electrons there is continually renewed, the optical radiation would decay within a century. But this is certainly not the case! The brightness of the Crab has hardly changed at all since the time of Charles Messier, who observed the nebula in the eighteenth century. Some mechanism for pumping in fresh relativistic electrons of very high energy is therefore operating continuously in this nebula. What could the mechanism be; what is its nature? This was the problem facing astrophysics more than a dozen years ago. It was solved in 1969, when the Crab was discovered to have a pulsar at its center (see Chapter 19).

The distinctive break in the spectrum of the Crab at $\nu_1 \approx 10^{13}$ Hz is in fact attributable to the energy lost by the relativistic electrons because of their synchrotron radiation. The frequency ν_1 is determined by the circumstance that relativistic electrons radiating at lower frequencies will survive in the Crab Nebula without significant energy loss for a longer time than the nebula itself has existed, whereas at higher frequencies the radiation comes from more energetic electrons whose lifetime is comparatively short. Were it not for the continuous pumping of such electrons into the nebula, they would not be there at all.

Since we know the frequency ν_1 of the break in the synchrotron spectrum and the age t_1 of the nebula, we can apply the theory to find the strength of its prevailing magnetic field, which acts continually to transmute the spectrum. The expression for the field strength has the simple form

$$H = 700 \, \nu_1^{-1/3} \, t_1^{-2/3} \text{ gauss,} \tag{17.2}$$

where $t_1 \approx 1000$ yr. Taking $\nu_1 \approx 10^{13}$ Hz we find $H \approx 4 \times 10^{-4}$ gauss, which is very close to the value adopted above. Note that H depends only weakly on the frequency of the break in the synchrotron spectrum of the Crab, which is uncertainly determined, by extrapolation, since no direct measurements are yet available for the brightness of the nebula in the far-infrared part of the spectrum. Because of severe radiative losses, the relativistic electrons of very high energy ($\approx 10^{14}$ eV) responsible for the x rays of the Crab will be able to survive only a few months. This circumstance evidently accounts for the comparatively small extent of the x-ray source in the central part of the nebula. After entering the nebula such electrons will simply be incapable of reaching the periphery, because that would take several years.

Relativistic particles moving along the intricate, closed loops of the magnetic field of the Crab Nebula will tend to become trapped in the nebula. If there are enough of them, the particles will press on the magnetic loops, distorting them and moving them apart. But the magnetic field itself will attempt to unravel and fill as large a volume as possible. This tendency is hindered by the fact that the magnetic field lines are bound to the gaseous filaments of the nebula, which thereby restrain the field and the relativistic particles moving in it from rapid, unlimited expansion.

But we are then compelled to conclude that the magnetic field together with its relativistic particles should continuously exert a pressure on the system of gaseous filaments. The pressure force should cause an accelerated motion of the filaments in the nebula, as is actually observed (see above). We can estimate the value of this acceleration from the disparity between the known age of the Crab (920 years) and the time inferred from the expansion of the filaments when the whole nebula was essentially a point, which should have been the case about 800 years ago if the expansion velocity were constant. In this manner we find an acceleration $g = 0.0016$ cm/s^2.

Knowing the pressure force of the magnetic field and relativistic particles and the acceleration imparted by that force, we can apply a simple equation from mechanics to find the mass $\mathfrak{M}$ of the filaments in the Crab Nebula:

$$\mathfrak{M}g = 4\pi R^2 \left(\frac{H^2}{8\pi} + P_r \right), \tag{17.3}$$

where R is the characteristic linear size of the nebula (about 1 pc), the cosmic-ray pressure P_r is itself approximately $H^2/8\pi$, and H should be taken at the periphery of the nebula. We may estimate that $H \approx 3 \times 10^{-4}$ gauss there. It follows that the mass of the filaments is $\mathfrak{M} \approx 5 \times 10^{32}$ g or 0.25 $\mathfrak{M}_\odot$. This value includes the mass of faint filaments that are not observed. It is quite close to the "broadside" estimate based on the density and combined volume of the gaseous filaments in the nebula. The point of importance for us is that the mass is decidedly smaller than that of the sun, indicating that it was a type I supernova that exploded in 1054. This theory for the secular acceleration of the filaments in the Crab was put forward in 1954 by the eminent Soviet astrophysicist Solomon B. Pikel'ner, whose untimely death occurred late in 1975.

As we have emphasized, in order to maintain the optical radiation and especially the x rays of the Crab Nebula for periods of centuries new batches of relativistic electrons must continually be injected. It is natural to suppose that along with these electrons, relativistic heavy particles should also be injected into the nebula—protons, α-particles, and so on—or in short, cosmic rays. Unlike the electrons, the heavy relativistic particles will not lose their energy to synchrotron radiation. If they were to enter the nebula in greater abundance than the relativistic electrons, they would accumulate there to such an extent that the pressure on the gaseous filaments would be very large. In that case the filaments ought to be accelerated far more strongly than is observed. We may therefore conclude that the source replenishing the Crab with relativistic particles supplies it either primarily with electrons or with approximately equal numbers of electrons and heavy nuclei. This conclusion is significant for the theory of the origin of cosmic rays, because a substantial fraction (if not all) the relativistic particles entering the Galaxy have been formed in supernova outbursts.

The other remnants of the historic type I supernovae that appeared in the Galaxy in 1006, 1181, 1572, and 1604 do not furnish such a wealth of information as the Crab Nebula. The reason is not just that they are more distant from us. Quite simply, the Crab has proved to be a far richer and more interesting object. Nonetheless, we shall give a brief description of the other supernova remnants.

There is no evidence for the existence of optically observable remnants of the 1006 supernova. However, at the site of the outburst in the southern constellation Lupus a rather weak, extended radio source has been detected. The angular size of this source is quite large: 25', or five times the size of the Crab Nebula. If we assume that the 1006 supernova had the same apparent magnitude as the "guest star" of 1054, then with allowance for interstellar absorption of light we find that the star exploded at a distance of about 3500 pc. At this distance the linear radius of the remnant ought to be about 12 pc. As we know the age of the remnant ($\approx$ 1000 year), we can determine the mean velocity of the expanding envelope, which is roughly 10,000 km/s. This velocity is much higher than in the Crab Nebula, and fully agrees with the ejection velocity of gas in type I supernovae, as determined from the width of the emission lines in their spectra (Chapter 15). It is worth noting that the 1006 supernova exploded fairly high above the galactic plane, where the interstellar gas density should be quite insignificant. Hence the envelope thrown off in the eruption of the star would have experienced little deceleration.

The OSO-7 satellite has recently disclosed a fairly weak x-ray source at the position of the 1006 supernova. Analysis of the x-ray spectrum of the source indicates quite a high temperature for the radiating plasma, about 45 million degrees. Probably the temperature is maintaining this level because the 1006 supernova exploded in a very tenuous medium, and its envelope has scarcely begun to slow down. This somewhat meager information delineates the extent of our current knowledge about the remnants of the supernova outburst of 1006.

At the site of the supernova observed in 1181 by Chinese and Japanese astronomers in the constellation Cassiopeia, there is a rather bright (and consequently young) radio source.

We know much more about the remnants of the outburst of Tycho's star, the supernova of 1572. Some very faint filamentary wisps of nebulosity have been observed at the site of this supernova. Observations spaced about 10 years apart reveal certain changes in the relative brightness of the filaments. But spectroscopic measurements do not show any appreciable radial velocities. As early as 1952 a radio source was discovered at the site of the explosion. From subsequent observations astronomers have established its structure, which is very remarkable. At radio wavelengths this source comprises a bright, very thin ring about 7' in diameter but with a thickness less than 0.01 of the radius. The radio spectrum of the nebula exhibits absorption lines of interstellar hydrogen at 21-cm wavelength, from which, by the same method described for Cassiopeia A (Chapter 16), one can estimate the distance of the source: about 5000 pc. This distance, again allowing for interstellar absorption, gives about -18^m as the absolute magnitude of the 1572 supernova, which is comparable to the absolute magnitude of type I supernovae.

The envelope of the nebula has a linear radius of about 5 pc. In view of the age of this supernova, the mean expansion velocity of its envelope should be 12,000 km/s. It is not understood why the filaments have such a low velocity in comparison. Perhaps the filaments are not remnants of the envelope originally ejected. The mass of that envelope ought to be very small. In fact, if we suppose that the total kinetic energy of the ejected envelope was 10^{50} erg (an order of magnitude greater than for the Crab Nebula) and that it had an initial velocity of about 20,000 km/s, then the mass of the envelope would have been hardly a few hundredths of a solar mass, too small to be preserved after 400 years. At the site of the 1572 supernova an x-ray source has been discovered, which we shall discuss in Chapter 20.

In the position in which Kepler's supernova appeared in 1604 there is a peculiar optical nebula with bright condensations, looking like a fan. Observations made 20 years apart indicate that the bright features are moving slowly through this nebula at a rate of 0".03 per year. Since the distance of the nebula, obtained in the same way as for Cassiopeia A and the 1572 supernova, is about 10,000 pc, the corresponding linear velocity in a plane perpendicular to the line of sight is about 1400 km/s, whereas the radial velocity derived from spectroscopic observations is only 230 km/s.

A fairly bright radio source was found some time ago at the site of the 1604 supernova. Its angular diameter is about 3′, which at a distance of 10,000 pc corresponds to a radius of about 5 pc. With this radius the envelope would have a mean velocity of 12,000 km/s, virtually the same as for the 1572 supernova. It is remarkable that the 1604 supernova exploded far away (about 1500 pc) from the galactic plane, where the interstellar gas density is very low. For how can we explain the comparatively slow dispersal rate of the gaseous filaments in the nebula? Surely the interstellar medium could not be retarding them. To solve this difficult problem we might have to assume that a nebula had previously been formed around the exploding star from matter flowing out of the star. That nebula could hardly have had much mass, however, because an old star had exploded with a mass only slightly exceeding the sun's.

To summarize, we still know too little about the nature of the remnants of the type I supernovae that appeared in 1006, 1181, 1572, and 1604. Apparently, though, the 1054 supernova differed strikingly from all of them in the low velocity of its ejected envelope and its comparatively large mass. In the final analysis this circumstance has been responsible for all the unique phenomena observed in the Crab Nebula. The 1006, 1572, and 1604 supernovae have the special property that the density of the interstellar medium around them is very low. Hence their envelopes, having experienced hardly any deceleration, have expanded to a radius of about 5 pc, causing a rapid decline in their radio luminosity. After a few thousand years they will have expanded so much that their radio surface brightness will be unobservable. On the other hand, the objects discussed in Chapter 16 (such as IC 443) exploded in a comparatively dense interstellar medium which, in protecting the remnants of the outburst from too rapid an expansion, preserved them and ensured their survival for tens of thousands of years.

18

Why Do Stars Explode?

Thus far we have considered only the consequences of supernova outbursts. Stellar explosions lead to the formation of extremely interesting nebulae with a most unusual character. These nebulae are literally stuffed with relativistic particles, or, simply stated, with primary cosmic rays. The cosmic rays should be generated, one way or another, comparatively early in the development of the nebular supernova remnants. Moreover, as shown by the example of the Crab Nebula, the stellar remnant of the explosion will in some cases remain a powerful source of cosmic rays, continuing to feed them into the nebula formed after the explosion. It is not yet clear to what extent this property of stellar remnants is universal, even in the particular case of type I supernovae.

The explosions of stars have important implications for the physics and dynamics of the interstellar medium. A tremendous disturbance begins to travel outward at a very high velocity, which gradually diminishes. After a few tens of thousands of years the explosion zone embraces a vast region of the interstellar medium measuring tens of parsecs across. The physical conditions in this zone differ strikingly from the undisturbed state. The zone contains an extremely hot plasma, heated to temperatures of several million degrees. In the region covered by this great disturbance the cosmic ray density and the magnetic field strength are much higher than usual. As it disperses through the surrounding interstellar medium, the disturbance enriches it with cosmic rays and produces changes in the chemical composition of the interstellar gas. We have already seen in Chapter 16 that the chemical composition of the swiftly moving filaments in Cassiopeia A is decidedly different from the ordinary. This observational fact alone shows that the explosion of a star serves as a crucible in which the "cooking" of complex nuclei takes place. In the language of metallurgists, supernova outbursts thereby carry out a process of "flotation," or enrichment of the interstellar medium with complex nuclei.

We hardly need stress the immense importance of the consequences that flow from this steadily operating process. When the universe was young—before the galaxies and stars had been formed—it comprised a fairly simple hydrogen–helium plasma, perhaps with a small admixture of deuterium. There were no heavy nuclei at that time. This circumstance is reflected in the chemical composition of the oldest generation of stars, the subdwarfs (Chapter 12). Most of the enrichment of the interstellar medium with heavy elements, one should understand, occurred very early in the formative stages of the galaxies. Along with the present subdwarfs a great many massive and supermassive first-generation stars were formed at that epoch—stars that were to explode as supernovae after tens of millions of years of evolution. The frequency of supernova outbursts was a hundred times greater then than now. For this reason the primary process of enriching the interstellar medium with heavy elements ended quite soon, within just a few hundred million years of the earliest history of our Galaxy (and similarly, of course, in other galaxies as well).

It is natural to ask how these important details in the chemical history of our stellar system are known. The story is a chronicle written in meteorites and the crust of the earth. Delicate chemical analysis has enabled us to find abundance ratios for the radioactive isotopes U^{238} (uranium 238), Pu^{244} (plutonium 244), Th^{235} (thorium 235), and the two iodine isotopes I^{127} and I^{129}. Since the half-lives of the nuclei of these isotopes are quite well established, we can use the measured relative abundances to find the age of the nuclei. In particular, measurements of the abundance ratio [Pu]/[U] in meteorite samples indicate that these superheavy nuclei were formed within a comparatively short span of time that occurred 8.5–10 billion years ago.

Very interesting results have been gathered from analysis of the abundance of the iodine isotopes and another substance found in meteorites, the heavy inert gas xenon, which is a stable decay product of the radioactive isotope I^{127}. This analysis demonstrates that the age of the iodine isotopes is only about half the age of the uranium, plutonium, and thorium isotopes. Otherwise the relatively short-lived I^{127} isotope would not have been preserved. On the other hand, examination of the xenon abundance in meteorite samples shows that within 180 million years of their formation the iodine isotopes had become part of the crystalline material of meteorites. As there can be no doubt that the meteorites were formed concurrently with the solar system some 5 billion years ago, we may conclude that the matter from which the solar system was formed had been enriched not long before by an exploding supernova. It is also noteworthy that the recently discovered differences in the chemical composition of the interstellar clouds (see Chapter 2) can be explained in a natural way by the effects of supernova outbursts.

Having digressed briefly into the appealing subject of the chemical history of the Galaxy, we return to the fundamental question of the causes of the stellar explosions observed as the supernova phenomenon. The radiation emitted by the remnants of these outbursts gives us an opportunity to estimate certain significant parameters of the explosions without which any scientific discussion of the problem would be impossible. Among these parameters are the mass of the envelope thrown off in the outburst, the kinetic energy of the envelope, its chemical composition, the extent to which supernova remnants are permeated by vast numbers of relativistic particles, and the energy spectrum of those particles. Moreover, studies of supernova outbursts in other galaxies by the techniques of modern astronomy, especially spectroscopy, can furnish the total amount of energy radiated— the most important property of an explosion. Observations of the same kind permit us to determine the initial velocity of the gases ejected in the explosion, so that we can estimate the specific energy of the outburst, that is, the amount of energy released per gram of material.

We should emphasize at the outset that modern science does not yet have a genuine theory of stellar explosions at its disposal. The problem has proved very difficult, as might have been expected. Even so, the situation should not be regarded as hopeless. A number of knotty questions facing a future theory have already been worked out, but the main point is that we have some understanding of what physical conditions in an evolving star should, as they undergo regular change, inevitably lead to a cosmic catastrophe.

Turning now to the theoretical ideas that have been developed to account for stellar explosions, let us first of all consider the possible sources of energy. The most natural candidate for such a source is nuclear energy.

We have already discussed this source in some detail to explain the quiet radiation of stars while they are still on the main sequence (see Chapter 8). As we emphasized in that context, the evolution of a star becomes much more complicated after the hydrogen nuclear fuel is depleted in its central regions. The equilibrium state of the star during the final stage in its evolution depends on its original mass, which is assumed to remain the same throughout the evolution. But this last assumption, as we have seen in Chapter 13, is not actually satisfied. For example, when real stars pass through the red giant phase their outer layers become detached and a white dwarf is formed from the inner layers.

Nevertheless, it is helpful to consider an idealized model star that continually maintains its mass and moreover does not rotate. Presumably this simplified treatment of the problem will clarify various significant properties of the closing phase of stellar evolution. Calculations show that if the mass of such an ideal star is below about 1.2 $\mathfrak{M}_\odot$, then the final product of the evolution will be a white dwarf, as described in Chapter 10. For stars with a mass greater than 1.2 $\mathfrak{M}_\odot$ but less than about 2.5 $\mathfrak{M}_\odot$, a configuration with degenerate gas can no longer be in equilibrium. In 1938 the American theoretical physicists J. Robert Oppenheimer and George M. Volkoff proved that after such a star exhausts its nuclear fuel supply it should collapse catastrophically and turn into a superdense object about 10 km across—a neutron star. We alluded to this situation in Chapter 12, where we also pointed out that stars with a mass exceeding some limit of about 2.5 $\mathfrak{M}_\odot$ should eventually collapse catastrophically to a point; these are the black holes that we shall discuss in detail in Chapter 24.

Depending, then, on the original mass of the idealized model star, theory predicts three types of final state for stars whose energy is exhausted:

1. white dwarfs;
2. neutron stars;
3. black holes.

Astronomers have known about the first of these for more than 60 years. Neutron stars were discovered only in 1967, after long, fruitless efforts. Finally, there are some grounds for believing that certain objects we know of actually represent black holes (Chapter 24). Thus even though the idealized model star is much simplified, it correctly predicts the existence of all three species of "dead" stars. The original theory, however, did not indicate any specific route by which the dead stars would be formed.

From all evidence, supernova outbursts are connected with the final step in the evolution of a star. The truth of this claim is apparent from the very peculiar chemical composition of the filaments in Cassiopeia A. We may accordingly expect supernova outbursts to be genetically related to the formation of neutron stars and black holes. Such a relationship has long been suspected, but only a decade ago was direct observational evidence obtained: neutron stars were discovered in supernova remnants.

It would be most natural to suppose that the colossal amount of energy released in supernova outbursts is nuclear in origin. However, simply in principle, by no means all the star's nuclear fuel can be responsible for the explosion. This is particularly the case for hydrogen, the basic nuclear fuel, whose thermonuclear reactions maintain the quiet radiation of stars on the main sequence. Even though the release of energy through total conversion of hydrogen into helium is very large (6×10^{18} erg/g), it takes place quite gradually. Hence an explosion—the very fast liberation of a great deal of energy—cannot occur in this case. The rate of thermonuclear reactions involving hydrogen is so gradual because chains of such reactions (Chapter 8) include the β-decay process as necessary links. Processes of this kind operate very slowly and simply cannot be speeded up, because they operate spontaneously. As an illustration, even at very high temperatures the reaction transmuting hydrogen into deuterium,

$$p + p \rightarrow D^2 + \beta^+ + \nu, \tag{18.1}$$

takes place sluggishly by β-decay.

But if the temperature is high enough the $3He^4 \rightarrow C^{12}$ reaction discussed in Chapter 8 will come into play, along with the ensuing reactions involving carbon nuclei and helium nuclei (α-particles),

$$C^{12} + He^4 \rightarrow O^{16} + \gamma, \qquad O^{16} + He^4 \rightarrow Ne^{20} + \gamma, \tag{18.2}$$

and these can produce a very large number of light carbon, oxygen, and neon nuclei. By the time the temperature has reached about 10^8 °K, the nuclei of these light elements can enter into reactions with protons, accompanied by a substantial and—the main thing—*rapid* release of energy, because such reactions do not operate by β-decay. In this manner every nucleus of a light element can successively combine with at least three or four protons, whereby an energy of about 10–20 MeV per nucleus will be liberated. For the heavier nuclei formed through successive attachment of protons, β-decay will be present and will seriously retard the reaction, so it will lose its explosive character. Nonetheless, even three or four successive combinations with protons will result in a fair degree of explosiveness. The whole question, though, is whether the star has enough nuclei of light elements for their explosion (and we have not yet said how this can happen!) to release the amount of energy necessary.

If the chemical composition of a star that ought to explode resembles that of the sun, then every gram of its material will contain about 5×10^{20} light nuclei. If for any reason an explosive reaction of the type described above involving light nuclei does take place, the specific energy release will be about 10^{16} erg/g. This is not so very much. For type II supernovae, the specific energy yield is at least 10 times as high. Imagine for a moment that our sun were to explode by such a reaction process; then about 10^{49} erg of energy would be released, but this amount would still be some 10 times less than the outburst of a type I supernova sets free. If we suppose that for some unknown reason the interior of the sun should be heated to a temperature of 10^8 °K, an explosion would most likely follow. But the gases would be dispersed at velocities no more than, say, 500 km/s, at least 10 times below the dispersal velocity observed in supernova outbursts (see Chapter 15).

If we want to explain the catastrophic release of energy during a supernova outburst in terms of nuclear reactions (and suitably explosive reactions can only occur with the nuclei of light elements), we would have to assume that the chemical composition of the interiors of exploding stars differs sharply from that inside the sun. The difference would take the form of an incomparably greater abundance of light elements (carbon, nitrogen, oxygen, neon) relative to hydrogen than occurs in the sun. For example, if in the sun there is only one atom of any of these elements for every thousand hydrogen atoms, then in a star about to explode there should be fully 2 to 3 percent as many light atoms as hydrogen atoms. But this star would have been formed at one time out of the interstellar medium, whose chemical composition is nearly the same as that of the solar atmosphere. Accordingly, in the evolutionary process the chemical composition of the star about to explode should undergo a very substantial change, thanks to various kinds of nuclear reactions. This change would, in a sense, prepare the star for the explosion, forming a potential "powder magazine" there, filled with dangerously explosive nuclear fuel.

At the extremely high temperatures that should inevitably develop when reactions involving light nuclei take place (we are talking about temperatures around a billion degrees), the matter will begin to acquire an explosive instability because of very rapid reactions such as $C^{12} + C^{12} \rightarrow Ne^{20} + He^4$ and $C^{12} + C^{12} \rightarrow Na^{23} + p$, as well as similar reactions for O^{16}, Ne^{20}, and other light elements. The characteristic time for such reactions is about 1 second, and the specific energy yield will reach 5×10^{17} erg/g. If, for instance, a mass of such material equal to 0.1 $\mathfrak{M}_\odot$ were to explode, about 10^{50} erg of energy would be released—already close to the energy liberated during the explosion of a type I supernova.

We may conclude, then, that the only potential nuclear fuel that might cause a star to explode is matter highly enriched with light elements. An ordinary cosmic mixture with a chemical composition like that of the sun cannot under any circumstances lead to the nuclear explosion of a star. The question thus far remains entirely open, however, as to how the conditions required for a nuclear explosion are prepared.

Finally, the prime origin of stellar explosions could be the release not of nuclear but of gravitational energy through catastrophic collapse. Probably both forms of energy are significant, although, as we have indicated above, the whole picture of the explosion of a star is still far from clear. Nevertheless, we shall now describe some theoretical developments which will undoubtedly be useful when a theory of stellar explosions is worked out in the future (perhaps before too long).

The British and American theoreticians Sir Fred Hoyle and William A. Fowler have considered an interesting model for a star on the eve of explosion (a "pre-supernova"). At the outset they have taken the case of a comparatively massive star of 30 $\mathfrak{M}_\odot$ whose material has not become mixed during its evolution. In such stars the matter in the central part will not be degenerate, as the density there will be comparatively low (Chapter 12).

These calculations evidently relate to the problem of type II supernova outbursts. In the closing evolutionary phase the temperature in the central regions of such a star (properly speaking, in the model star) will be exceedingly high, of the order of several billion degrees. At such a temperature all the hydrogen and helium will have been burned. Nuclear reactions will operate with great speed. The equilibrium state of the matter will show a predominance of nuclei of elements belonging to the iron group, for they are the most tightly bound and stable nuclei. The core of such a star is surrounded by a mantle whose temperature is considerably lower, say less than a billion degrees. The chemical composition of this envelope will be strikingly different from that of the core. Light elements will predominate in the mantle—oxygen, nitrogen, and neon, the potential nuclear fuel needed for the explosion of the star. Finally, the mantle is surrounded by an outer shell of hydrogen and helium. According to calculations based on this model, the central iron core would have a mass of 3 $\mathfrak{M}_\odot$; the oxygen mantle, 15 $\mathfrak{M}_\odot$; and all the rest would make up the fairly rarefied outer hydrogen–helium layer.

Conditions are ripe for a nuclear explosion when the iron core has evolved to the point at which a catastrophic contraction or collapse sets in. The characteristic time for such a collapse is approximately the free-fall time—about 1 second. During catastrophic contraction of the core, the mechanical equilibrium of the remainder of the star will also be destroyed, for the weight of the overlying layers will no longer be supported by gas pressure from beneath. Thus the outer layers of the star will also begin to fall in toward the center. After a very short time, again about 1 second, the kinetic energy of the infalling envelope will be converted into thermal energy, causing the material to become rapidly heated. In this manner conditions will develop to permit a nuclear explosion of the light elements in the mantle.

The essential fact, however, is that the catastrophic collapse of the core of the star should occur in less time than would be required for a calm reorganization of the envelope structure without an explosion. We have already examined this point rather fully in Chapter 6 in connection with the problem of how the mechanical equilibrium of a star would be violated by the instantaneous local release of a certain amount of energy. The time needed for calm reorganization of the structure of a star is determined by the velocity of sound passing through it. This velocity is given approximately by

$$v_s = (\bar{P}/\bar{\rho})^{1/2} \approx (g\,\mathfrak{M}/R)^{1/2} \tag{18.3}$$

(see Chapter 6). In our case, for a star with a mantle measuring 3×10^9 cm, the velocity $v_s \approx 10^9$ cm/s, and the contraction would pass through the star in $t_s \approx R/v_s \approx 3$ s. Now it is important to understand that if the temperature of the core material were to rise fast enough as the core contracts, then the collapse would not take place at catastrophic speed. At any given instant the star would be able to adapt its structure to the changing conditions, and no explosion would occur. We have touched on this matter in some detail when discussing the equilibrium of a star in Chapter 6.

Catastrophic collapse will happen only if there is a "refrigerator" in the core, removing from it the thermal energy released by the contraction process. This refrigerator, incidentally, should have an exceptionally high power, of the order of 10^{18} erg/g.

At least two possible types of refrigerators have been proposed. The first was suggested by Hoyle and Fowler. It entails a very intensive absorption of energy through the dissociation of iron nuclei into α-particles and neutrons. As the temperature rises, such a dissociation process will be inescapable, and it will be accompanied by the absorption of an enormous amount of latent heat of dissociation. Each iron nucleus will break down into 13 α-particles and four neutrons. The binding energy of the nucleons in an iron nucleus is 8.79 MeV, whereas the mean binding energy of a single nucleon in the mixture of α-particles and neutrons resulting from the dissociation is only 6.57 MeV. Hence in order to break the iron down into α-particles and neutrons, 2.22 MeV of energy must be used up, or 2×10^{18} erg/g. An excellent refrigerator, surely! It works in the following way: as soon as the temperature of the iron core has risen to a certain level, a further increase will be halted for a comparatively long while, because the gravitational energy released in the contraction will be expended in dissociating the iron nuclei. This pause in the heating of the contracting core will create favorable conditions for detonation of the "powder magazine" of the star, for the core will proceed to collapse catastrophically, while the envelope, being unable to readjust its structure calmly, will begin to cave in toward the center of the star, becoming heated rapidly as it does so. As a result, explosive reactions with the light elements making up the mantle will come into play.

That, in general outline, is Hoyle and Fowler's analysis of the way a massive star would explode. It is evident that before exploding the star must be far advanced in its evolution. Significantly, its interior will have changed radically in chemical composition. In particular, more than half the mass of the star, having formed a hydrogen–helium mixture at the beginning of the evolution, will have been converted into light elements. The immediate cause of the explosion of the star is the catastrophic contraction of its iron core in the presence of a refrigerator such as the latent heat of dissociation of iron into helium and neutrons. This kind of evolutionary path would evidently be typical for sufficiently massive stars. The theory described above should therefore correspond to type II supernova outbursts.

We wish to emphasize, however, that despite the valuable ideas incorporated in this theory, it definitely cannot be regarded as providing a complete description of the processes taking place when type II supernovae explode. For example, the theory says nothing about what happens if the material is heated to temperatures of several billion degrees and great numbers of neutrinos and antineutrinos begin to be formed there. These particles will escape from the star, carrying an immense amount of energy along with them.

The nuclear reactions leading to the production of neutrinos ν and antineutrinos $\bar{\nu}$ are as follows (this is called the Urca process):

$$e^+ + n \rightarrow p + \bar{\nu}, \qquad e^- + p \rightarrow n + \nu. \tag{18.4}$$

Beginning at a temperature of half a billion degrees, the neutrino radiation of massive stars will already exceed their photon radiation. As the temperature of the collapsing core rises, the power of the neutrino radiation of the star will increase vastly. It will grow especially after the iron in the central zone of the star has been dissociated, that is, during a later phase of the contraction. After this dissociation the iron refrigerator will cease to exist and a new, quite rapid rise in the temperature of the core will occur. When the core has been heated to 20 billion degrees (by which time its density will be about 10^{10} g/cm^3), the α-particles will begin to split up and a great many free (that is, not bound in nuclei) protons and neutrons will appear. These will cause a sharp increase in the rates of neutrino and antineutrino formation by the reactions 18.4. The newly formed particles will escape from the core, carrying off a tremendous amount of energy. We thereby have another, exceedingly powerful type of refrigerator.

The great energy of the neutrino radiation is derived from the gravitational energy of the collapsing core. The neutrinos and antineutrinos leaving the star will have an energy of about 10 MeV, considerably more than the energy of solar neutrinos (see Chapters 8 and 9). When the temperature of the collapsing core reaches 40 billion degrees and the density is about 3×10^{11} g/cm³, a new situation will develop: the core of the star will cease to be transparent to neutrinos. The neutrinos will now be absorbed by the protons and neutrons (the reactions 18.4, but reading from right to left). A new refrigerator will thereby turn on, raising the core temperature still more sharply, and severely retarding the contraction process. The core will evidently stop contracting when its density has reached about 3×10^{13} g/cm³ and its temperature exceeds 100 billion degrees. The envelope, no longer falling toward the center of the star, will heat up rapidly, and the powder keg (the light elements in the mantle) will explode.

Such is the general picture of the explosion of a massive star, if we allow for the processes of neutrino and antineutrino production in its hot, collapsing interior. It is also worth noting that a collapsing core can quickly cease its contraction for a completely different reason. For we have not yet taken into account the rotation of the collapsing star. The law of conservation of angular momentum, familiar from mechanics, tells us that as a star contracts its linear rotational velocity will increase rapidly. A situation can develop in which the enormous centrifugal forces resulting from this effect will halt the contraction of the stellar core, serving to stabilize it. The infalling mantle of the star will thus also be stopped; it will become strongly heated, and the conditions will be ready for a nuclear explosion.

Nor should one forget that the whole intricate physical behavior of a collapsing star prior to its explosion, as set forth above, all takes place within a negligibly short time, around one-tenth of a second. During this interval the catastrophically collapsing star emits a huge number of neutrinos. Calculations show that the total energy of these neutrinos reaches about 10^{52} erg, nearly a hundred times the kinetic energy of the envelope the star ejects! And it is approximately a thousand times the energy of all the neutrinos the sun has managed to radiate over its whole evolution of 5 billion years. At least 99 percent of the gravitational energy liberated during the catastrophic collapse of a star is carried away by neutrinos, and hardly 1 percent is converted into the forms of energy observed by astronomers.

It is interesting that neutrinos of high energy will interact more easily with matter; thus they can more easily be detected experimentally. If somewhere in the Galaxy a type II supernova were to explode at a distance of several thousand parsecs, and if we knew the time of the outburst to within a few hundredths of a second, then receiving apparatus already in place on the earth (such as that described in Chapter 9) would be capable of recording the explosion. A successful experiment of this kind would be extremely helpful for our understanding of the nature of stellar explosions. Thus far, however, we can only dream about such an experiment.

Up to now we have been discussing the causes of the explosion of massive stars observed as type II supernovae. Although, we must repeat, there is still no real theory for this phenomenon, considerable advances in dealing with the problem have been made in recent years. Matters are far worse regarding the theory of the explosion of comparatively low-mass stars, observed as the type I supernova phenomenon.

The distinctive structural property of stars of comparatively low mass during late phases of their evolution is that they contain a very dense degenerate core (Chapter 11). In this case the central layers of the star will collapse catastrophically because of the absorption of degenerate electrons by nuclei, which will happen when the evolving star reaches some sufficiently high density,[1] greater than 10^{11} g/cm^3. A great many neutrons will then be formed. Just as in the case of massive stars, the envelope, left unsupported by the pressure of the degenerate electrons, will cave in and begin to fall toward the center of the star. Matter will become strongly heated (up to 500 billion degrees) where the envelope collides with the collapsing core. Powerful neutrino radiation will thereby by generated by the Urca reactions 18.4, and the neutrinos will be absorbed by the envelope. In this fashion the envelope also will become strongly heated, and nuclear reactions involving light elements will produce an explosion. But there is much that remains unclear in this picture. For example, it is just as likely that neutrinos will be formed throughout the whole core, which should be hot enough for that purpose.

Comparatively low-mass stars may actually explode in a manner very different from the scheme described above. In particular, the magnetic field of the collapsing star may play a major role, as may the rotation of the core. During the contraction process the magnetic field may attain extremely high values, of the order of several billion gauss (the reasons for this behavior will be discussed in Chapter 20). Under certain conditions the magnetic field may transfer the gravitational energy released by the collapse outward into the envelope, strongly heating and thereby detonating it. In astrophysics generally, the magnetic field quite often acts as a drive belt, conveying a substantial amount of energy. The main question facing theories of type I supernovae—why only a small fraction of 1 percent of stars with the appropriate mass explode—still remains unanswered. Might there be a very narrow range of mass wherein a star awaits catastrophe at the end of its evolution? Or does the cause of an explosion perhaps lie in the rotational properties of the core? Could the magnetic field play the fatal role here? Alas, theory cannot yet answer these questions. At the present stage of its development

---

[1] Of course the star must have had time to complete its evolution during the period the Galaxy has been in existence; thus it should have a mass of at least 1.2–1.3 $\mathfrak{M}_\odot$.

the theory of the explosion of low-mass stars is restricted to analyses of stellar models only during their closing evolutionary phases.

A theory of stellar explosions should not only indicate the cause of the explosion in view of the prior evolution of the star, and not merely estimate the amount of explosive energy released, but should also explain the light curves of supernovae. Why, for instance, are the light curves of type I supernovae so similar to one another? And why are the light curves of type II supernovae so diverse? Theoreticians have found these questions very difficult indeed.

One procedure for coping with the problem has been to consider the propagation of a strong shock wave generated after detonation of the mantle in the extended outer envelope of a star in which the density decreases outward from the center. In such a configuration the properties of the shock wave will be determined by the energy of the explosion and by the density distribution law in the outer envelope. The luminosity of a supernova calculated theoretically by this technique has proved to be very dissimilar to the observed value; in particular, the theoretical outburst has been found to be too short in duration. To some extent this difficulty can be circumvented by assuming that the exploding star is a red supergiant of vast size, with a radius tens of thousands of times the sun's. The calculations then yield a light curve in better agreement with the observations. All the same, there is no reason to claim that theory is capable of correctly describing the light curves of type II supernovae (we are referring here to the case of quite massive stars).

To explain the characteristic light curves of type I supernovae, the rather curious hypothesis of radioactive decay has been advanced. The principle of this hypothesis is as follows. A large number of radioactive isotopes of various elements can be formed during the explosion of a star. As they decay with their natural periods, they will act as energy sources. The light curve of a type I supernova approximately 100 days after maximum is described by the simple exponential law

$$L = L_0 e^{-t/\tau}, \tag{18.5}$$

where L is the luminosity of the supernova and $\tau \approx 70$ days is the characteristic time required for the brightness to drop by a factor $e = 2.718$. If the luminosity of a supernova of this type is attributable to radioactive decay, then there should be a radioactive element with a half-life close to τ. The most suitable element is the radioactive isotope Cf^{254} of the transuranium element californium. The nucleus of this isotope splits spontaneously into fragments whose combined energy is about 200 MeV. This means that the decay of 10^{30} g of matter (less than 0.001 $\mathfrak{M}_\odot$ of Cf^{254}) will yield 10^{48} erg of energy, which is fully adequate to account for the radiation of type I supernovae 100 days after maximum. Californium was first synthesized when an American thermonuclear bomb was exploded in 1952 by irradiating uranium with a large number of neutrons.

The radioactive iron isotope Fe^{59} would also be suitable. For example, in order to explain the amount of energy observed to be released during the exponential (comparatively late) phase of a type I supernova, it would be enough for 0.1 $\mathfrak{M}_\odot$ of this iron isotope to be formed at the time of the explosion. It appears, however, that this much iron cannot be formed in the explosion of a star. Many transuranium elements (other than Cf^{254}) possess radioactive isotopes with suitable half-lives, ranging from 10 days to 1 year. Among these are Cf^{256}, Cf^{258}, the fermium isotope Fm^{260}, and isotopes with an atomic number $Z > 102$. Superheavy elements can, incidentally, be formed during the explosion of a star by successive attachment to heavy nuclei of neutrons, which should be formed in great abundance. But there is a limiting atomic weight for transuranium elements that could be formed in this manner. The reason for the limit is that beginning at some sufficiently high atomic weight, the next capture of a neutron by a nucleus would result in fission. This should happen somewhere around atomic weight 265.

Accordingly, if perhaps 5×10^{-4} $\mathfrak{M}_\odot$ of transuranium radioactive isotopes were formed during the explosion of a star of comparatively low mass, the observed light curves of type I supernovae could be explained both qualitatively and quantitatively. But unfortunately it is not yet clear whether so large an amount of transuranium elements can be produced in a stellar explosion. In view of the observed frequency of supernova outbursts (one every 100 years or so in the Galaxy), if we were to assume that ordinary superheavy nuclei should be formed during the explosion of a star as well as the transuranium elements, we would be confronted with another difficulty: the cosmic abundance of uranium ought to be 100 times that observed. All these circumstances undercut the justification for the transuranium hypothesis as a mechanism to explain the radiation of type I supernovae, although it cannot yet be discarded as definitely wrong.

One other highly efficient source of energy during a supernova outburst might be the activity of an erupting "pre-supernova" neutron star formed out of the core. This possibility will be considered in the next part of the book, in Chapter 20.

To close this chapter, let us return to the conclusion we reached in Chapter 15 regarding stars that explode as type I supernovae. We stated that although such stars would represent ancestors of white dwarfs and planetary nebulae, they would further need to have some rare sort of pathology. As indicated by our simple statistical arguments in that chapter, about 0.3 percent of all stars that evolve normally into white dwarfs ought to suffer from this malady. A newly formed white dwarf will apparently be so afflicted if it has an abnormally high mass, slightly in excess of the Chandrasekhar limit $\mathfrak{M}_C = 1.2\ \mathfrak{M}_\odot$, a quantity to which we alluded in Chapter 10. What is the trouble here?

Actually the theoretical value of $\mathfrak{M}_C$ refers to a cool white dwarf whose gravitation is balanced solely by the pressure of its degenerate gas. But a real white dwarf may be hot, with its interior at a temperature T of tens of millions of degrees; heavy nuclei will then form an ordinary gas whose pressure, hundreds of times lower, should also be taken into account. The extra pressure will raise the value of the critical mass by an increment

$$\Delta \mathfrak{M} = 2 \times 10^{-4}\ \overline{A}\ \frac{T}{10^7\ °\mathrm{K}}, \tag{18.6}$$

where $\overline{A} \approx 20\text{--}50$ is the mean atomic weight of the nuclei. Notice that as the white dwarf cools, the increment $\Delta \mathfrak{M}$ will diminish. Hence a white dwarf formed with a slight mass excess will, as it proceeds to evolve, inevitably collapse, and a type I supernova outburst will probably occur.

Stars in elliptical galaxies, where every supernova is of type I, are very ancient—about 10 billion years old. Thus at the present time only stars with a mass no more than 1 $\mathfrak{M}_\odot$ are coming off the main sequence there, whereas pre-supernovae of type I should have a mass of about 1.5 $\mathfrak{M}_\odot$ (allowing 1.2 $\mathfrak{M}_\odot$ for the creation of a white dwarf and perhaps 0.3 $\mathfrak{M}_\odot$ for a planetary nebula). We may infer that in such galaxies white dwarfs slightly above the critical mass would have been formed roughly 10^{10} years ago. Cooling slowly, these delayed-action bombs would just be arriving at their critical state in our day, when their mass excess $\Delta \mathfrak{M} \approx 3 \times 10^{-5}\ \mathfrak{M}_C$. On the other hand, in spiral galaxies such as our own, where star formation is continuously in progress, outbursts of these leftover white dwarfs will be accompanied, a hundred times as often, by explosions of relatively young white dwarfs (only a few hundred thousand years old) for which $\Delta \mathfrak{M} \approx 3 \times 10^{-3}\ \mathfrak{M}_C$.

If this interpretation is correct, then the type I supernovae in elliptical galaxies should all be as similar to one another as twins—displaying identical light curves—while in spiral galaxies their properties ought to be notably diverse. But that is exactly what the observations show!

Finally, an overmassive white dwarf might explode even while it is so young that the coeval planetary nebula has not yet had time to disperse (Chapter 13). Quite possibly this is how the Crab Nebula developed. Rather than representing the envelope of a supernova, the Crab would then constitute a planetary nebula that has acquired high velocity through interaction with the envelope. In this event the true remnant of the 1054 supernova, an object perhaps 20–30′ across with the Crab near its center, would as yet be undiscovered. At both radio and x-ray wavelengths its brightness should be insignificant. Discovery of this hypothetical object would mark a complete triumph of the ideas outlined above. Meanwhile we can only wait patiently, gathering and analyzing new observational material.

Modern astrophysics, at any rate, has taken an earnest approach toward interpreting the explosions of stars as the regular end of their evolution. And while many details in the grand picture we have drawn still remain shrouded in the fog of the unknown, one can only be astonished at how much has been learned in the past two decades.

Part IV

Stars Perish

To be called into a huge sphere and not to
be seen to move in 't, are the holes where
eyes should be, which pitifully
disaster the cheeks.

WILLIAM SHAKESPEARE, *Antony and Cleopatra*

*(A comparison of an outclassed man
to an unfit star, a sightless
face, an empty sky.)*

19

Neutron Stars and the Discovery of Pulsars

As we explained in Part II of this book, the terminal phase in the evolution of a star, which sets in after its supply of hydrogen nuclear fuel has largely been exhausted, depends substantially on the star's mass. We must include the proviso "substantially" here because a star's evolution can be affected not only by its original mass but also by the velocity and character of its rotation, the degree to which it is magnetized, its membership in a close binary system (Chapter 14), and perhaps other factors as well. Nevertheless, the original mass plays the decisive role. In the ideal case—a nonrotating, isolated star devoid of a magnetic field—theory predicts three different outcomes for the life of a star, depending on the initial value of its mass (see Part II).

1. If the core of the star was originally less massive than about 1.2 $\mathfrak{M}_\odot$, then after a comparatively brief red giant phase the star will turn into a white dwarf. Having cooled off for a few billion years, it will become a cold "black dwarf" or, figuratively speaking, a dead cosmic body, the corpse of a star. We have discussed this process in some detail in Chapter 13.

2. If the original mass of the stellar core exceeded about 1.2 $\mathfrak{M}_\odot$ but was lower than 2.4 $\mathfrak{M}_\odot$, then upon depletion of a large part of the nuclear

fuel there will be a catastrophe. The inner layers of the star will cave in toward the center through the action of gravity, which will no longer be counteracted by gas pressure. At practically the same time an explosion will blow off the outer layers of the star at an enormous velocity, around 10,000 km/s. This phenomenon will be observed as a supernova outburst (see Part III). Collapsing at the free-fall speed, the inner layers of the star will contract by a hundred thousand times in scarcely a few seconds. The star will thereby shrink to a volume 10^{15} times smaller; its mean density will rise by just as large a factor until it surpasses the nuclear density, and its linear size will diminish until it is only about 10 km across.

Once it has reached this size and density, the star will become stabilized and will virtually cease to contract any further. A new equilibrium configuration will develop, but under conditions qualitatively distinct from the equilibrium of an ordinary star. The physical properties of this superdense matter, whose pressure must balance the gravitational attractive force of the collapsed star, are most unusual. In many respects they resemble the properties of matter in the atomic nucleus—a mixture of strongly interacting protons and neutrons. Such an object is analogous to a macroscopic "drop" of nuclear material. The main distinction between this aggregate and nuclear material is that a collapsed star is massive enough for gravitational interaction among its elements to be of fundamental significance, whereas gravitation is unimportant for nuclei with their negligible mass. It is entirely understandable why, ever since the 1930s, theoreticians have called a star formed through gravitational collapse a neutron star.

Thus supernova explosions are accompanied by the formation of neutron stars, a distinctly new type of cosmic object whose existence was predicted by theoreticians some time ago.

3. In the event that the collapsing star has a mass above a certain critical limit (about 2.5–3 $\mathfrak{M}_\odot$), its unlimited contraction through the influence of gravity can no longer be stopped by anything. A neutron star cannot be formed as a stable configuration. The force of gravity, unchecked and uncompensated, will contract the material of the collapsing star to an indefinite degree, and its size will become arbitrarily small. The star will contract to a point, but . . . but here the paradoxical rules of the general theory of relativity come to the fore. Because of the gigantic value of the gravitational potential, the effects of general relativity theory, which under normal cosmic conditions are wholly negligible in magnitude, here take control. This situation is bound up with the absorbing problem of black holes, which has inevitably become a central problem of astronomy today; we shall return to it in Chapter 24. For the present we shall examine the no less interesting problem of neutron stars.

Of the three kinds of products that can result from the terminal phase of stellar evolution—white dwarfs, neutron stars, and black holes—the first to be detected by astronomical observations were the white dwarfs (see Part II). It is worth stressing again that in this case practice was well in advance of theory. White dwarfs were, so to speak, empirically discovered before astronomers understood what such a star is and why it radiates. Physicists arrived at the concept of a degenerate gas quite a while after white dwarfs were discovered. Of course there is nothing surprising about the degeneracy: any piece of metal—a substance known to humanity ever since the bronze age—contains electrons in a degenerate state. But to see and study this phenomenon is not the same thing as to understand it, and there are different levels of understanding, too. The fact remains, though, that white dwarfs were first perceived, and only later understood.

Neutron stars represent an altogether different situation. They were discovered by theoreticians "at the tip of the pen" a third of a century before they were genuinely discovered by astronomers. In our twentieth century that is a very long time span! The reason for this delay in the discovery of real neutron stars is quite natural, for it was clear from the outset that these objects would be very difficult to detect by astronomical observation. If a cosmic body is only 10 km across, then even if it were as close as the nearest stars (about 10 light years) it could not be distinguished with the most powerful telescopes. In fact, if the surface temperature of such a body (a model neutron star) were the same as on the sun, or 6000°K, then its absolute magnitude would be about 30^m, and its apparent magnitude would be as faint as 27^m. But the limiting magnitude of the faintest astronomical objects accessible to the largest modern optical telescopes is about 23^m. If the surface of the neutron star were as hot as that of the hottest stars known to optical astronomy (about 100,000°K), it would still be undetectable, even if it were assumed to be improbably close to us! Actually the closest neutron stars are most likely dozens of light years away. It is evident, then, that efforts to discover neutron stars by optical means were doomed to failure.

Over the past decade, however, a striking improvement in our capability for observing neutron stars has taken place. The development of x-ray astronomy has marked a new era in the search for ways to detect neutron stars and thereby to prove that they actually exist. We shall describe the achievements of x-ray astronomy in Chapter 23. At this juncture we wish to point out only that immediately after the first cosmic x-ray sources were discovered, the suspicion arose that they were the long-awaited neutron stars. An initial look at the situation gave persuasive support to such a conjecture.

In fact, a neutron star newly formed through gravitational collapse should have an exceptionally high temperature, of the order of several billion degrees (see Chapter 18). To simplify the calculations, suppose that the surface temperature of such a star is equal to 10^9 °K. Then we would expect the object to radiate like a black body at that temperature, and according to the Wien displacement law $\lambda_m T \approx 0.3$ the position of maximum radiation would correspond to a wavelength $\lambda_m \approx 3 \times 10^{-10}$ cm = 0.03 Å. The photons of this radiation would have an energy $h\nu \approx 400$ keV, which falls in the range of very hard x rays. According to the Stefan–Boltzmann law, a unit area of 1 cm² on the surface of such an object will emit x rays of the monstrous power $\sigma T^4 = F \approx 10^{32}$ erg cm^{-2} s^{-1}, and the whole superhot neutron star would radiate at a power $L = 4\pi R^2 F \approx 10^{45}$ erg/s, where $R = 10^6$ cm is the radius of the neutron star.

This value of L is exorbitantly large. Suffice it to say that the Galaxy, consisting of a hundred billion stars of all kinds, radiates only about 10^{44} erg/s throughout the entire spectrum. Even if the temperature T were 10^8 °K, quite a modest value for the temperature of a newborn neutron star, then L would be of order 10^{41} erg/s, still a very large amount. In this case the maximum radiation would fall at an x-ray photon energy $h\nu \approx 40$ keV. Most of the detectors of cosmic x rays now available operate in this or a softer energy range. Now suppose that a hot neutron star is located at the other end of the Galaxy, at a distance $r = 60,000$ light years = 6×10^{22} cm. Then the flux density of the x rays received from it would be $F \approx 2 \times 10^{-6}$ erg cm^{-2} s^{-1}, which is a hundred times the intensity of the brightest source, Scorpius X-1, in the hard x-ray range (see Chapter 23). But modern techniques of x-ray astronomy can measure sources that are thousands of times weaker.

It might seem that all is well, especially since the spectrum of many x-ray sources can be interpreted satisfactorily as the radiation of a black body heated to a temperature of a few tens of millions of degrees. For a moment astronomers had the impression that the mysterious neutron stars, the pride of twentieth-century theoretical thinking, had finally been discovered. Alas, nature has once again proved far more complex and rich than had been suggested by the primitive ideas of scientists. Calculations have shown conclusively that originally hot neutron stars will cool off with catastrophic speed.

For example, in matter at very high temperatures a perfectly specific cooling mechanism will operate, involving the annihilation of electron and positron pairs with the formation of neutrinos and antineutrinos:

$$e^- + e^+ \rightarrow \nu + \bar{\nu}. \tag{19.1}$$

At lower temperatures, neutrinos will be generated through the reactions

$$n + n \rightarrow n + p + e^- + \bar{\nu}, \qquad n + p + e^- \rightarrow n + n + \nu. \tag{19.2}$$

Formed in great abundance, the neutrinos will escape from the neutron star without hindrance, carrying away an enormous amount of energy and thereby cooling the star very quickly. Calculations show that within just a month after it has been created, a neutron star will have a surface temperature below 10^8 °K. Indeed, the cooling of a neutron star can take place even faster because the stellar material will evidently be in a superfluid state. Newborn neutron stars will therefore cool too rapidly to be identified as the sources of the x rays observed.[1]

So again, like the bluebird of the fable, neutron stars—real ones, not their pale mathematical reflections described by the theoretician's pen— literally slipped from the hand. And suddenly something completely unexpected happened: neutron stars were discovered after all! But they were by no means discovered where they were being sought, or by those who were looking for them. In February 1968, like a bolt from the blue, the highly reputable scientific journal *Nature* carried a paper by the well-recognized British radio astronomer Antony Hewish and his colleagues, announcing the discovery of *pulsars.* It is worth telling more fully just how this discovery—nearly the most important one in twentieth-century astronomy— took place.

At the famed Cavendish Laboratory of Cambridge University, a program had been under way since 1964 to observe scintillations (rapid, irregular variations) in the flux of radio waves from discrete cosmic sources. Scintillations occur when radio waves pass through condensations in the plasma of the outer solar corona and the adjoining regions of the interplanetary medium; these regions of inhomogeneous density cause irregular diffraction of the radio waves and thus the variability. Scintillations could be observed only if the angular size of the radio sources was very small, less than 0".5. The phenomenon has a most familiar optical analogy: the twinkling of stars in the atmosphere. We also notice that planets do not twinkle but shine with a calm glow whose brightness remains constant for short periods. The reason is that planets have a quite large angular size, up to tens of seconds of arc, while the angular size of stars is vanishingly small. As we have said, a similar behavior is observed in the radio range.

The most notable objects to undergo radio scintillations are the *quasars,* extremely distant extragalactic objects whose angular sizes are measured in thousandths of a second of arc. Quasars had been discovered the year before and at that time were on the mind of a great many astronomers. Hewish decided to employ the scintillation technique to distinguish quasars from

---

[1]It is true that certain young neutron stars (less than 10^5 yr old, with a surface temperature above 10^6 °K) might be recognized from their thermal x rays. Such objects would be comparatively rare, however.

other observed sources of cosmic radio waves. Following his plan, a radio telescope of very large dimensions for the time was constructed just for this purpose; in one direction it extended 470 m, and it was designed to operate at a wavelength of 3.7 m. Although the telescope was sizable, it was a rather crude affair because of the long wavelength of the radio signals being received. Moreover, it was stationary and could record radio waves from sources only when they were crossing the meridian. By a simple retuning of the electrical circuitry without any mechanical adjustment of its elements, the radio telescope could be aimed at various zenith distances, so that sources at different declinations in the sky could be investigated. It is interesting to note that this radio telescope was built in only a few months, mainly by students at Cambridge University under the guidance of a handful of engineers. The whole installation cost but £10,000, which may be a world record and certainly fits the best tradition of the Cavendish Laboratory, where James Clerk Maxwell, Lord Rayleigh, J. J. Thomson, and Ernest Rutherford had worked and made memorable discoveries. Alas, modern observational astronomy as a rule needs immeasurably greater resources, and the story we shall now recount is indeed unique.

Scintillations grow in amplitude as the source approaches the sun in the sky; this effect results from the rising density of interplanetary plasma near the sun. By observing the same source at different seasons of the year, when the source is at different angular distances from the sun, one can record substantial changes in the amount of scintillation in its radio emission. It is for this reason that scintillations of radio sources are observed only in the daytime, when the angular distance of the sources from the sun is small. No appreciable scintillations can be detected at night.

Since the scintillations in the radio flux take place very rapidly, they must be studied with special recording equipment having a very short time constant, in all events shorter than the characteristic time for changes in the flux because of scintillations. This equipment differs fundamentally from the receiving apparatus ordinarily used in radio astronomy, where the time constant (the signal integration time) is generally quite long, at least a few seconds and often considerably longer. Those long integration times are dictated by the need to squeeze out the maximum sensitivity, which is essential if sources with as weak a flux density as possible are to be recorded. Such receivers are routine equipment for radio astronomers, but naturally they are incapable of detecting very fast variations in the flux.

How, then, were pulsars discovered with this radio telescope? In the summer of 1967, Jocelyn Bell, 24, a graduate student of Professor Hewish, found an unfamiliar source of some kind that was scintillating at *night*—a behavior unlike anything ever noticed before! Repeated observations confirmed that every day, at a fixed instant of sidereal time, this remarkable source actually does cross the meridian, thereby demonstrating that it is real. By November 1967 the time constant of the receiver had been made several times as short, and an astonishing phenomenon was discovered: the variations in the flux from the enigmatic source were not taking place at random, as is true of ordinary scintillations of sources caused by the irregular diffraction of radio waves when they encounter inhomogeneities in the interplanetary medium. Instead, these variations were strictly periodic. Very short pulses of radio emission, lasting about 50 milliseconds, were observed to follow one another at perfectly constant intervals of approximately 1 second. The pulses differed in amplitude, however. Two other sources of the same type were also discovered on that occasion. Figure 19.1 shows some early records of the radiation of three pulsars, obtained at a frequency of 81 MHz (which corresponds to the 3.7-m wavelength).

Here was something for the Cambridge radio astronomers to ponder! For they were confronted with an altogether extraordinary phenomenon. What could be said in the first few weeks about the nature of these puzzling sources? At first, the strict periodicity of the radio signals unavoidably aroused the notion that they might be of artificial origin. They could, in particular, represent terrestrial sources of some unknown purpose; they might be signals transmitted from artificial satellites or space probes; or, strange to imagine, they might be coming from extraterrestrial civilizations. This last possibility was discussed in all seriousness at Cambridge, and it was apparently the reason why the authors of the report on this remarkable discovery decided not to publish their results until they had clarified the nature of the mysterious objects. An unprecedented case in the history of twentieth-century astronomy!

Figure 19.1 Signals of three of the first pulsars discovered, as recorded at the Mullard Radio Observatory, Cambridge University, at 81-MHz frequency early in 1968.

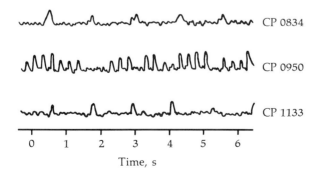

Only after it became obvious that the mysterious sources were located far outside the solar system (how this was shown we shall see presently), and thus that they represented a hitherto unknown class of astronomical objects, did the British radio astronomers publish their discovery—nearly half a year after Miss Bell detected the first puzzling source. The newly discovered sources at once acquired the very apt name *pulsars.* Pulsars are radio sources whose emission is concentrated in separate pulses that follow each other after a perfectly definite time interval.

The discovery of pulsars literally rocked astronomy. The author of this book cannot forget the summer of 1968, when we avidly awaited the latest issues of *Nature,* which was publishing all the latest news from the "pulsar front." Above all it was important to obtain as much factual information about these enigmatic objects as possible. Theoreticians, however, were far from able to comprehend everything about this phenomenon all at once. The striking, almost absolutely accurate periodicity of the pulses inevitably attracted their interest. The brevity of the periods between pulses was especially noteworthy. For example, the period of the first pulsar discovered, which was soon designated[2] CP 1919, is 1.337 s, a far shorter value than the pulsation or rotation period of any cosmic object then known to astronomy. To judge from the fact that each pulse was extremely short (typically a few hundredths of a second), it was apparent that the linear size of the emission region was very small, in any event smaller than a hundredth of a light second (a few thousand kilometers). Since the observed repetition period of pulsars varies smoothly with the seasons of the year by one ten-thousandth of its value, pulsars are evidently located far beyond the solar system. This smooth annual variation in the period is explained in a natural way by the earth's orbital motion around the sun and the accompanying Doppler effect, which changes the value of the period depending on where the earth is situated in its orbit.

Although their positions were known quite accurately, the pulsars could not at first be identified with any known class of astronomical objects. Very likely the first successful identification was that made in Australia in the summer of 1968. The large cross-shaped radio telescope operating there at meter wavelengths was perhaps the best instrument in the world to use in a search for new pulsars. Nearly half of all the pulsars discovered by 1970 had been found with this radio telescope, located in desert country near the village of Molonglo, close to Canberra. Of all the pulsars discovered with this radio telescope, probably the most interesting is the object called PSR 0833 — 45. The position of this pulsar ($\alpha = 8^{\mathrm{h}}33^{\mathrm{m}}$, $\delta = -45°$) is close to

---

[2]The letters CP stand for "Cambridge pulsar," and the number 1919 means that the pulsar is located at 19 hours 19 minutes right ascension.

that of a very interesting nebula found in the southern constellation Vela. The nebula, whose angular size is about 1°, is a source of nonthermal radio emission and constitutes the remnant of a supernova outburst. It was in this supernova remnant that γ rays were discovered (see Chapter 16). The pulsar PSR 0833 — 45 is located inside the radio nebula, although not at its very center.

A remarkable property of the pulsar in Vela is its exceptionally short period—just 0.089 s! For nearly half a year it was the champion in this respect among all known pulsars, but by the end of 1968 it had yielded its place to another, even more interesting object (see below).

The discovery of a pulsar in a radio nebula, a supernova remnant, was no accident. The Australian radio astronomer Malcolm I. Large, who was directing the search for new pulsars at Molonglo, had been working from the very outset on the possibility that pulsars and supernova outbursts might be genetically related. Since the nebula in Vela, the brightest super-nova remnant in the southern sky, is also comparatively close to us, it is perfectly natural that the attention of investigators should have been con-centrated on this most interesting object.

The idea of a connection between pulsars and supernova remnants received its most effective proof in November 1968, when probably the most interesting pulsar of all was discovered in the Crab Nebula.

I shall never forget my feelings when I found out about this discovery. I was on the telephone with William E. Howard, then Deputy Director of the U.S. National Radio Astronomy Observatory. The conversation concerned details of my forthcoming visit to the United States, to this observatory[3]—and suddenly, without warning, such stunning news! For many years I had been occupied with various problems involving the Crab Nebula, perhaps the most remarkable object in the sky. And here, if you please: along with all the other marvels found in this nebula, it has a pulsar. And what a pulsar!

It did not, however, become clear right away what sort of object this pulsar was. Two young radio astronomers, David H. Staelin and Edward C. Reifenstein, had discovered the pulsar. They had detected pulses of radio emission coming from the Crab Nebula area but could not determine the period. The most surprising thing was the American radio astronomers' insistence (and with very good reason) that *two* pulsars had been discovered in the vicinity of the Crab Nebula. What a wild notion! It couldn't fit through any gates, as the saying goes.

At this very time I was in the United States, and I recall having made a wager with my American colleagues. I claimed that there could be only one pulsar in the Crab Nebula, while they, chuckling and pointing at the records

---

[3]Having been awarded the Karl G. Jansky Lectureship.—R. B. R.

of the pulses, maintained there were two! The stakes were just a matter of principle: one dollar against one ruble, that's all.

Well, no sooner had my three-week mission to the United States ended than everything was cleared up. Using the giant radio telescope of the Arecibo Observatory in Puerto Rico, 300 m in diameter, the American radio astronomer John M. Comella showed that his colleagues at the National Radio Astronomy Observatory had indeed discovered two pulsars, one with the record short period of 0.033 s, the other with a record long period, 3.7 s. Soon afterward, however, the short-period pulsar was found to be located at the very center of the Crab, while the long-period one is 1°5 away. (Remember that the Crab Nebula is only 5′ across.) Moreover, the earliest observations of the pulsar in the Crab had a very low angular resolution, precluding accurate measurement of the positions of the newly discovered pulsars. Thus even though the long-period pulsar is comparatively close to the Crab, the two objects are not genetically related (see below). So it seems to me that I have serious grounds for claiming that I won the bet, notwithstanding that there really were two pulsars. I still haven't lost hope of collecting my dollar—which, to be sure, has depreciated by a third since that time.

The discovery of a pulsar in the Crab Nebula with a period of $\frac{1}{30}$ s has been of considerable benefit to our understanding of the nature of these objects. Ever since pulsars were discovered the prime question has concerned their astonishingly rigid periodicity. It soon became apparent that, for example, in the first Cambridge pulsars the period changes by less than 10^{-14} of its value in each cycle! Only the very best quartz clocks run at such a remarkably constant rate. What is responsible for the natural clockwork associated with pulsars?

Two such mechanisms are known to astronomers: the pulsation of stars and the rotation of stars. Let us consider the first mechanism. The phenomenon of pulsation in stars has been familiar for many decades. Such pulsations are observed in their most distinct form in Cepheids. Several empirical laws relate the various parameters describing these stars. For instance, the relation between period and mean density has the form $P \propto \bar{\rho}^{-1/2}$, where P is the period of a Cepheid and $\bar{\rho}$ is its mean density. This law follows directly from the basic equation of a pendulum, familiar to every schoolchild: $P = 2\pi(l/g)^{1/2}$, where P is the period of the pendulum, l is its length, and g is the gravitational acceleration. We can apply this equation to a pulsating star. In this case the gravity $g \approx G\mathfrak{M}/R^2$ and $l \approx R$, where $\mathfrak{M}$ is the mass of the star, R is its radius, and G is the gravitational constant. Substituting the quantities g and l into the pendulum equation, we obtain

$$P = 2\pi G^{-1/2} \left(\frac{R^3}{\mathfrak{M}}\right)^{1/2} \propto \bar{\rho}^{-1/2}, \tag{19.3}$$

because the mean density of the star is naturally given by

$$\bar{\rho} \approx \frac{\mathfrak{M}}{4\pi R^3/3}. \tag{19.4}$$

With Cepheids the characteristic pulsation periods are measured in days. According to Equation 19.4 their mean densities are very low, $\bar{\rho} \approx 10^{-7}$ g/cm^3, a thousand times lower than the density of air; for Cepheids are giant stars of high luminosity with vast photospheres. It now becomes clear that if we were to attribute the clockwork of pulsars to stellar pulsations, then the corresponding stars would have to be very dense. The simple relation $P \propto \bar{\rho}^{-1/2}$ at once tells us that the mean density $\bar{\rho}$ of such stars ought to be around 10^3–10^4 g/cm^3. But we know that these are typical mean densities for white dwarfs (Chapter 10). Thus it might seem that the pulsar phenomenon can be explained by the pulsations of white dwarfs.

Alas! Accurate theoretical calculations have shown that the natural oscillation period of a white dwarf cannot be shorter than a few seconds. Yet the pulsar in the Crab Nebula has a pulsation period of only $\frac{1}{30}$ s. Admittedly, by resorting again to theoretical devices involving a far-reaching revision of the equation of state for white dwarf material and of the models of white dwarfs, and by taking the effects of general relativity theory into account, one can decrease the limiting pulsation period of these stars to 3 s. But this value is still unacceptably long.

Some theoreticians have tried to circumvent this perplexing state of affairs in the following way. A real body (such as a pulsating star) is known to oscillate not only at its fundamental frequency (in our case, at a frequency given by a modification of the pendulum law) but also at the higher harmonics of that frequency, that is, at frequencies two, three, or in general n times higher than the fundamental frequency. In particular, on this hypothesis one would regard the pulsar phenomenon as a manifestation of white dwarfs pulsating at a very high harmonic. Not a few theoretical artifices would be needed, however, to "suppress" the effects due to pulsations at the lower harmonics, for those pulsations would ordinarily be far stronger. In any event it is hard to comprehend why some white dwarf would pulsate only at, say, its fifth harmonic. This very artificial theory has been in existence for just a short time; we mention it only to give a glimpse into the inquiring atmosphere in which theoreticians seeking to interpret some puzzling new phenomenon spend their creative lives.

In principle, one could attempt to explain the clockwork operating in pulsars by the pulsations of *neutron stars,* which had never been detected previously even though they had literally shouted their presence. But because of the extremely high mean density they are expected to have, their pulsation period should be shorter than 10^{-3} s, too short a value for pulsars. Thus it has not proved possible to account for the pulsar phenomenon in terms of the pulsation of any type of star.

After the efforts to explain pulsars by stellar pulsations had failed, it was natural for astronomers to turn their attention to explaining the phenomenon by the rotation of some class of stars. This mechanism seemed fairly promising, since the rotation of a massive starlike body whose surface is radiating nonuniformly could perfectly well account for the remarkable constancy of the periods of pulsars. But how could a cosmic body have a rotation period so short that it would spin around its axis in only 1 s, or even $\frac{1}{30}$ s in some cases? When pulsars were discovered, the shortest rotation periods known to astronomers were rather longer than an hour (in the eclipsing binary system WZ Sagittarii the orbital period, which for close binary systems is equal to the period of axial rotation, is 81 min).

It is perfectly evident that periods as short as those encountered in pulsars can only be achieved through the rotation of cosmic objects of very small size (compared to ordinary stars). On the other hand, there is a limit to the angular rotational velocity, specified by a balance between the centrifugal force exerted on each element of the star and the gravitational force attracting that element toward the center of the star. Mathematically, this condition takes the form

$$\frac{v^2}{R} = \Omega^2 R < \frac{G \mathfrak{M}}{R^2} , \tag{19.5}$$

where v is the equatorial velocity of the spinning star, $\Omega = 2\pi/P$ is its angular velocity, and $\mathfrak{M}$ and R, as usual, denote the mass and radius of the star. If $\Omega^2 R \approx G\mathfrak{M}/R^2$, then before flattening into a disk the star will be torn to pieces. This relation implies that a typical white dwarf having a mass $\mathfrak{M} = 1\,\mathfrak{M}_\odot$ and a radius $R = 7000$ km cannot rotate faster than about once every 10 seconds.[4] Thus rapidly rotating white dwarfs are not capable of producing pulsars.

One other hypothesis put forward when pulsars were discovered may be worth mentioning. It sought to explain them as very close binary systems, each of whose components is a very tiny, extremely dense star. If both components are white dwarfs and they are nearly in contact with each other, then it was shown that the minimum period ought to be 1.7 s. But one might suppose that the components of such a binary system are still more compact

---

[4]Lately it has been learned that former novae belonging to systems of close dwarf stars rotate with periods of the order of tens or hundreds of seconds (see Chapter 14).

than white dwarfs—that they are neutron stars. Yet this hypothesis will not work either! A system comprising two very close neutron stars would be an enormously powerful emitter of gravitational waves (Chapter 24). Because of the energy losses associated with this type of radiation, the neutron stars would fall in toward each other after a few years and collapse. Before that happens, as the components come together their period will decrease rapidly, which is in sharp conflict with the observations (see Chapter 20).

Finally, we might describe one further original idea discussed at that time. This interpretation is a version of the preceding one, but the binary system would comprise a planet of low mass revolving in a very tight orbit around a neutron star. Such a system, it turns out, would emit hardly any gravitational waves at all, and in this sense it would be perfectly stable. The temptation to introduce a planet revolving about a star inside its magnetosphere was aroused particularly by an interesting phenomenon in our solar system. Jupiter has a satellite, Io, which travels around this largest planet of the solar system within its magnetosphere. Io has a major influence on the strong radio emission of Jupiter, which exhibits a periodicity matching the orbital period of the satellite. Although this was certainly a fresh and interesting idea for explaining pulsars, it was soon shown to be untenable: gravitational forces exerted by the neutron star would tear apart so close a planet, in much the same way as a planet was presumably disrupted into the fragments now observed as the particles comprising the rings of Saturn.

All the possibilities have now been considered except one: that the clockwork of pulsars represents the axial rotation of neutron stars. In other words, pulsars are very rapidly spinning neutron stars.

20

Pulsars and
Supernova Remnants

Properly speaking, the idea that pulsars are rapidly rotating neutron stars was by no means a surprising conclusion. The way was paved for it, in a sense, by the whole development of astrophysics during the decade before pulsars were discovered. By 1967 astronomers no longer doubted that a neutron star could be formed through a supernova outburst. Moreover, several facts emerging from a simple analysis of observational evidence pointed directly to such a process.

Much was said in Part III of this book about the Crab Nebula, that laboratory of modern astrophysics. In particular, we explained that if left to itself this nebula would have ceased to radiate after a century or so, whereas it has actually persisted for nearly 1000 years. As we indicated, the Crab Nebula is radiating because the relativistic electrons confined in it are being decelerated in its magnetic field. The mere fact that the Crab nevertheless continues to radiate in the optical region of the spectrum can mean only one thing: some source is constantly pumping new relativistic electrons into the nebula.

Direct observations have clearly shown where this source is located: at the very center of the Crab Nebula near the faint 16th-magnitude star that Walter Baade, one of the greatest observational astronomers of our century, intuitively considered the optical remnant of the star that exploded in July of 1054. Baade also discovered a remarkable variability in a small area of the nebula immediately adjacent to this little star, as we mentioned in Chapter 17. According to Baade's measurements, individual wisps of nebulosity vary at a rate no less than 30,000 km/s, or $\frac{1}{10}$ the velocity of light. Such behavior is a distinct sign that the processes responsible for the variability involve an enormous quantity of fast particles being ejected from some small region.

Finally, radio observations carried out by British researchers during a lunar occultation of the Crab led directly to the discovery of a very small source (about 0″.5 across) located at its center (see Chapter 17). But today we know that this tiny radio source is actually the same as the *pulsar* discovered only four years later. Its finite although diminutive angular size is not real, being just a consequence of the scattering of the radio waves emitted by what is virtually a point source. There have been many occasions in the history of science when long before a discovery was made it had been perceived, but not understood. Astronomers, we shall find, had been observing the pulsar in the Crab for at least a century without suspecting what a wonder it was!

To return, however, to pulsars. As a matter of fact their discovery had literally knocked at the door of astronomy. In a theoretical sense, John H. Piddington and especially the Soviet astronomer Nikolai S. Kardashov had been on the verge of discovering the pulsar in the Crab Nebula in 1964. Kardashov was studying the difficult problem of the origin of the magnetic field in the Crab. We pointed out in Chapter 17 that this field is remarkably regular. It certainly does not constitute a tangle of magnetic lines of force wound up at random, the configuration that one might expect. Moreover, the field itself is quite strong, falling off only very gradually from the center of the nebula toward the periphery.

In order to explain these surprising properties of the field in the Crab, Kardashov proposed the following elegant model. When a supernova explodes, the inner parts of the star will contract in a catastrophic collapse. Even though the star will become a hundred thousand times smaller, two important quantities ought to remain unchanged, or nearly so. The first is the angular momentum, and the second is the magnetic flux. The angular momentum of a celestial body may be defined as the product of its equatorial velocity of rotation about its axis by its mass and its radius:

$$K = v\mathfrak{M}R. \tag{20.1}$$

The law that angular momentum is conserved despite any processes that may be taking place in a physical body is one of the fundamental laws of mechanics. Admittedly, we have made the proviso above that the angular

317

momentum should be "nearly" conserved. What does this mean? It merely means that some of the material of the exploding star, along with some of its original angular momentum, will be thrown outward into interstellar space. But in any event a substantial part if not most of the initial angular momentum of the star will indeed be conserved.

Now the mass of the star (excluding the part ejected in the explosion) will remain unchanged during the catastrophic collapse process. As we have said, however, the radius will shrink by a hundred thousand times. Thus the condition that angular momentum be conserved implies that the equatorial velocity of the collapsing star should increase by just as large a factor as its radius has decreased. By the final stage of the collapse, when the neutron star has been formed, its equatorial rotational velocity may have risen to a stupendous value, even close to the velocity of light!

It is worth observing here that if the rotational velocity of the star was comparatively high before the explosion, and if comparatively little mass (and angular momentum) was thrown off during the explosion, then long before the collapsing star has reached the neutron star phase its contraction will have been halted by centrifugal forces, which will be comparable to the force of gravitational attraction that has caused the star to collapse. This circumstance, of course, places important restrictions on the supernova outburst process itself.

Thus a star newly formed through gravitational collapse should start out with an enormous equatorial velocity. On the other hand, this velocity is clearly related to the rotation period P by

$$vP = 2\pi R, \tag{20.2}$$

whence

$$P = \frac{2\pi R}{v} \propto R^2, \tag{20.3}$$

since $v \propto 1/R$. For example, if in the course of catastrophic collapse the radius of a star shrinks by 100,000 times, its rotation period will become 10 billion times shorter! To illustrate, if our sun, which is rotating very slowly about its axis in a period of about 27 days, were suddenly to turn into a neutron star, it would then begin to rotate many thousands of times every second!

A simple law of mechanics therefore tells us that neutron stars should be spinning very swiftly. The fact that the predicted periods of neutron stars are considerably shorter than the observed periods of pulsars should not disturb us; an explanation is given below.

In just as natural a way we find that neutron stars formed through collapse should be strongly magnetized. This result is a direct corollary of the law of conservation of magnetic flux, which may be written as follows (see also Chapter 16):

$$HR^2 = \text{constant};\qquad\qquad(20.4)$$

thus as a star contracts the magnetic field on its surface should be inversely proportional to the square of its radius. In graphic language, the conservation law signifies the indestructibility of the magnetic lines of force that are firmly coupled to the material of the collapsing star conducting electricity.

Now let us see what this law implies. Suppose that prior to the explosion the magnetic field on the surface of the star was very weak, say around 1 gauss. (The field strength on the earth's surface is of this order.) We must then conclude that the magnetic field of the neutron star after the gravitational contraction will have the huge intensity of 10 billion gauss! Nothing of the sort has ever been achieved in any physical laboratory in the world. In a restricted space, and for a brief time measured in microseconds, physicists are able to create magnetic fields with a strength of a few million gauss. But they would scarcely dream of magnetic fields as strong as those anticipated on neutron stars.

It was Kardashov's idea that the regular magnetic field in the Crab Nebula has resulted from a twisting of the field of the neutron star in the plasma surrounding it, which fills a certain rotating disk connected magnetically to the collapsing star. Calculations by Kardashov showed that a field of the required strength could in fact have been wound up in this manner during the time the Crab has been in existence.

Kardashov's work was aimed not so much at predicting remarkable properties for the neutron stars formed in supernova outbursts as at explaining the quasiregular magnetic field of the Crab Nebula. Nevertheless, it underlined very clearly two fundamental properties of neutron stars that result from gravitational collapse: their great speed of axial rotation and the exceedingly high strength of their magnetic field. In a paper published in 1967, about a year before the discovery of pulsars was first announced, the young Italian astrophysicist Franco Pacini also called attention to these two basic properties of neutron stars. The Italian scientist placed special stress on the simple fact that the kinetic energy of a rotating neutron star will be converted into electromagnetic energy, and for this reason its rotation period will steadily lengthen. Essentially the same prediction of a continual loss of kinetic rotational energy by a neutron star had been made by Kardashov three years earlier. In his view the energy of the magnetic field in the Crab Nebula would be drawn from the kinetic rotational energy of the neutron star, which accordingly would tend to slow down. Pacini's service, however, was in expressing this simple concept in a clear, distinct form. This role too has been of much significance in the history of science.

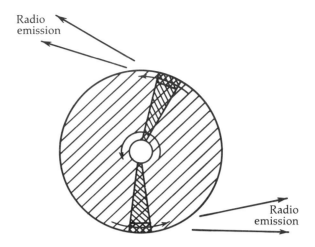

Figure 20.1 Gold's model of a pulsar.

Perhaps the first to claim that the newly discovered pulsars are rapidly spinning neutron stars was the noted British-American astronomer Thomas Gold. To Gold we owe the simple pulsar model shown schematically in Figure 20.1. An important feature of this model is that the magnetic axis of the neutron star, which at the same time is the axis of a cone within which for some reason strong radio emission is concentrated, does not coincide with its rotation axis. Accordingly, when the observer is in a favorable position relative to the neutron star, he will receive from it radio pulses separated by equal time intervals.

Hardly had Gold described this model when a similar interpretation of the phenomenon was given by Pacini, who well deserved the opportunity to refer to his previous theoretical analysis, published before the pulsars had been discovered. That paper, as we have seen, contains the important theoretical prediction that the rotation periods of all neutron stars should lengthen with time. The validity of this prediction was very soon confirmed by direct observation. Early in 1969, for example, using the giant radio telescope in Puerto Rico, the American radio astronomers D. W. Richards and J. M. Comella found that the period of the Crab Nebula pulsar is regularly increasing by 36 nanoseconds a day.[1] Soon afterward a steady, although considerably smaller, lengthening of the period was also observed in other pulsars.

---

[1] A nanosecond is a billionth of a second.

One might ask how period changes of such a tiny magnitude could be measured at all. This question has a simple answer: if the pulsation period were strictly constant, then the exact phase could be predicted a long while in advance. For instance, one might forecast that after exactly one year, or 31,556,925.61 s, a given pulsar should send us a regular pulse of radio waves. By comparing the times when pulses are actually observed to the times that would be expected if the period remained constant, one can establish small fluctuations in the period of a pulsar to great precision. In general, the accuracy will improve as a pulsar continues to be monitored; it will be higher for pulsars of relatively short period and, of course, for those whose period changes more rapidly.[2]

In all pulsars without exception, the period is increasing. This fact decisively corroborates the concept that pulsars are rapidly spinning neutron stars. Moreover, the two pulsars of shortest period, NP 0532 in the Crab and PSR 0833 − 45, are located inside radio nebulae that are remnants of supernova outbursts.

A perfectly natural question (or actually two questions) arises in this latter connection: Why is a pulsar by no means observed in every radio-emitting supernova remnant, and why are most pulsars not situated within radio nebulae? Let us take the first question. Indeed, in most of the radio nebulae known, pulsars have not been found. For example, no pulsar has been discovered in the brightest radio source in the sky (apart from the sun), Cassiopeia A, practically the youngest supernova remnant (see Part III). The same is true of the famous system of filamentary nebulae in the constellation Cygnus and the remnants of the historic supernovae of 1572 (Tycho's star) and 1604 (Kepler's star). As for the 1572 supernova, we have our own ideas (see below). The standard catalog of the Australian radio astronomer David K. Milne contains more than 100 radio nebulae, but pulsars have been detected in only three or four. There is a simple reason: the radio emission of pulsars is not isotropic (that is, not of the same intensity in all directions), but is concentrated within a certain cone whose axis is inclined relative to the rotation axis of the pulsar (see Figure 20.1). One should also keep in mind, of course, that for distant radio nebulae the radiation flux received from pulsars will be small.

Several other weak pulsars in even more distant nebulae will quite possibly be discovered in the near future. Nonetheless, the effect of di-

---

[2]The fact that the rotation periods of newborn pulsars should be very short ($\approx 0\overset{s}{.}001$) means that these objects were formed in just a few tens of minutes during the catastrophic collapse of the cores of certain very rapidly rotating stars. But white dwarfs (or equivalently, the cores of red giants; see Chapter 13) are rotating comparatively slowly (see note 3 at the end of Chapter 10). Might this situation account for the type I supernova phenomenon itself?

rectivity in pulsar radiation should play an essential role in any explanation of why pulsars are missing from supernova remnants. We are very lucky, for example, that along with the many remarkable properties of the Crab Nebula mentioned in Part III, it also happens to contain a pulsar oriented in an especially felicitous way with respect to observers on the earth.

The absence of radio nebulae around the great majority of pulsars can be explained even more simply. Most of the pulsars presently known are at least 1 million years old, while the age of even the most ancient supernova remnants observed as radio nebulae is only a tenth of that or less. In Chapter 16 we estimated the age of these nebulae. But how is the age of pulsars determined?

Their age can be established quite reliably, perhaps with even more confidence than the age of radio nebulae. We have explained that the period of every known pulsar is steadily lengthening. Thus the young, recently formed neutron stars that radio astronomers observe as pulsars should be spinning considerably faster than old objects whose rotation has had a chance to slow down somewhat. Accordingly, if we know the period of a pulsar and the amount it increases per unit time, we can calculate the age of the object. Let $\dot{P}$ denote the change experienced by the period P of a pulsar in one second. Then if the value of $\dot{P}$ were constant throughout the whole prior evolution of the pulsar, its age in seconds would clearly be expressed by the simple equation

$$t_1 \approx \frac{P}{\dot{P}}. \tag{20.5}$$

But at the beginning of a pulsar's life its deceleration should have been particularly strong, and the value of $\dot{P}$ would naturally have been substantially greater than that typical of an old pulsar. We must therefore conclude that pulsars should actually be younger than given by the expression 20.5. The theory of pulsar deceleration to be discussed below gives the equation

$$t_1 \approx \frac{1}{2}\frac{P}{\dot{P}} \tag{20.6}$$

for the age of a pulsar.

Figure 20.2 shows schematically how the period of a pulsar depends on its age, to clarify what we have said. If we apply Equation 20.6 to pulsars located inside radio nebulae, we obtain striking results. For example, in the case of NP 0532, the pulsar in the Crab Nebula, $P = 0\overset{s}{.}033$ and $\dot{P} = 36.5$ ns/day $= 4.2 \times 10^{-13}$ s/s. Hence Equation 20.6 gives $t_1 = 3.9 \times 10^{10}$ s $= 1240$ yr, one-third longer than the historical age of the Crab. The age of PSR 0611 + 22, which is located near the radio nebula IC 443 and which has

322

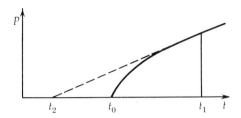

Figure 20.2 Age dependence of the period of a pulsar (schematic).

$P = 0\overset{s}{.}335$ and $\dot{P} = 5.2$ ns/day $= 6.0 \times 10^{-14}$ s/s, is found to be about 90,000 yr. Finally, PSR 0833 − 45, with $P = 0\overset{s}{.}089$ and $\dot{P} = 11$ ns/day, should be less than 12,000 yr old.

As for all the other pulsars—those not found in radio nebulae produced by supernova outbursts—their ages as determined from Equation 20.6 are generally 1 million years or longer. For instance, one of the first pulsars discovered, CP 1133, would be 5 million years old. And the oldest pulsars would have ages in excess of 1 billion years.

It is now perfectly apparent why the overwhelming majority of pulsars are not surrounded by radio nebulae. After some millions of years these nebulae will have dissolved completely in the ambient interstellar medium so they can no longer be observed. The radio nebulae formed at the sites of supernovae are comparatively ephemeral structures, whereas neutron stars must be extremely durable. Only the youngest neutron stars would be embedded in radio nebulae that have not yet had time to disperse.

The difference in longevity between pulsars and supernova remnants undoubtedly represents the main reason for the absence of radio nebulae around the great majority of pulsars. In principle, however, there could be another very curious reason. Pulsars are traveling through space at fairly sizable velocities. This is true, for example, of the Crab Nebula pulsar (see below). A method discussed in Chapter 21, involving only radio astronomy techniques, can yield the velocity of some pulsars relative to the velocity of clouds moving in the interstellar medium. Since the clouds will not be traveling very fast, this method gives us the tangential velocity components of the pulsars themselves. By 1977 the proper motion of 10 pulsars had been determined. They are found to have a very high tangential velocity,[3] in the range of 100–550 km/s.

---

[3]For example, the proper motion of CP 1133, one of the closest pulsars, was measured not long ago. It was found to be about 0″.6/yr; hence, estimating 130 pc as the distance of the pulsar (see Chapter 21 for an explanation of how this is done), we obtain a tangential velocity of 360 km/s.

If the velocity of a newly formed pulsar is about 400–500 km/s, then during the life of the radio nebula genetically associated with it (say 10^5 yr) the pulsar will withdraw to a distance of 40–50 pc from its birthplace, or two to three times the radius of the oldest radio nebulae observed. Thus in principle a situation can occur in which a pulsar has moved rather far away from a radio nebula that has not yet melted into the interstellar medium. In this event a genetic relationship between a pulsar and a radio nebula would be established not by coincidence in space and age, but by age alone. Thus far, despite several attempts, no such pairs have been discovered. But in the future the search for paired nebulae and pulsars may be crowned with success.

Now the question arises: Why should newly formed neutron stars have such high space velocities? The answer evidently lies in the special circumstances under which neutron stars are produced. All the evidence, observational as well as theoretical, indicates that neutron stars are formed in the course of supernova explosions. It is very difficult to imagine, however, that such an explosion would be perfectly symmetrical. The presence, say, of a magnetic field whose axis differs from the rotation axis will necessarily cause matter to be ejected asymmetrically from the star, even if only slightly so. What will be the result? When a star explodes, matter is ejected at velocities up to about 10,000 km/s (Part III), and 10 percent or more of the mass of the star is flung out. Clearly, then, even if only about 10 percent more matter is ejected on one side than on the other, the body remaining after the explosion (the neutron star) will, by the law of momentum conservation, acquire a recoil velocity of at least 100 km/s; probably the velocity will be appreciably higher. It is interesting that the inference of high velocities for the neutron stars formed through gravitational collapse was made theoretically before this result was demonstrated by observations. Another way in which high velocities can be imparted to newly formed pulsars will be set forth in Chapter 22.

Altogether the pulsars in the Galaxy ought to number tens of thousands. Only a few of them are observable; today no more than about 200 pulsars are known. Since the average age of pulsars, as determined from the lengthening of their periods, is about 3×10^6 yr, new pulsars are formed at the rate of about one every 30 years, a value reasonably close to the frequency of supernova outbursts.

Pulsars, as we have seen, have very high space velocities, surely in excess of 100 km/s. But if this is so, they should wander a long way from their birthplace after some billions of years. Most pulsars have such a great forward speed that they ought to leave the Galaxy eventually. It follows that old pulsars should form a vast quasispherical halo around our Galaxy, hundreds of thousands of light years across. Yet nothing of the kind is observed.

Pulsars show much the same space distribution as old stars in the galactic disk. They are concentrated toward the galactic plane within a layer about 500 pc thick. And at distances beyond 15,000 pc from the galactic center, pulsars are altogether absent.

How can we reconcile the swift motions of pulsars through space with their comparatively flat space distribution? The answer probably lies with the magnetic field of pulsars. One or two million years after a pulsar is formed, its field will have decayed considerably, and the pulsar will cease to emit radio waves. Hence the great majority of observable pulsars are actually no more than 2 to 3 million years old. The law 20.6 therefore gives a good age determination only for young pulsars.

Suppose that perhaps 10^9 ancient, extinct pulsars comprise a giant halo around the Galaxy, and that such objects also populate intergalactic space. An important question arises: Are these extinct pulsars truly dead cosmic objects, or do they retain activity of some kind that might, if only in principle, allow us to observe them? Apparently a certain amount of residual activity does persist. For example, on rare occasions (say, once every few hundred thousand years) a "dead" pulsar may well undergo a starquake comparable to those observed in the young pulsars NP 0532 and PSR 0833 − 45 (see Chapter 21). Such starquakes might be accompanied by eruptions: cracks could develop in the crust of the neutron star, and from these fissures might issue streams of material from the interior, rich in free neutrons. The process might generate strong bursts of γ rays. Is it possible that some of the γ-ray pulses detected recently (Chapter 23) are coming from defunct pulsars?

This illustration suggests the unexpected consequences that can flow from what would seem to be a simple observational fact—the high velocities of pulsars traveling through interstellar space.

Of all the pulsars presently known, perhaps the most interesting is NP 0532, near the center of the Crab Nebula. All its properties are exceptional ones: it has the shortest period, which is lengthening the most rapidly, and hence it is the youngest of all known objects of this class. But its most notable feature is probably that it is the only pulsar observed to emit not radio but optical radiation.

We mentioned in Chapter 17 that two petty stars of about 16^m are found in the central part of the Crab (see Figure 17.5). These stars are separated by less than 5″. The northerly star is of no special interest: it is considerably closer to us than the Crab and is not genetically related to it, but simply happens to be projected against the nebula. But the southerly star is a completely different matter. Its spectrum, obtained as early as 1942, proved to be highly peculiar. Its brightness in the ultraviolet region indicated that the surface of this star is very hot. The most remarkable feature of the spectrum, however, was the absence of any lines, either emission or absorption lines. These unusual properties of the southerly star in the Crab served as a basis for Baade's proposal that the star constitutes the remnant of the supernova that exploded in 1054. Subsequently, as described in Chapter 17, Baade investigated the surprisingly rapid variability of the structure in the central part of the Crab adjacent to the southerly star. This behavior suggested that the activity of the southerly star is ongoing—an important new argument supporting Baade's hypothesis.

Immediately after Staelin and Reifenstein had discovered the pulsar in the Crab Nebula and Comella, at the Arecibo Observatory in Puerto Rico, had measured its exceptionally short period, the thought arose that this pulsar might also be emitting pulses in the optical wavelength range. Since the pulsation period was known from the radio observations, the problem facing optical astronomers was considerably simpler. The first optical pulses from NP 0532 were detected at the beginning of 1969 by the American astronomers W. J. Cocke, M. J. Disney, and D. J. Taylor. They used the comparatively modest 36-inch telescope at the Steward Observatory of the University of Arizona. At nearly the same time optical pulses with a $0\overset{s}{.}033$ period were recorded by two other groups of researchers.

It may have been suspected even earlier that the pulsar NP 0532 was somehow connected with the southerly faint star in the central part of the Crab. This possibility was supported by the proximity of the two objects as well as the unusual properties of the southerly star that we have outlined. The optical observations confirmed this idea strikingly. Direct photoelectric observations of the brightness of the star, carried out with a short integration time, revealed an extraordinary phenomenon: the brightness does not remain constant (as astronomers had tacitly assumed for years), but varies with a strict periodicity, as illustrated by Figure 20.3. With extreme precision the period has been found to be equal to that of the radio pulsar NP 0532. Just as in the radio range, the main pulse in optical light is accompanied by an interpulse located approximately (but not exactly) midway between successive main pulses.

Thus the southerly faint star in the Crab Nebula, familiar to astronomers for more than 100 years, proved to be not a star at all but a *pulsar!* Were it not for the development of radio astronomy, which ultimately led to the discovery of pulsars, no one would ever have had so deranged a notion as to look for such remarkably short periodicity in the optical radiation of an object known for so long! This example graphically demonstrates the interaction and intimate relationship between optical and radio astronomy, two mighty branches of the same old tree. There are not a few instances of this kind: we need only recall the discovery of quasars. In such cases the role of

326

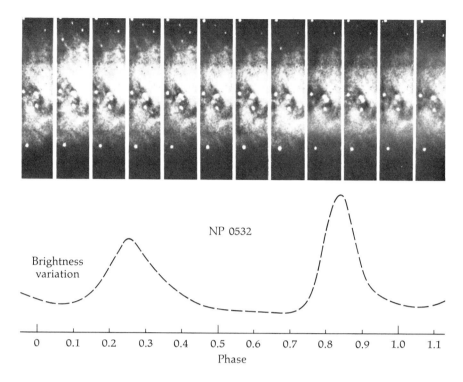

Figure 20.3   A light curve for the pulsar in the Crab Nebula. The
photographs show the central part of the nebula at times
corresponding to each phase (Kitt Peak National Observatory).

the guide pointing out a hitherto unknown phenomenon of nature has generally been played by radio astronomy.

The strange character of the southern star in the Crab is very plainly
shown by the photographs in Figure 20.4. These photographs were taken
using the familiar stroboscope principle. The light beam passing through
the telescope is focused on a television camera, and the beam is interrupted
by a rotating shutter with a period exactly the same as that of NP 0532. If the
shutter is open in phase with the flashes of optical radiation from the pulsar,
then the maximum radiation of the pulsar will be recorded every time by the
camera. But if the shutter is rotating out of phase with the pulsar, the camera
will receive only the very faint glow emitted by the pulsar during the intervals between pulses. On the other hand, the images of the ordinary stars
nearby will of course be completely independent of the phase of the rotating
shutter. There are three stars in the lower picture of Figure 20.4, but the

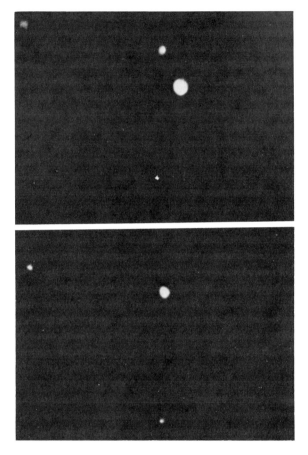

Figure 20.4 Stroboscopic observations of the visible light of the
Crab Nebula pulsar (Lick Observatory).

upper picture shows a fourth one, the brightest of all. And this is indeed the
southern star in the Crab Nebula, photographed when the shutter in front
of the television camera was in phase with the flashes of the pulsar. In
modern astronomy it would be hard to find any more striking proof of the
unusual nature of a celestial object long known.

Several months after the optical radiation of the pulsar NP 0532 had
been recognized, Hale V. Bradt and his colleagues discovered that the object
is emitting x rays. It soon became clear that earlier x-ray observations of
the Crab Nebula contained a pulsating component whose contribution
amounted to about 6 percent of the total x-ray emission of the nebula, and
the pulsation period was 0^s033, the same as that of NP 0532. It goes without

328

saying that if a radio pulsar with this period had not been discovered, the pulsating component in the x rays of the Crab could hardly have been noticed. Since that time many studies of NP 0532 have been made in the x-ray range. In 1970 the pulsar was found to be emitting pulsed γ rays with photon energies up to 100 MeV. And in 1972, very hard γ-ray photons with energies as high as 10^{12} eV were discovered. These were recorded by a very interesting technique based on the Čerenkov effect, which depends on the electrons formed in the earth's atmosphere when it absorbs a beam of hard γ rays arriving from the Crab Nebula.

On a single scale Figure 20.5 displays "light curves" of NP 0532 in various parts of the spectrum. The lowest radio frequency at which the radiation of this pulsar has been recorded is about 3×10^7 Hz (a wavelength of 10 m), while the highest γ-ray frequency is more than 3×10^{25} Hz, so the ratio of the frequencies at either end of the vast band of electromagnetic waves emitted by NP 0532 is 10^{18} to 1, a fantastic value. Comparing now the light curves at radio, optical, and x-ray frequencies, we notice that the pulse always begins sooner at radio frequencies (especially low ones) than at high frequencies. At low radio frequencies some additional radiation

Figure 20.5 Brightness curves for the Crab Nebula pulsar in four spectral regions.

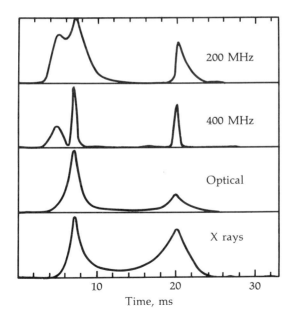

is emitted ahead of the primary maximum and distinctly separate from it in time. This subsidiary peak in the radiation is called the precursor. The intensity of the precursor falls off rapidly toward high radio frequencies, which shows that its radio spectrum is very steep, considerably steeper than the radio spectrum of the main pulse and the interpulse.

The optical range shows no trace of a precursor. From the optical light curve of NP 0532 we see that the intensity of the radiation does not drop to zero between pulses; it maintains a small but perfectly real level. Accurate photoelectric measurements demonstrate that between the main pulse and the interpulse the intensity never falls below a level of about 1 percent of the peak intensity in the main pulse. But in the x-ray range this intensity level between pulses reaches 10 percent. Accordingly, the spectrum of this component of the pulsar radiation (which is nearly isotropic) is much harder than the spectrum of the bulk of the radiation, concentrated in the pulses. Another curious feature is that the x-ray intensity of the pulse and the interpulse are nearly the same, whereas in optical light the main pulse is three times as intense as the interpulse.

The results gathered by observing NP 0532 in the near-infrared range are of special interest. Here the intensity of NP 0532 diminishes toward low frequencies, whereas the flux of infrared radiation emitted by the Crab Nebula itself rises as the frequency decreases. Figure 20.6 schematically illustrates the behavior of the spectra of the Crab Nebula and the pulsar in it throughout the entire range of electromagnetic radiation, from radio waves to γ rays. This diagram shows above all that the high-frequency (optical and x-ray) spectrum of NP 0532 is definitely not a simple continuation of the radio spectrum. One other distinction between the radio and optical emission of this pulsar is also important. Although the optical intensity of NP 0532 is definitely constant over appreciable spans of time (in particular, NP 0532 maintains the same stellar magnitude to within a fraction of 1 percent), its radio pulses are subject to large intensity fluctuations. Fairly often the radio pulses are observed to become much stronger than usual, by more than a hundred times. In fact, this is why Staelin and Reifenstein were able to discover the pulsar in the first place; their receiving equipment would not have been capable of recording the very short periodicity when the pulses from NP 0532 were at their ordinary strength. But instead, from time to time they reliably detected extremely powerful transient pulses of radio waves from this pulsar. It is worth noting in this regard that giant pulses are also observed in the radio emission of several other pulsars. This difference in the behavior of the radio and high-frequency emission of NP 0532 clearly suggests that different mechanisms are responsible for these types of radiation.

The bulk of the electromagnetic radiation of NP 0532 is concentrated in the high-frequency part of the spectrum. Knowing the flux density of the

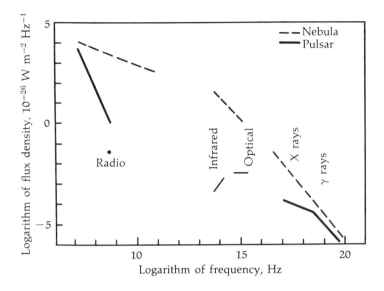

Figure 20.6   Spectra of the Crab Nebula and its pulsar.

radiation arriving from this pulsar throughout the whole frequency range (this quantity will be denoted by F, as usual) as well as the distance r of the Crab Nebula, which is about 2000 pc, we can easily find the power radiated by the pulsar:

$$L = 4\pi r^2 F \approx 10^{36} \text{ erg/s},$$

which is hundreds of times the radiant power of the sun. Most of this power is contributed by x rays.

As early as the 1930s it was established that the southerly star at the center of the Crab possesses a small but perfectly measurable proper motion. Measurements by John C. Duncan indicated an annual displacement along the two coordinates in the sky of $\mu_\alpha = 0''.010$ and $\mu_\delta = 0''.002$. Since that time observations by a number of astronomers, in particular at the Pulkovo Observatory near Leningrad, have yielded improved values for the proper motion of this star. Knowing the proper motion and the distance of the object, we can readily determine the projection of its space velocity on a plane perpendicular to the line of sight. If we again take $r \approx 2000$ pc, we find a projected velocity of about 120 km/s. Thus the pulsar in the Crab Nebula, like other pulsars, has a high space velocity. We have alluded to this point above.

These are the basic results gained from observations of NP 0532 at various frequencies. Certain other observational results—and what is especially interesting, some theoretical attempts to explain the nature of this pulsar—will be considered below. But first let us examine two other pulsars that also appear to be emitting high-frequency radiation. Only one of these objects is generally recognized as a pulsar: the remarkable PSR 0833 − 45 associated with the Vela radio nebula, which we have already mentioned. The second object, located in the nebula at the site of Tycho's supernova of 1572, has not yet been certified as a pulsar. But in the opinion of the author of this book, a pulsar is being observed there, although not at all in the usual fashion.

We are referring to the rather bright x-ray source that has been found to match this nebula. One can show theoretically that the x rays observed cannot possibly be emitted by the nebula itself, as happens for other supernova remnants such as the Crab Nebula and Cassiopeia A. How are we then to account for this x-ray source, since no radio pulsar is observed there?

The phenomenon can be understood, however, if we look carefully at the x-ray light curve of the pulsar NP 0532 (Figure 20.5). It is clear from the curve that for the pulsar in the Crab the level of x rays between pulses amounts to about 10 percent of the peak intensity reached during the pulses. Since pulses occupy a rather small fraction of a period (about 10 percent), we may at once infer that averaged over a long time span the almost isotropic x-ray emission between pulses is practically equal to the average value of the x rays emitted in the pulses themselves. But according to the pulsed radiation model (Figure 20.1), a pulsar can be observed only when the rotation axis of the neutron star is favorably oriented with respect to the observer. On the other hand, the radiation between pulses can be observed whatever the observer's position relative to the rotation axis of the neutron star, which is why we say it is nearly isotropic. Thus the x rays from the source in the Tycho radio nebula evidently belong to a young (400 years old) pulsar there whose axis is unfavorably oriented with respect to the earth. This circumstance precludes us from observing radio pulses.

Perhaps before too long a very faint star may be discovered at the center of the Tycho nebula. It would be none other than the pulsar, observed through the side lobes of its radiation beam pattern. This little star ought to have quite unusual properties. For example, we would expect it to emit linearly polarized radiation, and possibly to have a periodicity representing some sort of dependence of the radiation between pulses upon the phase of the period. (In Figure 20.5 such a dependence is distinctly apparent: the intensity is considerably higher between the main pulse and the interpulse than between the interpulse and the main pulse.) The same type of periodic effect might conceivably be exhibited by the x rays as well. We might further mention that in the softer part of the x-ray spectrum an important role could be played by thermal radiation from the hot surface of the neutron star, which in 400 years should have cooled off to 5 million degrees.

Efforts to find optical objects at the sites of the most intense pulsars have generally failed to give positive results thus far. But at the position of PSR 0833 — 45, astronomers working with the new 153-inch Anglo-Australian telescope on Siding Spring Mountain inland from Newcastle, New South Wales, have recently announced the discovery of pulsating optical radiation from an object of 25^m.2. This magnitude represents a fantastically low level of light.

In the x-ray range, observations made in 1972 in the region of the Vela X nebula revealed, along with the extended x-ray source having a comparatively soft spectrum (probably of thermal origin; see Chapter 16), a "point" source whose position coincides (within the observational error) with the pulsar PSR 0833 — 45 located there. But the most interesting circumstance concerns the finding that the x rays from this source are pulsating with a period of 0^s.089, exactly the same as the period of the radio pulsar.

Our understanding of the situation is complicated by the fact that the phase of the x-ray pulses does not coincide with that of the radio pulses but lags behind by about one-third of a period. A possible explanation for this unexpected phenomenon might be provided by the interpulse recorded for PSR 0833 — 45; although it does not appear at all at radio frequencies, it is very strong in the x-ray region of the spectrum. A similar situation is observed for the pulsar in the Crab Nebula. As we have seen, the relative intensity of the interpulse is considerably higher in x rays than at optical and radio frequencies. For PSR 0833 — 45 the situation is even more pronounced: at the time when a pulse of radio waves is received, x rays are not observed at all. Thus we must explain not only why no radio emission occurs during the bursts of x rays, but also why there are no x rays during the bursts of radio waves. Another feature is that the radio profile of PSR 0833 — 45 has an unusually smooth and simple form for a pulsar (Figure 20.7).

Satellite-borne detectors of γ rays in the energy range $E > 35$ MeV have recorded two fairly broad pulses from PSR 0833 — 45, which are out of phase with the radio pulses. The newly detected faint optical radiation of this pulsar exhibits a rather similar behavior. Finally, it is noteworthy that pulsating radiation has also been received from this object in the ultrahard γ-ray range ($E > 10^{11}$ eV). Many more observations will be needed before we can hope to comprehend the intricate problems posed by this very interesting pulsar.

The fact that hard radiation has been detected only in two or three young pulsars impels us to conclude that the duration of the x-ray phase

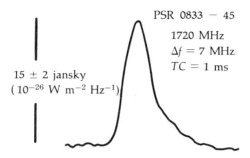

Figure 20.7   A radio profile of the pulsar in the Vela nebula.

in pulsars is short compared to the radio emission phase. But the latter phase is quite limited too; for most pulsars it is around 1 million years (see above). Thus we have arrived in a purely empirical way at the concept that the "activity" manifested in radio emission, optical radiation, and x rays should steadily wane as a neutron star evolves.

21

Pulsars as Radio Sources

For pulsars, perhaps the most difficult things to establish are two properties basic to any normal source of radio waves: the flux and the spectrum. The difficulties result primarily from the nature of pulsars themselves. As we have explained, the radio emission of pulsars varies with time in a most complicated fashion. In particular, these variations take place very rapidly: for example, two successive pulses may have strikingly different profiles (a profile is analogous to a light curve in the optical range). In other words, within a time of the order of a second (the typical length of a pulsar period), substantial changes in the flux density may occur.

One distinctive characteristic of pulsars as radio sources is their negligible angular size. Because of this they experience scintillations unlike any other sources known to radio astronomy. And although the spectrum of pulsar radio emission appears to remain quite stable, the behavior of the scintillations depends significantly on the frequency of the radiation. As a consequence, the spectrum undergoes remarkably strong distortions when the radiation of a pulsar passes through the interstellar medium. For instance, the radio flux density at a given frequency may drop to zero in a few minutes because of scintillations, while at a slightly different frequency the flux may decrease at some other time. Adding to this circumstance the fact that the distortions due to scintillations show rapid time variability, we see that the true radio spectrum of a pulsar is not easy to determine.

In order to exclude the influence of scintillations one must begin by averaging the observations over hundreds of periods. A further difficulty enters here, however, because after so many periods the true radiation (before distortion by scintillations) of a pulsar may undergo considerable change. In this manner one obtains synthetic profiles and spectra, smoothed with respect to time. Then by comparing various synthetic profiles established for the same pulsar, one can seek to identify cycles of variability on time scales ranging from minutes to years. Naturally we do not yet have much information on the long-term variability in the basic parameters of pulsar radiation, for the first pulsars were discovered less than a decade ago.

Figure 21.1 displays synthetic spectra for several pulsars as derived at the famed Jodrell Bank radio astronomy observatory of the University of Manchester. As a rule the spectral flux density declines steeply toward higher frequencies. But some pulsars, such as PSR 0329 + 54, exhibit a fairly flat maximum near 400 MHz. Just as for other sources of cosmic radio waves, the pulsar spectrum can usually be fit by a power law of the form $F_\nu \propto \nu^{-\alpha}$, where F_ν is the averaged spectral flux density; for most pulsars the

Figure 21.1   Synthetic spectra of six pulsars.

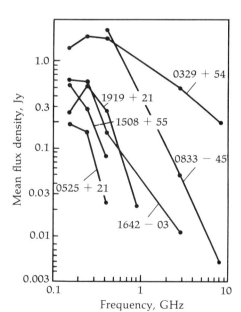

spectral index α lies between 1 and 2. The spectral index does not show any dependence on the period of the pulsar.

The profiles of pulsar radio emission are often marked by a great deal of fine structure. For example, the pulsar in the Crab Nebula exhibits individual features in its radio profile that are far narrower than those in its optical light curve. Although these features fluctuate quite rapidly from period to period, the synthetic profile of NP 0532 is quite similar to the light curve of its high-frequency emission in the optical range.

The radio emission of pulsars is strongly polarized. Individual pulses occasionally show nearly 100 percent linear polarization. We observe the very interesting behavior that even within a single pulse, in just a small fraction of a period, the polarization parameters (such as the direction in which the electric vector vibrates) may vary sharply. This important phenomenon will be considered below; meanwhile we will add that as a general rule pulsars exhibit elliptical polarization.

As we have indicated several times, to a first approximation the periods of pulsars are very stable. However, prolonged series of observations do reveal some very interesting variations in pulsar periods. To begin with, such observations have established that the period of every pulsar without exception is steadily lengthening. As a result we can estimate the age of a pulsar with some confidence (Chapter 20). Because of the Doppler effect induced by the earth's orbital motion at a velocity of about 30 km/s, the periods of pulsars vary over the course of a year—by 10^{-4} of their value for pulsars located comparatively close to the ecliptic, and by a smaller but perfectly measurable amount for other pulsars. Whenever the period of a pulsar is determined to the high precision now attainable, the Doppler effect is excluded and the value given represents the "heliocentric" period— the period that would be measured by an imaginary observer located on the sun.

An extremely interesting phenomenon was discovered in 1969 by the Australian radio astronomers V. Radhakrishnan and Richard N. Manchester. They noticed an abrupt decrease in the period of the pulsar PSR 0833 − 45, which occurred sometime between February 24 and March 3 (no observations were made in the interim; see Figure 21.2). The period diminished by a rather substantial amount, $\approx$ 200 ns. Since this pulsar has a period of $0\overset{s}{.}089$, the break in the period is fully 2×10^{-6} of its value. It is also interesting that after this abrupt shortening, the period proceeded to lengthen again 1 percent faster than before the break. Two and one-half years afterward, late in 1971, the break in the period of PSR 0833 − 45 was repeated almost exactly.

Such strange behavior can be caused only by a genuine, discrete change in the rotation period of the neutron star. The change in the rotation period

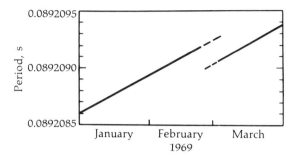

Figure 21.2   A starquake observed in PSR 0833 — 45.

should in turn result from an abrupt decrease in the moment of inertia of the star, because of complicated internal processes of some kind, such as as a change in the character of the coupling between the interior of the neutron star and its surface layers. The phenomenon of an abrupt break in the period of a pulsar has received the graphic and apt name of a *starquake*. On a far smaller scale than for PSR 0833 — 45, several starquakes have been observed for the pulsar in the Crab Nebula. Analysis of starquakes opens up the only opportunity of its kind to investigate the interiors of neutron stars, just as the analysis of seismic phenomena on the earth is a most important technique for studying the internal regions of our planet.

The great precision with which pulsar periods and their variations can now be determined permits us to draw one other important conclusion regarding the nature of pulsars. Suppose that a pulsar is a component of a binary system. Then the value of its period should fluctuate regularly according to its orbital motion in the system. But periodic changes of this kind have hardly ever been observed in the pulsation period of any pulsar. This simple fact carries the significant implication that pulsars, or the neutron stars they represent, are not components of multiple star systems. Such a conclusion is surprising in itself, for duplicity is very widespread among stars. As indicated in Chapter 14, at least 50 percent of all stars belong to binary systems, and the percentage is appreciably higher for massive young stars. Yet among the 200 pulsars known at the present time, only one seems to show evidence of membership in a binary star system (see below). No type of star possessing such a propensity for being solitary had previously been known to astronomy.

The lack of companions to pulsars can, however, be understood in terms of prevailing ideas about the formation of neutron stars. Above all it is quite likely that the explosion of one component of a binary system as a supernova will cause the pair to break up. This will happen provided that the distance separating the binary components is great enough and that the evolution of the two components proceeds more or less independently. Furthermore, when an explosion occurs the greater part of the mass of the star ought to be cast into interstellar space at quite a high velocity. If, however, an explosion were to occur in a close binary system, in which the distance between the components is small, the situation could be altogether different. In this case, as we saw in Chapter 14, it is the less massive star that will explode. The pair then will not be disrupted, even if most of the exploding star is ejected into interstellar space.

But if most binary systems are comparatively close, why are pulsars not observed to be components of those systems? The young Soviet astrophysicist Viktorii F. Shvartsman has put forward a very interesting hypothesis to explain this enigma. He believes that in a binary system, especially if the pair is close, gas will continually leave the normal component and fall onto the neutron star—a process called accretion. This process will serve to suppress the radio waves emitted by the neutron star and to extinguish the pulsar associated with it. When the pulsar is young and active, accretion will be incapable of muffling the radio emission of the neutron star. But the number of such very young pulsars can literally be counted on the fingers of one hand. Most pulsars are actually rather old, and if they belong to binary systems their radiation will be suppressed (see also Chapter 23).

An exceptional case is PSR 1913 + 16, a very weak pulsar discovered at the Arecibo Observatory in the summer of 1974. It is a member of a close binary system with an orbital period of 7^h45^m. The distance between the components is only a little more than the radius of the sun. The second component should be either a white dwarf or an even more compact object that demonstrably does not fill its Roche lobe. Hence there is no accretion in this system, and accordingly the pulsar can be observed.

The accretion of gas onto a neutron star belonging to a binary system can itself lead to interesting and important consequences. We shall have much to say about this situation in Chapter 23. One should recognize, however, that the question of why pulsars are so rarely double has not yet been finally settled. Theoreticians still have much food for thought here.

Synthetic pulsar profiles show wide diversity. Although, as we have mentioned, these profiles vary considerably with time, for a given pulsar their basic features remain unchanged and can serve as a fingerprint. For example, there are pulsars whose profiles consist of one simple pulse, such as PSR 0833 − 45, mentioned often above. And there are pulsars whose synthetic profiles comprise two or sometimes three subpulses. The differences in behavior are clear from Figure 21.3, which shows synthetic profiles for a number of pulsars.

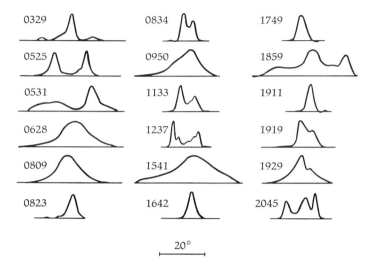

0329

0525

0531

0628

0809

0823

0834

0950

1133

1237

1541

1642

1749

1859

1911

1919

1929

2045

20°

Figure 21.3 Synthetic pulse profiles for 18 pulsars.

Pulsar radiation is observed within a time interval (called the window) amounting to about $\frac{1}{30}$ of a period. Figure 21.4 shows how the width of the window depends on the period of the pulsar. It is convenient to measure the window width in angular units, with 360° representing the whole period. According to Figure 21.4, the points corresponding to various pulsars are distinctly grouped about a line representing a window 9° wide.

Although the window width remains nearly constant for a given pulsar, individual fine details in the profile (the true rather than the synthetic profile) can migrate from place to place within the window. In some pulsars these displacements are surprisingly regular in character; the subpulses seem to drift around in the window. This phenomenon was first observed for PSR 1919 + 21, and it has now been found in many other pulsars. Figure 21.5 is a diagram of this interesting effect. For such pulsars one can specify a second period, defined as the time interval required for the profile to repeat itself. Usually the second period P_2 is several times as long as the main period P_1 determined by the rotation of the neutron star. But the period P_2 is by no means obeyed with the same high precision as the main period P_1.

Of the several types of variation experienced by the pulse profiles of pulsars, perhaps the most puzzling is the complete cessation of the radio emission for a considerable number of periods. Thus, the radiation of PSR 1237 + 25 will suddenly turn off for a few minutes, after which it turns on

340

again without the slightest break in the period. Sometimes PSR 0809 + 74 will turn off for several periods. Such behavior probably indicates that for some reason the rotating neutron star has abruptly stopped emitting radio waves. In this regard, we wish to stress that the details of the fundamental radio emission process of pulsars are still far from well understood. We shall return to this problem presently.

Although the nature of the radio emission of pulsars continues to be quite obscure and mysterious, the radiation itself has opened up rich new opportunities for studying the interstellar medium. From the outset astronomers have appraised the remarkable pulsed character of this radio emission at its true worth. Another very useful property is that in many cases the radio waves are linearly polarized. These features of pulsar radio emission enable it to be used as a highly effective probe for investigation of the interstellar medium.

The effect by which pulses of radio waves emitted by pulsars are dispersed in the interstellar medium has been especially widely applied. This very interesting phenomenon warrants more detailed examination. Are all electromagnetic waves propagated through the interstellar medium at the same velocity? The question is significant, for even a slight difference in the propagation velocity of electromagnetic waves of varying length might, in principle, produce a perfectly measurable effect: the difference in the arrival time of pulses at different wavelengths could steadily build up over vast

Figure 21.4 Relation between window width and period of pulsars.

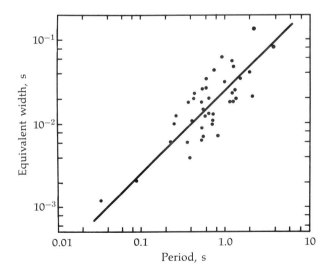

341

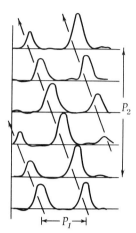

Figure 21.5 Drift of pulses within the window.

interstellar distances. At the turn of this century the creative Russian astronomer Gavriil A. Tikhov sought to detect such an effect in eclipsing binary
stars; if the effect existed, the times of stellar eclipse ought to depend on the
color of the light (say, on whether measurements are made in the blue or
the red). In his day, however, ideas about the nature of the interstellar
medium were not even in a rudimentary state; there were simply none at all.
After several years the interstellar calcium lines were discovered by Johannes
Hartmann (Chapter 2), and the study of the interstellar medium had begun.
But now we understand well why Tikhov's attempt to detect interstellar
dispersion of light was groundless. For the interstellar medium is so tenuous
that even over a path of 1000 light years, ordinary dispersion would cause
a pulse of red light to lead a simultaneously emitted pulse of blue light by
only a tiny fraction of a second.

The discovery of cosmic radio waves fundamentally altered the old
problem of measuring the dispersion of electromagnetic waves in the
interstellar medium. We can always regard the medium as a plasma (even
in H I zones, where hydrogen is not ionized; see Chapter 2). The theory
of the propagation and dispersion of radio waves in plasma is a well-
developed branch of macroscopic physics. Here we shall merely give the
expression for the refractive index of electromagnetic waves in plasma containing no magnetic field:

$$n = \left(1 - \frac{e^2 N_e}{\pi m_e \nu^2}\right)^{1/2} = \left(1 - 8.1 \times 10^7 \frac{N_e}{\nu^2}\right)^{1/2}. \tag{21.1}$$

Here N_e is the number density of free electrons in the plasma and ν is the
frequency of the radiation.

342

As Equation 21.1 indicates, for radio waves the refractive index of plasma is less than unity. An elementary physics course tells us that electro-magnetic waves are propagated through a medium with a refractive index n at a velocity $v_{ph} = c/n$, where $c = 3 \times 10^{10}$ cm/s is the velocity of light in a vacuum. Since n is less than unity, $v_{ph} > c$, and we would seem to have a contradiction to the special relativity principle. But there is actually no contradiction here. For v_{ph} is the *phase* velocity corresponding to waves of a perfectly definite frequency. The relativity principle states that signals cannot be transmitted at a speed faster than light. However, no signal at all can be transmitted by a monochromatic wave (a wave with an exact fre-quency). For this purpose a group of waves differing slightly in frequency is necessary. Such a group of waves (or wave packet) will be propagated through a medium at a certain *group* velocity that differs from the phase velocity. If waves are propagated through sufficiently rarefied plasma, their group velocity will be given by the expression

$$v_g = c \cdot n = c \left(1 - \frac{e^2 N_e}{\pi m_e v^2}\right)^{1/2} < c, \tag{21.2}$$

and the propagation time of the group of waves will be

$$t = R/v_g. \tag{21.3}$$

From Equations 21.2 and 21.3 we find that the difference between the propagation times for a group of waves in a medium (plasma) and in a vacuum—that is, the *lag* of the group—will be

$$\tau = \frac{R}{v_g} - \frac{R}{c} = \frac{e^2}{2\pi m_e c} \frac{N_e R}{v^2} = 4.15 \times 10^3 \frac{D}{v^2}, \tag{21.4}$$

where the quantity $D = N_e R$ represents the number of free electrons in a cylinder whose base has an area of 1 cm^2 and whose length is R. In Equation 21.4 the frequency v is expressed in megahertz, and R is expressed in parsecs. The quantity D is called the dispersion measure.

Now suppose that we measure the arrival time of pulses at two fre-quencies differing from each other by the slight amount Δv. A pulse of radio waves containing a whole set of frequencies will have been emitted at a certain instant, and were it not for the dispersion caused by the interstellar medium, the pulse would be observed at all frequencies simultaneously. But since dispersion is present, the pulse will be observed at higher frequencies before it is observed at lower ones. The difference Δt in the times when a pulse is observed at frequencies differing by the amount Δv will be given by

$$\Delta t = 1.25 \times 10^4 \, D \frac{\Delta v}{v^3} \text{ s.} \tag{21.5}$$

To gain a feeling of whether this quantity is large or small, let us make a numerical calculation. Suppose we are observing near a frequency $\nu = 100$ MHz, and that $D = 100$ cm^{-3} pc. Then if $\Delta\nu = 1$ MHz, Equation 21.5 states that Δt is no less than 1 second! This is a very large value, especially when we remember that 1 s is a typical period for a pulsar. This example clearly shows that interstellar dispersion of pulsar radio signals will very seriously distort the structure of the pulses observed. In some cases, unless special measures are taken (such as narrowing the frequency band accepted by the radio receiver), dispersion can wash out the pulses and make them unobservable. We have alluded previously to the distortion of pulsar observations by interstellar dispersion.

With the techniques of modern radio astronomy, the dispersion measure D for any pulsar can be determined to high accuracy, up to one part in 10^5. The precision is high enough to permit the variation in the value of D to be measured in some cases. The variations in the dispersion measure for the pulsar in the Crab Nebula are of special interest. In this case $D = 57$ cm^{-3} pc $= 1.75 \times 10^{20}$ cm^{-2}. However, at times when the central part of the Crab Nebula is active and rapidly moving wisps are formed (see Figure 17.10), the value of D will change by $\Delta D \approx 10^{16}$ cm^{-2}. Such increases in the dispersion measure generally last a few weeks, after which D reverts to its original value. There can be no doubt that these variations in D are caused by the passage of radio waves through moving plasma clouds in the central part of the Crab.

If the density of free electrons in the interstellar medium were known with full confidence, then the value of D measured for some pulsar would at once enable us to determine its distance. Actually, however, the situation is much less favorable. A complicating factor is that the free electron density N_e fluctuates over quite a wide range in various regions of the interstellar medium (see Chapter 2).

Calculations show that in H I zones, which occupy a large part of the interstellar medium, $\overline{N}_e \approx 3 \times 10^{-2}$ cm^{-3}. This value of N_e may be taken as the average electron density in the interstellar medium, most of which corresponds to zones of nonionized hydrogen, or H I. Today it is customary to use this value of $\overline{N}_e$ to estimate the distances of pulsars from their empirical dispersion measure, although such a method can yield large errors in certain cases. For example, if a faint, optically unobservable H II zone is projected by chance against a pulsar, one could greatly overestimate the distance of the pulsar by using the observed dispersion measure.

Distances of individual pulsars can be determined most reliably if the 21-cm line of interstellar hydrogen is present as an absorption feature in their radio spectrum. One can then apply the common radio astronomy method which presupposes that the interstellar hydrogen is concentrated

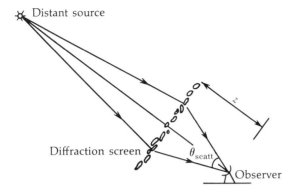

Figure 21.6 Scintillation of radio waves received from a pulsar.

in the arms of spiral structure in the Galaxy (Chapter 3). Thus far, however, this method remains seriously limited because it requires that the pulsar have a high radio flux density. Only for very few pulsars has the distance been established by this method.

Up to now we have neglected the effect of the small irregularities in the interstellar plasma, which produce strong scattering of radio waves. A good deal has been said about the important phenomenon of scintillation of pulsar radio emission. Now let us consider this process somewhat more deeply.

In principle, scintillation results from interference in the radiation of a pulsar. Because of scattering by plasma irregularities, the observer simultaneously receives a bundle of rays whose optical paths differ. The situation is shown schematically in Figure 21.6. For simplicity, this diagram considers all the irregularities in the interstellar medium as being located at some definite distance z from the observer, playing the role of a diffraction screen. Actually, of course, they pervade the whole space between the source and the observer. Since the path difference from ray to ray depends strongly on the wavelength, interference between the rays will cause the observer to perceive sizable intensity fluctuations in adjoining spectral regions. Moreover, any motion of the observer or the source relative to the clouds of inhomogeneous interstellar matter will also serve to change the optical path of the rays, inducing the random brightness fluctuations of the source observed as scintillations. This is why an analysis of scintillations can yield the relative velocity of the source and the irregularities, as described in Chapter 20. As the frequency of the radiation increases, the effects of scattering by irregularities in the interstellar plasma diminish, and ultimately when the frequency is high enough they disappear altogether.

Now let us return to Figure 21.6, in which θ denotes the angle within which the observer receives radiation scattered by irregularities in the interstellar medium. Next let a denote the size of the irregularities, ΔN_e the excess electron density in them, and L the thickness of the region in which the irregularities are concentrated. One can show that the relation

$$\theta = \left(\frac{L}{a}\right)^{1/2} \Delta N_e r_0 \lambda_0{}^2 \tag{21.6}$$

holds, where $r_0 \approx e^2/mc^2 = 2.8 \times 10^{-13}$ cm is the classical electron radius.

By applying simple diffraction theory such as that depicted schematically in Figure 21.6 to real scintillations of radio emission from pulsars, one finds that the size a of the irregularities is of order 10^{11} cm, while the electron density ΔN_e in them is around 10^{-4} cm^{-3}. These very puny ripples in the interstellar plasma evidently arise when it is disturbed by streams of charged cosmic ray particles.

Another consequence of the diffraction of pulsar radio waves by irregularities in the interstellar medium is a greater duration for pulses at low frequencies. The reason is that not all rays reaching the observer within the angle θ have the same group lag. Because of this difference, which can

Figure 21.7   Spreading of pulses at three frequencies.

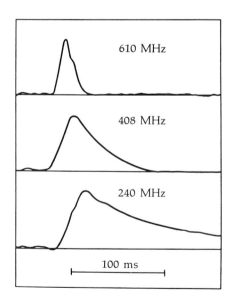

610 MHz

408 MHz

240 MHz

100 ms

reach several milliseconds, the pulse tends to spread out at low frequencies, so it is observable for a longer time interval. Figure 21.7 shows this type of pulse spreading at several frequencies for the pulsar PSR 1946 + 35.

Since the radio emission of pulsars is polarized and interstellar plasma is magnetized, we would expect the polarization properties of the radiation to change as it passes through the medium. A most interesting sort of interaction between linearly polarized radiation and magnetized plasma is called Faraday rotation of the plane of polarization. The polarization plane of an electromagnetic wave whose wavelength is λ meters will rotate through an angle of φ radians, given by the equation

$$\varphi = 0.81 \, \lambda^2 \overline{N_e H_{||}} l, \tag{21.7}$$

where $\overline{H}_{||}$, expressed in microgauss (μG), is the component of the interstellar magnetic field parallel to the direction of wave propagation, and the distance l from the radio source to the observer is expressed in parsecs. By comparing the direction of the electric vector in the wave at two frequencies, one can establish directly from the observations the product $R = \varphi/\lambda^2 = 0.81 \, \overline{N_e H_{||}}$ rad/m², which is called the rotation measure. On the other hand, the observations also supply the dispersion measure $D = \overline{N_e} l$ for the same pulsar. Hence one can directly measure the mean value of the longitudinal component of the interstellar field vector:

$$\overline{H}_{||} = 1.2 \frac{R}{D} \, \mu G. \tag{21.8}$$

In this manner dozens of values of $\overline{H}_{||}$ have now been measured, corresponding to the directions toward various pulsars. In typical cases $\overline{H}_{||}$ is found to be about 2–3 μG. Of all existing methods for measuring the strength of interstellar magnetic fields (including the Zeeman effect in the 21-cm line, and studies of the small amount of optical polarization of starlight produced by interstellar dust grains), this one is the most trustworthy and clearly visualized.

If the sign of the magnetic field changes, then the direction in which the polarization plane rotates will also change. Since the interstellar magnetic field is not entirely chaotic but is organized to some extent (for instance, its lines of force tend to stretch along the spiral arms of the Galaxy), we would expect the sense of the Faraday rotation to be consistent over large areas of the sky. Observational results confirming this behavior are illustrated in Figure 21.8. The filled circles signify that the mean intensity of the interstellar magnetic field in the direction of a given pulsar is directed toward the observer, and open circles, away from the observer. The size of

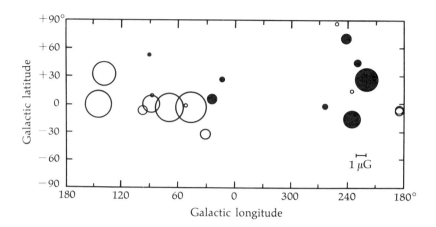

Figure 21.8 Distribution of the interstellar magnetic field over the sky, obtained by analyzing the Faraday rotation of pulsar radio emission (R. N. Manchester).

the circles is proportional to the interstellar field strength. As a matter of fact, Figure 21.8 gives a good idea of the capabilities of pulsar radio astronomy today. And thus pulsars, quite independently of their own nature, have furnished astronomers with a powerful method for investigating the interstellar medium.

22

Theories about Pulsars

Now that we have devoted several chapters to the basic observational facts concerning pulsars, it is fitting to discuss the theories proposed in explanation. In some measure we have already elucidated the theoretical aspect of the pulsar problem when we identified them with rapidly rotating neutron stars. This identification is undoubtedly an outstanding achievement of modern astronomy. But it signifies little, of course, to prove that pulsars are rapidly spinning neutron stars. We must still comprehend the reason for the fantastic power of their electromagnetic radiation. This is a question of fundamental importance. While astronomers have been aware theoretically of the neutron star phenomenon for four decades, the implausibly high activity of these objects was completely unexpected.

In Chapter 19 we said that more than 30 years after the need for neutron stars to exist had been substantiated theoretically, and shortly before the discovery of pulsars, Piddington, Kardashov, and soon Pacini came to the conclusion that neutron stars should rotate rapidly and be strongly magnetized. But how could such an object manage to emit radio waves? This problem has turned out to be far from simple.

We must admit that today there is no generally accepted quantitative theory for the radio emission of pulsars. That does not mean that in the nearly 10 years since pulsars were discovered, theoreticians have done nothing in this area. There has surely been progress, and many facets of the problem have been clarified to some extent. But the problem has proved an exceptionally difficult one indeed. The physical conditions in pulsars are too extreme. They cannot, for example, be modeled in the laboratory at the present time. The monstrous strength of their magnetic field, which we have mentioned above, radically alters the way that radio waves are propagated in plasma. Moreover, the physical structure of the interior of a neutron star presents a major theoretical problem. Although we have discussed a crude model, it remains very important yet extraordinarily difficult to understand the cause of the starquakes, and even the axial rotation of a superfluid body raises not a few questions. For such material lacks viscosity and cohesion between its particles, so one would think that it could not rotate like a solid body.

Thus the physics of neutron stars poses more than enough unsolved problems. Accordingly, in reviewing the present status of the theory of pulsars we shall take up only the simplest and best accepted principles. But even here scientists lack a unanimity of views, and the observational evidence has no definitive interpretation.

First of all we should say at least a few words about the expected theoretical properties of neutron stars. The possibility that neutron stars could exist as stable configurations in a state of equilibrium under the action of gravity and pressure forces was proposed as early as 1934 by Zwicky and Baade, who suggested that these objects might be formed in supernova outbursts. Long afterward it was still completely unclear whether neutron stars were indeed produced or whether they only represented an elegant mathematical construct.

But theoreticians meanwhile continued to investigate the superdense state of stellar matter. It had long since become apparent that the hypothetical neutron stars could not be homogeneous configurations; in other words, the physical state of such a star should change from its periphery in toward its center. Nor could one regard the material of a neutron star as consisting only of very densely packed neutrons. At all depths protons and electrons should accompany the neutrons as impurities. Heavy nuclei should predominate near the surface, while in the centermost regions superheavy elementary particles should exist—the hyperons, which are highly unstable under the conditions of laboratory experiments. It has been found that in the outer layers of a neutron star these nuclei should form a crystal lattice, so that the periphery of a neutron star should constitute a solid crust. Meanwhile the inner layers ought to be in a superfluid state.

In developing models for neutron stars theoreticians have encountered major difficulties, arising mainly from our inadequate knowledge of the

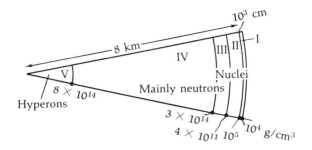

Figure 22.1   Internal structure of a neutron star.

nature of the nuclear forces acting between the particles that comprise a neutron star. Nevertheless, a semiqualitative neutron star model has been constructed. Figure 22.1 shows schematically how matter in a neutron star is presently believed to be stratified. Near the surface the material consists primarily of very densely packed iron nuclei. A comparatively small number of nuclei of helium and other light elements are also found there, as well as a very dense degenerate electron gas like that in the interior of white dwarfs (Chapter 10). The presence of electrons is required to compensate for the positive space charge of the nuclei.

As the interior of the neutron star is penetrated, the density rises and the electrons tend to be pressed into the nuclei. As a result neutron-rich nuclei are formed, heavier than the iron nuclei. When the matter reaches a density of about 3×10^{11} g/cm^3 these heavy nuclei will cease to be stable. They will begin to discard their neutrons, and with further progress into the interior the matter will gradually become a mixture of very densely packed neutrons, with the heavy nuclei now appearing as a comparatively minor impurity. Eventually, at a density of about 5×10^{13} g/cm^3, the heavy nuclei will disappear altogether.

At extremely high densities (approaching that in a nucleus itself), the chief constituent of the matter will be very tightly packed neutrons, with a comparatively small admixture of protons and electrons. When the density is about 3×10^{14} g/cm^3 the charged elementary particles, protons and electrons, will number only a few percent of the neutrons. Finally, near the very center of the neutron star, hyperons will appear and begin to play a substantial role (especially the sigma-minus hyperon, designated by the symbol Σ^-), as well as mu mesons, which along with the neutrons, electrons, and protons will be the predominant particles there. It should be remembered, however, that at the present time the physical conditions in the innermost zone of a neutron star are especially poorly understood. Our

351

knowledge of the manner in which these particles interact under such strange conditions is still too incomplete.

Because of our inadequate understanding of the physical conditions near the center of a neutron star, the models constructed for such stars are still far from perfect in the way they express a fundamental law: the theoretical relation between the radius of a neutron star and its mass. Nevertheless, theoreticians have succeeded in obtaining results of a sort. For instance, it has been found that the smaller the mass of a neutron star, the larger its radius will be.

In this regard we must point out that theory has not yet established the range of permissible values for the mass of a neutron star, although most specialists consider that the mass should be comparatively small, from 0.15 to 1.5 $\mathfrak{M}_\odot$. This important question is still a long way from being clearly understood.

In Figure 22.2 the solid curve gives the dependence of the radius on the mass according to one model that has been proposed for a neutron star. For masses above 1 $\mathfrak{M}_\odot$, different models imply different $R(\mathfrak{M})$ relations, a consequence of our unsatisfactory knowledge of the equation of state of matter at densities above 10^{15} g/cm³. A comparison with the mass–radius relation for white dwarfs shown in Figure 10.1 is instructive here; remember that $R_\odot = 7.0 \times 10^{10}$ cm.

The starquakes mentioned in Chapter 21 (the abrupt changes in the period of PSR 0833 − 45 and NP 0532, the Crab Nebula pulsar) are of great interest. For the first pulsar two such breaks have been observed, separated by an interval of more than two years; the relative change in the period has reached 10^{-8}. In NP 0532, breaks in the period are considerably more rare. Most likely the starquakes result from a discrete change in the moment of inertia of the rotating neutron star. Such a change could occur if the neutron star were to shrink its radius by about 0.01 cm. How are we to understand such a phenomenon?

Since a neutron star is rotating so rapidly, its equilibrium configuration should be a figure close to an ellipsoid of revolution. But because of the continuous slowing of the rotation, the parameters of this ellipsoid should change: it should become increasingly less flat. But the rigidity of the solid crust of the neutron star prevents a smooth change in its figure. For this reason elastic stresses will accumulate in the crust; when its limit of resistance has been reached the crust will suffer a sudden deformation with the character of a shear. Thus by studying starquakes we can gain a deeper understanding of the mechanical properties of neutron stars.

Figure 22.2 shows another theoretically computed parameter of a neutron star: its moment of inertia I, which describes its kinetic energy of rotation, $E = I\Omega^2/2$, where Ω is the angular velocity. For neutron stars of per-

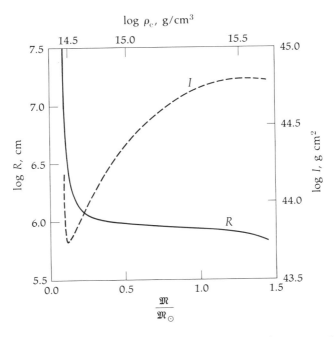

Figure 22.2 Theoretical dependence of the radius and moment of inertia of a neutron star on its mass.

missible mass, $0.15 < \mathfrak{M} < 1\text{-}2 \ \mathfrak{M}_\odot$, the moment of inertia will be in the range $7 \times 10^{43} < I < 7 \times 10^{44} \ \text{g cm}^2$.

From the observed lengthening of the period P of a pulsar with time, we can obtain the rate $\dot{P}$ at which the period increases (Chapter 20); the angular velocity will correspondingly change at a rate $\dot{\Omega} = -2\pi\dot{P}/P^2$. The rate of change in the kinetic rotational energy of the pulsar will be

$$\dot{E} = I\Omega\dot{\Omega} = -4\pi^2 \frac{I}{P^3}\dot{P}. \tag{22.1}$$

In pulsar physics the first task is to understand why rotating neutron stars slow down and thus continually liberate energy. The simplest explanation of this behavior would be that they are strongly magnetized. Then a neutron star rotating in a vacuum would emit magnetic dipole radiation at a frequency equal to the rotation frequency. The power of such radiation is given by the expression

$$L_m = \tfrac{2}{3}c^{-3} H_0^2 R^6 \Omega^4 \sin^2 \theta, \tag{22.2}$$

353

where θ is the angle between the magnetic axis and the rotation axis and H_0 is the magnetic field strength on the surface of the pulsar. In the case of NP 0532, for instance, $\Omega = 190 \text{ s}^{-1}$ and $\dot{\Omega} = 2.4 \times 10^{-9} \text{ s}^{-2}$, so that the pulsar is losing energy at a rate $-\dot{E} = 3 \times 10^{38}$ erg/s. If magnetic dipole radiation is responsible for the deceleration, then by setting $-\dot{E} = L_{\text{m}}$ we find that $H_0 \approx 3 \times 10^{12}$ gauss.

Another cause of the deceleration of this pulsar might be its emission of gravitational waves (Chapter 24). For such a mechanism to operate we need only assume that the figure of the neutron star is slightly asymmetrical, forming a triaxial ellipsoid. In this event the power of the gravitational radiation emitted by the rotating neutron star will be

$$L_g = 6.5 \, c^{-5} \, \varepsilon_e^2 I^2 \Omega^6, \tag{22.3}$$

where I is the moment of inertia and ε_e is the eccentricity of the equatorial ellipse of the neutron star. It turns out that the gravitational waves would have a frequency equal to twice the rotation frequency.

As Equation 22.3 indicates, the dependence of the power of the gravitational radiation on the angular velocity Ω is considerably stronger than in the case of magnetic dipole radiation. Hence an appreciable effect would be anticipated only for pulsars that are spinning very fast, such as NP 0532 (see also Chapter 24). For a magnetic dipole whose center is displaced from the center of the neutron star, the radiation will be asymmetric. The star will then acquire a recoil momentum along the direction of the rotation axis. Might this circumstance account for the high velocities at which pulsars travel?

If the deceleration of rotating neutron stars were caused by their magnetic dipole radiation, then we should theoretically have $\dot{\Omega} \propto \Omega^3$. But a statistical analysis of many pulsars with known Ω and $\dot{\Omega}$ gives an empirical law $\dot{\Omega} \propto \Omega^{3.4}$. The pulsar NP 0532, for which Ω and $\dot{\Omega}$ have been observed with particular care, conforms instead to a law of the form $\dot{\Omega} \propto \Omega^{2.7}$. We may conclude that the simple model considered above for the deceleration of magnetized rotating neutron stars is not adequate. Above all, the assumption that the neutron star is surrounded by a vacuum is certainly wrong. And as a result the problem becomes substantially more complicated.

A magnetized, rotating conductor will induce an electric field in the space around it. The component of this field perpendicular to the conductor's surface will tend to pull out electrons and ions from it. For a real neutron star the strength of the electric field can reach enormous values. Moreover, the surface temperature of a neutron star is quite high. For these reasons the space surrounding a neutron star will be filled with a great many charged particles which, moving along the magnetic lines of force, will rotate together with the neutron star about its axis at the same angular velocity. This solid body rotation should prevail out to some critical distance from the rotation axis—that is, within a cylinder. The radius $R_1 = c/\Omega$ of the cylinder is determined by the condition that the velocity of solid body rotation on its surface be equal to the speed of light.[1] However, if the plasma surrounding the neutron star is dense enough, the region of solid body rotation will be smaller, being then determined by the condition that the plasma have equal magnetic- and kinetic-energy densities.

Thus a magnetized, rotating neutron star should be enveloped in a fairly dense magnetosphere, which represents an extension of the star in an electrodynamic sense. Calculations show that the electric charges in the magnetosphere of a neutron star should be separated; that is, substantial space charge should exist there. The charge density is given by the expression

$$n_- - n_+ = \frac{\Omega H}{2\pi e c}. \tag{22.4}$$

At the surface of NP 0532, for example, where $H \approx 3 \times 10^{12}$ gauss and $\Omega = 190 \text{ s}^{-1}$, the charge density should be $n_- - n_+ \approx 10^{13} \text{ cm}^{-3}$, a rather large amount. Of course the total plasma density near the surface of this pulsar should be far higher.

Charged particles that have first been accelerated by the electric field to relativistic energies will flow out of the magnetosphere to infinite distance along open lines of force, and charges of differing sign will follow different lines. Figure 22.3 shows a diagram of a pulsar magnetosphere for the simplest case, in which the magnetic and rotation axes coincide. Even if the axes are not coincident the magnetosphere will have the same general structure.

Near the light cylinder crossed by the magnetic lines of force as they extend away to infinite distance, the lines will be seriously deformed by the streams of relativistic plasma escaping from the magnetosphere of the pulsar. The energy flux of the particles and magnetic field crossing the light cylinder may be estimated roughly from the expression

$$\frac{dE}{dt} \approx \left(\gamma\rho c^2 + \frac{H^2}{8\pi}\right)cR_1^2 \approx cH^2R_1^2, \tag{22.5}$$

where $\gamma = (1 - v^2/c^2)^{-1/2}$, and $\gamma\rho c^2 = \varepsilon_p$ is the energy density of the relativistic particles. In the case of a dipole magnetic field, $H = H_0 R^3/R_1^3$, so that

$$\frac{dE}{dt} \approx \frac{H_0^2 \Omega^4 R^6}{c^3}, \tag{22.6}$$

---

[1]This cylinder is accordingly called the light cylinder.

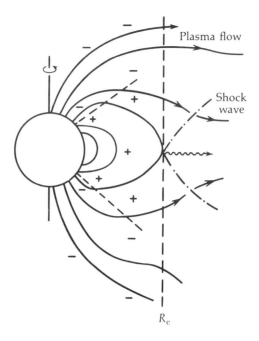

Figure 22.3   Structure of the magnetosphere of a pulsar. R_c is the radius of the light cylinder (M. Roderman).

and we arrive at an expression practically the same as Equation 22.2 for the power of magnetic dipole radiation in a vacuum. But now, of course, the physical meaning is different: most of the energy will leave the pulsar magnetosphere as a stream of relativistic particles.

The theory enables us to find only the total energy flux of these particles. Without further assumptions we cannot estimate the number of particles escaping from the pulsar magnetosphere and the average energy of each particle, let alone the energy spectrum of the particles. Specifically, let us consider the pulsar NP 0532. Equation 22.4 in conjunction with an R^{-3} law for the variation of $n_- - n_+$ sets a lower limit on the flux of charged particles through the surface of the light cylinder for this pulsar:

$$N \geq cR_1^2 |n_- - n_+|_{R=R_1} \approx \frac{H_0 \Omega^2 R^3}{ec} \approx 10^{34} \text{ s}^{-1}. \tag{22.7}$$

On the other hand, the source of the radiant energy of the whole Crab Nebula, whose power is of order 10^{38} erg/s, must undoubtedly be the corpuscular radiation of the pulsar NP 0532. Noting further that the relativistic electrons in the nebula have an energy of 10^{10}–10^{14} eV (1 eV = 1.6 × 10^{-12} erg), we conclude that the flux of charged particles across the light cylinder will be 10^{36}–10^{40} erg, and their numerical density there will be roughly 10^9–10^{13} cm^{-3}, a sizable value.

Thus the electrodynamics of magnetized, rotating neutron stars has the logical consequence that a powerful extended magnetosphere with a substantial separation of charge will inevitably be formed around them. Just as inescapable is the conclusion that there must be a pulsar wind—streams of charged particles flowing out of the magnetosphere along infinite lines of force. This conclusion is by no means trivial. Before the discovery of pulsars it was tacitly assumed that the atmospheres of neutron stars would be altogether negligible in extent. For example, even if the temperature of such an atmosphere were 1 million degrees, the characteristic scale height of the atmosphere, defined by the usual barometric formula $h = kT/m_H g$ = $kTR^2/Gm_H \mathfrak{M}$, would be only about 1 cm ($g = G\mathfrak{M}/R^2$ is the acceleration of gravity at the surface of the neutron star). So tiny a value of h would mean that the density of the neutron star atmosphere would drop practically to zero at a distance of a few tens of centimeters.

But the enormous gravitational potential of neutron stars will cause very deep potential wells to develop around them, and interstellar gas ought to collect in these wells. With certain simplifying assumptions, the density distribution in the neighborhood of a neutron star will be described by a law of the form

$$n = n_\infty \left(1 + \frac{Gm_H \mathfrak{M}}{kTr}\right)^{1/2},$$
(22.8)

where n_∞ is the density of the undisturbed interstellar gas and T is its temperature. If $r = R = 10^6$ cm and $T \approx 10^4$ °K, Equation 22.8 yields $n \approx 10^4$ n_∞. The existence of a pulsar wind fundamentally alters this situation. In particular, processes may occur in such a magnetosphere that are accompanied by nonequilibrium radio emission of very great power.

Let us now examine the most general properties of this radio emission. We are concerned here with the nature of the radiation window, the synthetic pulse profile, and the characteristic changes in the polarization, all of which were described in Chapter 21. As for the window, it is essentially explained by the beacon effect (see Figure 20.1). The comparison is an accurate one. A beam of radiation from a bright spot firmly coupled to the rotating neutron star describes a giant cone in space. When the beam passes across the observer, he records a pulse of radio waves. The beam is characterized geometrically by its directional pattern. In limiting cases the pattern may be pencil- or fan-shaped (Figure 22.4). A pencil beam has approximately the same angular size in all directions. For a fan beam the angular size is

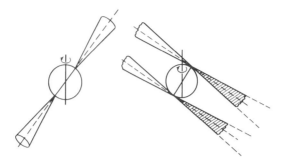

Figure 22.4 A pencil beam (left) and a fan beam of radiation.

comparatively small in one coordinate, but it may be a full 360° along the perpendicular coordinate. From the standpoint of emission mechanisms, both pencil and fan beams are possible.

Observations of the profile of pulses indicate that the window is about 9° wide (Chapter 21). As a rule, however, the profiles have quite an intricate structure; in particular, they are composed of considerably narrower subpulses. The question arises, then: what produces these subpulses? In principle, there are two ways to account for this effect.

1. The radiation emanates from a very limited area firmly coupled to the rotating neutron star. In this event the subpulses would correspond to a radiation beam with a complex, dissected pattern.

2. The radiation of each subpulse comes from a definite spot on the rotating neutron star, and each has a simple rather than a dissected beam pattern.

The presence of several subpulses would mean that the radiating spots are spread over a fairly large area on the rotating neutron star. Evidence for the second alternative is provided by the fact that the subpulses tend to maintain their individuality for several rotation periods, and that they have a well-defined polarization. Thus the characteristic element of the radio emission of a pulsar would be the subpulse, whose radiation would have a simple beam with a definite state of polarization. The observed profile of a pulsar would be made up of a sequence of such subpulses passing through the pulsar window.

If this interpretation is correct, the relative duration of each subpulse would be determined by the size of the region in which the radiating spots are located. As for the interpulses observed in some pulsars, they would be attributed to spots located in an entirely different (almost diametrically opposite) area on the rotating neutron star. The comparatively weak emission between pulses observed in the Crab Nebula pulsar and some others would arise from emission regions distributed over a large volume in the neighborhood of the neutron star.

What agency could keep the radiating spots in strictly definite places around a rotating neutron star? Evidently the only such factor would be the very strong magnetic field. It is also quite natural to ascribe the observed polarization and its variation as the subpulses cross the pulsar window to the varying orientation of the magnetic field with respect to the observer as the neutron star rotates.

The very large, striking changes in the direction of the polarization vector for the optical radiation of the Crab Nebula pulsar (Figure 22.5) and the radio emission of PSR 0833 − 45 (Figure 22.6) can be explained most simply by a pencil beam pattern. In both cases, essentially one subpulse passes through the pulsar window; hence the radiation comes from comparatively few emission spots. Probably these spots are associated with the magnetic poles of the corresponding neutron stars. Calculations that we shall not reproduce here demonstrate that the rotation of the magnetic axis vector can fully explain the rapid variations observed in the direction of polarization. When applied to the Crab Nebula pulsar, this theory gives interesting additional information. The rate of change in the direction of linear optical

Figure 22.5   Variations in the optical polarization vector for the Crab Nebula pulsar.

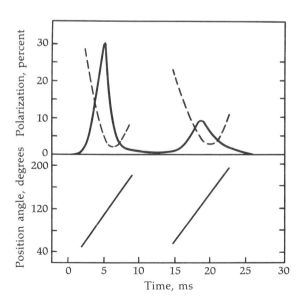

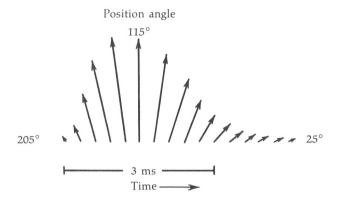

Position angle

115°

205°   25°

├────────── 3 ms ──────────┤

Time ──────▶

Figure 22.6 Variations in the radio polarization vector for the Vela pulsar during a single pulse.

polarization of this pulsar implies that its magnetic axis is nearly perpendicular to its rotation axis, and the minimum angle between the magnetic axis and the line of sight is only 2°. Curiously enough, the radio emission of the Crab Nebula pulsar does not show changes in the direction of polarization so regular as those of the optical radiation.

Beams of highly directive radiation, then, are tied to the spinning neutron star and revolve through space. Just how are they formed? Strange though it may seem, this clear question does not yet have a definitive answer. Various groups of investigators hold different views.

For example, the first astronomer to comprehend the nature of pulsars, Thomas Gold, attributes the formation of a radiation beam to the magnetosphere of the neutron star, which is rigidly coupled to the star and rotates along with it. Gold's hypothesis is clear from the diagram of Figure 20.1. The pulsar magnetosphere contains ionized gas (plasma) which rotates together with the magnetosphere at the same angular velocity as the neutron star. As the surface of the light cylinder is approached from the inside, the rotational velocity will approach the speed of light. Then, because of the effects predicted by special relativity theory, the plasma electrons rotating about the axis of the neutron star will emit radiation at frequencies close to $\Omega(E/mc^2)^2$, where $E = \gamma mc^2 = mc^2(1 - v^2/c^2)^{-1/2}$ is the energy of electrons rotating at a velocity v sufficiently close to c. Relativistic charged particles always radiate within a certain cone whose axis coincides with their instantaneous velocity vector and whose vertex angle $\theta = mc^2/E$. One can always imagine that the plasma is not distributed uniformly near the surface of the light cylinder, but is concentrated in separate clouds. This circumstance would account for the presence of the radiation beams.

360

A somewhat similar kinematic picture of the radio emission of pulsars has been developed by other authors, particularly Vladimir V. Zheleznyakov and F. Graham Smith. According to this interpretation, in the pulsar magnetosphere near the light cylinder a plasma-filled region exists which serves as a source of nonequilibrium, almost isotropic radio waves. As the magnetosphere rotates rigidly along with this cloud that it contains, the Doppler effect will strongly raise the frequency of the radiation whenever the beam is pointed toward the observer. From the viewpoint of the external observer, then, such radiation will be sharply directive.

Unquestionably the kinematic theories of pulsar radio emission afford an unconstrained explanation of why the radiation has the same directional pattern at different frequencies. This behavior is especially noteworthy for the pulsar in the Crab Nebula, whose x-ray pulses are practically simultaneous with the pulses of radio emission. To account for the polarization of pulsar radiation and its characteristic time variations, however, would be very difficult in terms of a kinematic mechanism alone.

A completely different radiation geometry has been proposed by Radhakrishnan and D. J. Cooke. In their interpretation, the radiation need not have anything at all to do with the surface of the light cylinder. Relativistic electrons again serve as the radiating agent, but they are intrinsically relativistic, not (as in Gold's model) just because they are rotating at nearly the speed of light, being located near the surface of the light cylinder. The Radhakrishnan–Cooke model would have the relativistic electrons moving in the vicinity of the magnetic poles of the neutron star and radiating through a certain modification of the synchrotron mechanism. Unlike the ordinary synchrotron mechanism, where the relativistic electrons travel in spirals around the magnetic lines of force and the angle between the velocity and field vectors is quite large, here the electrons would move almost exactly along the field lines, radiating only because the lines are curved. In this sense, of course, Gold's mechanism too is a synchrotron one.

The application of this generalized synchrotron mechanism to the radio emission of pulsars is certainly of interest, as it explains in a rather natural way the most important property of that radiation, its directivity. A more detailed analysis discloses serious difficulties, however. The width of the beams, which governs the duration of the subpulses, should depend on the frequency of the radiation, even if just weakly. But nothing of the kind is observed; the subpulses are equally long at all frequencies. There are also serious discrepancies between the observed polarization properties of pulsar radio emission and those predicted by various modifications of the synchrotron mechanism.

To summarize, we may say that a generally accepted theory of pulsar radio emission does not yet exist, although individual parts of the picture have evidently been filled in. The whole intricate behavior of the radio emission of pulsars should result from the joint action of many different factors: the strong magnetic field, the collective interaction of the charged particles with the field, and, of course, the motion of plasma at relativistic velocity near the inner boundary of the light cylinder.

Up to now we have been discussing principally the geometry of the radiation. We turn next to the physical processes that might be responsible for it. The question of the power of pulsar radiation is of special importance. The power can be evaluated if one measures the radiation flux within a pulse throughout the whole radio range, knows the ratio of the pulse length to the period of the pulsar, and specifies the directional pattern of the radiation (such as a pencil or fan beam). In this manner a very wide span of values is obtained for the power of various pulses: 10^{27}–10^{31} erg/s, which is equivalent to 10^{14}–10^{18} megawatts. It is interesting to compare this power with the combined output of all the radio and television transmitters in the world; if all were operating simultaneously they would radiate only a few thousand megawatts. We see that the natural cosmic radio transmitters are immeasurably more powerful than the artificial ones on the earth.

The Crab Nebula pulsar stands by itself. Along with its radio emission at an average power of about 10^{31} erg/s (on occasion it can be 100 times higher for brief intervals), this pulsar also radiates in the optical, x-ray, and γ-ray ranges. Its optical radiation, which is noteworthy for its remarkable stability, reaches a power of about 10^{34} erg/s, or 2.5 times the luminosity of the sun. But x rays and γ rays carry the greatest amount of power emitted by this pulsar. "Hard" photons of this radiation escape from the pulsar at a rate of around 3×10^{37} erg/s, which is ten thousand times the power radiated by the sun at optical frequencies and close to the radiant power of giant stars at frequencies in the infrared, optical, and ultraviolet ranges.

Another interesting estimate concerns the power that a pulsar radiates per unit volume in its emission region. In view of the geometry of pulsars, we may infer that the depth of the emission region cannot exceed the radius of the light cylinder. On the other hand, the duration of the pulses implies that the projection of this region onto the surface of the neutron star should have a linear size of a few tenths the radius of the star. It follows, in particular, that the volume of the emission region for the Crab Nebula pulsar is no more than 10^{23} cm^3. The power emitted per unit volume by the pulsar in the x-ray and γ-ray ranges therefore exceeds 10^{14} erg cm^{-3} s^{-1}. This is a fantastically high value, billions of times greater than the rate at which thermonuclear energy is generated per unit volume in the interior of a typical star.

An important quantity describing the intensity of radiation is its brightness temperature (Chapter 4). While the optical radiation of the Crab

362

Nebula pulsar has a brightness temperature of nearly 10 billion degrees, and while in the x-ray range the value is "only" a hundred thousand degrees, the brightness temperature at radio frequencies attains the incredible value of some 10^{28} degrees!

The exorbitantly strong radio emission of the pulsar in the Crab Nebula (and indeed of all other pulsars) completely precludes explanation by the combined effect of charged particles radiating independently, as would be the case for synchrotron or thermal radiation. Astronomers encountered a comparable situation in the 1940s, when radio observations revealed giant bursts of solar radio waves. Studies of this interesting phenomenon necessitated the conclusion that the bursts arise from organized, coherent motions of electrons within comparatively large volumes of plasma. When elementary charges take part in organized motion of this kind, the amplitudes of the elementary electromagnetic waves they emit are additive. Hence the intensity of the electromagnetic wave resulting from the whole system of radiating charges (which is proportional to the square of the resultant amplitude) will be an immense number of times greater than the sum of the intensities of the elementary waves emitted by the individual charges.

Under laboratory conditions a good example of radio emission from electric charges in organized (coherent) motion is provided by ordinary transmitting antennas. For instance, a transmitter with an effective surface area of 10^4 m^2 can radiate several megawatts of power. If the whole surface surrounding the magnetosphere of the Crab Nebula pulsar were covered with such transmitting antennas, the radiant power would be only about 10^9 megawatts, or 10 billion times less than the power of the radio emission of the pulsar. Even if the entire volume of its magnetosphere were densely filled with transmitting antennas of this kind, the radiant power would be a hundred thousand times below the value observed! This illustration demonstrates how insignificant the works of human hands and brains are compared to the spontaneous processes of nature.

An exceedingly high brightness temperature can also be developed if the radio emission mechanism is a certain variety of the maser amplification process. In Chapter 4 we have learned about the natural cosmic masers that have so unexpectedly turned out to be associated with the process of star formation. In this case maser amplification can occur only for a narrow frequency interval corresponding to the radio lines of the OH and H$_2$O molecules. But under certain conditions the maser amplification process can operate over a very broad band of the spectrum, with no connection to any particular radio line.

It is worth emphasizing once more how extraordinary are the conditions whereby radio waves are generated and propagated. For example, in pulsar radiation the electric field strength in the region of generation reaches several billion volts per meter. The inference follows at once that by itself this intrinsic electric field accelerates to relativistic energies the electrons that generate it. In this complicated situation such elementary concepts as the refractive index and superposition of waves lose their customary meaning.

To sum up, the radio emission of pulsars may be said to represent a coherent process that operates under very unusual conditions. In order for such radiation to be produced at all, a pulsar magnetosphere must contain a sufficient number of free electrons. The need for free electrons to exist in the magnetosphere follows from the basic laws of electrodynamics (see above). Moreover, quite a strong flow of plasma emanates from the surface of the neutron star; we have called this phenomenon the pulsar wind. The particle density can be rather substantial in such a stream. We have already discussed this matter more fully in connection with the pulsar in the Crab Nebula. Unfortunately, at the present level of theoretical development it is not clear in what part of a pulsar magnetosphere the radio waves are generated. It might be in a layer near the magnetic poles immediately adjacent to the surface of the neutron star, or at the distant periphery of the magnetosphere near the inner surface of the light cylinder. A great deal of effort is warranted on the part of both observers and theoreticians if the nature of the pulsar radio emission associated with the activity of neutron stars is to be understood.

The high-frequency radiation of the Crab Nebula pulsar NP 0532 undoubtedly differs radically in origin from the radio emission. This circumstance is apparent just from the general spectrum of the electromagnetic radiation of this pulsar displayed in Figure 20.6. The radio- and high-frequency spectra do not match up at all; the high-frequency radiation (which comprises the great bulk of the energy) in no sense represents a continuation of the radio emission of this pulsar. Another fundamental difference between the two types of radiation is that the high-frequency emission is very constant—the pulse profile remaining unchanged with time—while the radio emission is variable and differently polarized. All evidence suggests that whereas the radio emission of NP 0532, like that of other pulsars, is due to some still uncertain coherent mechanism, the high-frequency radiation is instead a sum of the radiation emitted by individual relativistic electrons moving in the pulsar magnetosphere, and thus is synchrotron radiation. This finding greatly simplifies the task of theoretical interpretation of the properties observed in the optical and x-ray emission of NP 0532.

The starting point for a theory of the high-frequency radiation of the Crab Nebula pulsar should, in our opinion, be the striking fact that the spectrum falls off steeply in the near-infrared range. What causes this be-

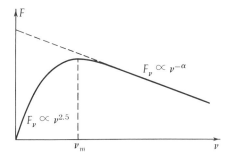

Figure 22.7   Spectrum of a source with synchrotron self-absorption.

havior? In 1970 the author of this book put forward arguments indicating that the falloff should be attributed to synchrotron self-absorption, while the high-frequency radiation of NP 0532 would be due to the ordinary synchrotron mechanism responsible for the bulk of the radiation of the Crab Nebula over the whole range of electromagnetic waves (see Chapter 17). In the phenomenon of synchrotron self-absorption, for a very high density of relativistic particles a cosmic object will cease to be transparent to its own synchrotron radiation. This opacity will set in at a certain frequency and will grow progressively toward lower frequencies. As a result, the intensity will seem to cut off at low frequencies, and the synchrotron spectrum of the source will take on the form shown in Figure 22.7. For comparison, the dashed line in the figure indicates the spectrum the same source would have if it were transparent—that is, if synchrotron self-absorption were absent. Notice that the high-frequency spectrum of the Crab Nebula pulsar (Figure 20.6) has the same character as in a source in which synchrotron self-absorption is important.

The frequency ν_m at which the synchrotron self-absorption effect becomes significant will be relatively high for a compact source, particularly for a source having a high density of relativistic electrons.

The angular size of a source of synchrotron radiation whose spectrum exhibits synchrotron self-absorption is given by the expression

$$\theta = 9 \times 10^{21} \, \Phi(\alpha) \, \nu_m^{-5/4} [F(\nu_m)]^{1/2} \, H_\perp^{1/4}, \tag{22.9}$$

where $\Phi(\alpha)$ is a dimensionless factor of order unity. This equation, a very important one for radio astronomy, was first derived in 1963 by the Soviet radio astronomer Vyacheslav I. Slysh. At that time it was especially useful for analyzing the nature of the newly discovered quasars, whose radio spectra often show the synchrotron self-absorption phenomenon at frequencies in the range of hundreds or thousands of megahertz (corresponding to the decimeter or centimeter wavelength ranges). The fact that in the case of the Crab Nebula pulsar $\nu_m \approx 10^{14}$ s^{-1} (100 million megahertz) is very surprising in itself. It implies above all that the pulsar represents a super-compact source.

Analysis shows that in the region where the optical and x-ray synchrotron radiation is emitted by the Crab Nebula pulsar, $H_\perp \approx 3000$ gauss, while the total amplitude of the magnetic field vector there is about 10^6 gauss. Accordingly, the emission region is located somewhere near the light cylinder, whose radius for NP 0532 is about 1500 km, or 100–200 times the radius of the neutron star. For this reason the field is several million times weaker in the vicinity of the light cylinder than at the surface of the neutron star.

Applying the theory of synchrotron radiation, one can calculate the density of relativistic electrons in the magnetosphere of NP 0532 and their flux in the Crab Nebula. This flux turns out to be just enough to supply the nebula continuously with the energy it needs to compensate for its powerful radiation. Thus the synchrotron theory offers a natural explanation for the optical and x-ray emission of quite the most singular object in the sky.

23

X-Ray Stars

During the postwar years the vigorous development of space astronomy, like that of radio astronomy, brought about a revolution in our science. Perhaps the most impressive achievement of astronomy conducted in space has been the outstanding success of x-ray astronomy. The earliest x-ray observations of the sun were made immediately after the war, in 1946, using photon counters lifted above the atmosphere by small rockets. But fully 16 years elapsed before observing techniques progressed far enough that the first x-ray source located far beyond the solar system could be detected. The low resolving power typical of x-ray astronomy (roughly 10° at that time) did not immediately permit reasonably accurate determination of the position of the new x-ray source in the sky. The suggestion was even made that this quite intense source might be situated at the center of the Galaxy.

It soon became apparent, however, that the source has no relation to the galactic center. It lies nearly 20° away, in the constellation Scorpius. This circumstance implies that the source is comparatively close to the sun; for the galactic disk—the region containing the great majority of stars—is no more than about 500 pc thick, while the radius of the disk extends out to 15,000 pc. Since the galactic latitude of the x-ray source in Scorpius is about 20°, its most probable distance should be only (250 pc)/sin 20° ≈ 750 pc.

(Simple methods of this kind for roughly estimating the distances of un-known galactic sources are very widespread in astronomy.)

The newly discovered x-ray source was named Scorpius X-1. Similar nomenclature for new sources was adopted during the early years of ener-getic development in radio astronomy. The brightest radio sources were named after the constellations in which they were found: Cygnus A, Cas-siopeia A, Taurus A, Virgo A. These names have been retained to the present, although now every astronomer knows that Taurus A is the Crab Nebula and Virgo A is the giant spheroidal galaxy NGC 4486. Shortly after the discovery of Scorpius X-1, x rays were recorded from the Crab Nebula (as discussed in Chapter 17) and from two new sources in the con-stellation Cygnus, which were at once named Cygnus X-1 and Cygnus X-2.

In the eight years after Scorpius X-1 was discovered, developments in x-ray astronomy were none too rapid. Observations were made with rockets, but launches were infrequent and more or less at random (an exception was the lunar occultation of the x-ray source in the Crab Nebula, described in Chapter 17). Even so, valuable information about the nature of x-ray sources was acquired during this period. Of special importance was the enormous amount of x-ray flux received from the Scorpius X-1 source. For x-ray photons in the 1–10 keV energy range (wavelengths $\lambda \approx$ 10–1 Å), the flux density is of order 3×10^{-7} erg cm^{-2} s^{-1}. A seventh-magnitude star contributes approximately the same (bolometric) flux density. Not until 1966 did advances in the technology of x-ray observations permit the Scorpius source to be placed accurately to within a few minutes of arc in the sky. As a result the enigmatic object could at once be identified with a rather faint star of 13^m to which no one had previously paid any attention. By this time observers had recognized that the x-ray flux from Scorpius X-1 was quite strongly variable: the day-to-day flux variations reach many tens of percent. The optical star identified with this source also changes its bright-ness irregularly (from about 12^m to 13^m), but these fluctuations have vir-tually no correlation with the changes in the x-ray flux (see below, however).

The spectrum of the Scorpius X-1 source has been measured repeatedly, and in the 1–20 keV range it follows closely the exponential law

$$F_E \propto e^{-E/kT}, \tag{23.1}$$

where E is the energy of the x-ray photons and T is a parameter having the significance of a temperature. The value of T is some tens of millions of degrees. Very hot plasma of temperature T would have such a spectrum, and this plasma should be transparent to its own x rays. In addition to the varia-tions in the x-ray flux, changes are simultaneously observed in the spectrum, which nonetheless maintains its exponential behavior. During these changes the temperature characterizing the spectrum varies from 25 to 100 million degrees! At high energies ($E \geq 50$ keV), however, the spectrum of Scorpius X-1 contains an appreciable amount of excess radiation that definitely does not merely continue the radiation of hot plasma in this spectral range.

Figure 23.1 shows the optical spectrum of the inconspicuous star identified as Scorpius X-1. In the near-infrared region the spectral flux density of the radiation increases with the frequency, while in the visible and ultraviolet the curve stays almost horizontal. An important point on the spectral curve is contributed by observations in the ultraviolet near 1500 Å, obtained by the techniques of space astronomy. This point lies on the extension of the horizontal part of the curve in Figure 23.1.

Superimposed on this bright continuous spectrum are some rather faint lines representing the Balmer hydrogen series, helium, and ionized carbon and oxygen atoms. The intensity of these lines, like their radial velocities, is strongly variable. In a few hours, for example, the radial velocities may vary by many hundreds of kilometers per second, meanwhile changing their sign. This behavior means that the ionized gas clouds radiating the lines are moving at comparably high velocities, sometimes toward and sometimes away from the observer. It is interesting to find that the radial velocities of

Figure 23.1 Optical and infrared spectrum of the Scorpius X-1 source (Hale Observatories).

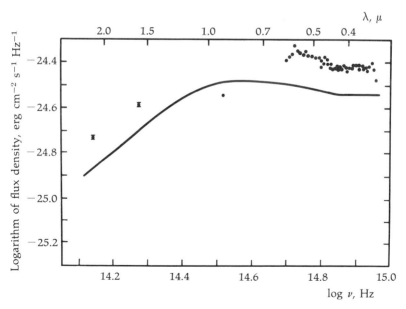

lines belonging to various elements are different and often have opposite signs. All told, the evidence implies that violent processes accompanied by the ejection of fairly large masses of gas are taking place in the neighborhood of the Scorpius X-1 source.

For the most part the optical continuous spectrum of the Scorpius X-1 source is probably an extension of the x-ray spectrum. Thus both the x-ray and the optical radiation of this source represent ordinary thermal emission by a very hot gas whose temperature is some tens of millions of degrees. But since the absorption coefficient of such gas rises sharply toward lower frequencies, in the near-infrared and red spectral regions the gas will cease to be transparent to its own radiation. At these frequencies the gas should therefore radiate like a black body.

In the range of frequencies satisfying the condition $h\nu < kT$, the intensity should depend on frequency according to the classical Rayleigh–Jeans law

$$F_\nu = \frac{2\pi kT}{c^2} \nu^2 \frac{R^2}{r^2}, \qquad\qquad (23.2)$$

where R is the radius of the emission region and r is the distance of the source. The empirical spectrum of Figure 23.1 shows, in fact, that $F_\nu \propto T\nu^2$, as should be the case for the Rayleigh–Jeans law. Knowing that $T \approx (3–5) \times 10^7\,°K$ and taking $r \approx 500$ pc as a rough estimate, we readily obtain an approximate radius $R \approx 10^9$ cm for the emission region; thus the x-ray source should be only about 10,000 km across! It must be a very compact object. From the distance r we have adopted (which can hardly be in error by more than a factor of two in either direction, because of the high galactic latitude of the source), the power of the x rays emitted by Scorpius X-1 (its x-ray luminosity) may be estimated as $L_x \approx 10^{37}$ erg/s, or 2000–3000 times the total bolometric luminosity of the sun!

Given the size of a source and the kinetic temperature of the plasma filling it, and having a theory for its radiation (a very good and trustworthy theory is available), we can estimate without difficulty the number density of particles (electrons and ions) in the plasma. For Scorpius X-1 the density turns out to be about 10^{16} cm^{-3}, quite a high value, close to the particle density in the upper layers of the solar photosphere. Finally, from the size and density of the source we can at once determine the total mass of gas emitting the x-ray photons observed to be arriving from the source. On an astronomical scale the mass is entirely negligible, only about 10^{20} g, or a hundred million times lower than the mass of the earth. This dense cloud of plasma contains a thermal energy supply of about 10^{36} erg. Accordingly, if left to itself the plasma cloud ought to radiate away all its energy in just a tenth of a second!

370

But this source has now been under observation for 15 years, and it will survive for many thousands of years, at least. Thus there should be some very powerful agent that pumps energy continuously into the hot plasma. A hot plasma cloud that is being permanently heated in some fashion and is reemitting x-ray photons—this is only a rather minor detail in an altogether exceptional cosmic body that cannot be observed directly.

As we shall see, analyses of the x-ray and optical radiation of Scorpius X-1 have enabled several important conclusions to be drawn about its nature and have revealed completely unexpected properties, hitherto unknown to astronomy. In its general characteristics this source is not unique. Much the same properties have been found in another source, Cygnus X-2, which has been identified with a peculiar 15^m star.

No sooner had galactic x-ray stars been discovered, of course, than the theoreticians began to ponder their nature, and especially the origin of their enormously energetic x rays. That's the way theoreticians are: although at first the information on x-ray stars was completely inadequate (even now, to put it mildly, the data are anything but plentiful), there was no lack of hypotheses and theories. But one should not be too severe toward the theoreticians—they are true to human nature. They wanted very much indeed, and so understandably, to comprehend the essential principles of these remarkable objects.

At that time a feeling was in the air that the discovery of neutron stars was just around the corner. (Remember that a few years were to pass before pulsars were discovered.) The first suggestion for interpreting the nature of neutron stars was forthright. The spectrum of the radiation observed did not preclude the possibility of a thermal origin, that is, a description by the Planck law with a temperature of the order of 10^7 °K. But the idea that the x-ray sources might represent hot neutron stars was quickly found to be untenable (see Chapter 19).

A new era in x-ray astronomy was inaugurated in December 1970, when from a launch site in Kenya the Americans placed the specialized x-ray satellite *Uhuru* in an equatorial orbit (see the Introduction). Before this satellite was launched about 35 cosmic x-ray sources were known, but after *Uhuru* had been operating for two years the number of known x-ray sources had risen to nearly 200. The satellite recorded practically all sources whose flux density exceeds 0.001 that of Scorpius X-1 (for x-ray photons in the 2–20 keV energy range).

The sources observed fall into two classes. Sources of the first class have a galactic latitude lower than 20°; the others have a latitude higher than 20°. As a rule the strongest sources belong to the first class. We may infer from this circumstance that the two classes of sources actually represent objects of completely different kinds. For if all x-ray sources represented a single type of object and were located within our Galaxy, then the sources observed in high galactic latitudes would on the average be much closer to us. But in that event they ought to appear brighter. We observe such a picture in optical astronomy: the brightest stars are not concentrated at all toward the Milky Way, while the faint, telescopic stars are concentrated very strongly there.

The high-latitude x-ray sources are distributed isotropically over the sky. Some of them have been identified with extragalactic objects—individual galaxies or clusters of remote galaxies. It follows that, in many cases at least, the high-latitude x-ray sources are very distant extragalactic objects. As for the bright sources located in low galactic latitudes (in the Milky Way belt), the great majority belong to our Galaxy. Altogether about 100 such sources have been found. Of these about 10 have been identified with supernova remnants; we discussed them in Chapter 16. However, the bulk of the galactic x-ray sources observed should belong to an entirely special class of objects of stellar character, more or less resembling the sources Scorpius X-1 and Cygnus X-1. We shall henceforth call such sources "x-ray stars."

In addition to their concentration toward the galactic equator, x-ray stars exhibit a well-defined concentration toward the galactic center: a longitudinal sector extending 60° to either side of the galactic center contains more than half of them. We therefore arrive at the conclusion that these sources have an average distance from us equal to the sun's distance from the galactic center—about 10,000 pc. The same result follows from a spectral analysis of the x-ray sources located in the vicinity of the constellation Sagittarius, which lies in the direction of the galactic center. Such sources often have a spectrum with a cutoff on the low-energy side. The cutoff results from absorption of x rays by the interstellar gas; and in order for the spectrum to drop off steeply at a photon energy of about 3 keV, the observed value, just as many interstellar atoms are needed as are actually located between the sun and the galactic center.

Knowing the distance of such a source (typically 10,000 pc) and its x-ray flux density (which is observed directly), we may infer that the object has an x-ray power up to 10^{38} erg/s, tens of thousands of times the bolometric luminosity of the sun. This important conclusion is supported by observations of five x-ray sources in the galaxies closest to us, the Magellanic Clouds, whose distance is known to be 60,000 pc. On the other hand, statistical analysis of the observational data indicates that hardly any objects among the x-ray sources have a radiant power lower than 10^{36} erg/s. Were this not the case, then along with the comparatively bright sources an appreciably larger number of faint ones would be observed in the Galaxy.

Thus the x-ray stars of the Galaxy comprise a population fairly restricted in both power (10^{36}–10^{38} erg/s) and number (about 100 objects).

In fact, most of the x-ray stars now existing in the Galaxy have already been observed. Confirmation of the scarcity of x-ray stars is provided by observations of x rays coming from the Andromeda Nebula, which is recorded as a very weak source. This "nebula" is, of course, a giant spiral stellar system, comparable in many respects to our Galaxy. Since the Andromeda Nebula is about 600,000 pc distant from us, its measured flux density gives us the combined power of all the x-ray sources it contains—about 2×10^{39} erg/s. X-ray sources have an average power of about 10^{37} erg/s; thus we may at once conclude that the Andromeda Nebula has roughly a hundred such x-ray stars.

It is evident, then, that x-ray stars represent an exceedingly rare phenomenon. In our Galaxy, as in the Andromeda Nebula, there is only one x-ray star for every billion "ordinary" optical stars. Indeed, it would be hard to cite any other population among the diversified objects in the Galaxy that is so rare. Perhaps only globular clusters are comparable in scarcity to x-ray stars. But globular clusters are huge aggregates of matter comprising hundreds of thousands of very old stars, whereas x-ray stars are very compact objects, undoubtedly connected with the terminal phase of stellar evolution.

One important property of the radiation of x-ray stars is its variability. As a rule, variations are recorded in the x-ray flux. These flux variations show a wide variety of behavior. Some sources exhibit very rapid, irregular flux variations. For example, in one of the brightest sources, Cygnus X-1, major changes in the flux can occur in a time interval shorter than 0.001 s! This fact in itself implies that the linear size of such a source should be smaller than 0.001 light second, or 300 km (one-twentieth the radius of the earth). Actually, the size appears to be considerably smaller still.

The flux variations in the brightest source, Scorpius X-1, have been studied quite thoroughly. In this source the changes observed in the flux are not so rapid as those occurring in Cygnus X-1. The level of radiation may remain more or less constant for several days. Then both the optical and the x-ray flux will change by up to 20 percent, but the variations in optical and x-ray emission are not correlated. Superimposed on this quiescent radiation are individual flares lasting a few hours. The flares extend over both the optical and the x-ray range. During a flare the flux may change two or three times, and the mean flux is twice as great in this active phase as in the quiescent phase. No periodicity can be discerned in the flares.

Simultaneous optical and x-ray observations demonstrate that during flares the x rays of Scorpius X-1 become harder (the proportion of high-energy photons in the spectrum increases), while the optical radiation becomes bluer. If we accept the model indicated above, according to which the source is regarded as a compact plasma cloud, we must suppose that

373

at the time of a flare the temperature and mean density of the plasma rise. Because of this circumstance the plasma will become opaque for shorter wavelengths, so by the Rayleigh–Jeans law its optical spectrum will grow bluer.

Perhaps the most noteworthy finding made with *Uhuru* has been the discovery of a strict periodicity in the x-ray flux variations of some sources. This discovery, as we shall see presently, has provided the key to understanding the nature of x-ray stars and to intelligent quantitative analysis. Prior to *Uhuru* the investigation of these objects amounted to an unsystematic collecting of observational data. The essence of the discovery was as follows.

Studies of the flux variations in the medium-intensity source Centaurus X-3 had shown that the source emits radiation at both high and low levels. When the radiation is at the low level the flux is 10 times weaker. These two radiation states alternate with the remarkably exact periodicity of 2.08707 days. During this period the source emits a low level of radiation for about $0^d.5$ (Figure 23.2). An explanation of such a strict periodicity poses no difficulty for astronomers. The x-ray source Centaurus X-3 belongs to a binary system whose orbit plane is inclined at a small angle to the line of sight. As the x-ray component of this binary system moves in its orbit, it periodically goes behind the normal, optical component, which thereby eclipses it. As a result the x-ray flux reaching the earth drops sharply. When the

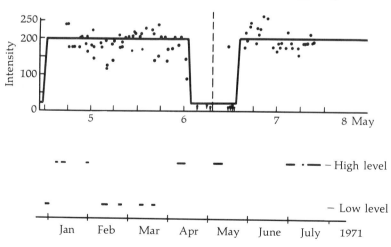

Figure 23.2 An x-ray light curve for the Centaurus X-3 source, recorded by the *Uhuru* satellite in the 2–6 keV energy range.

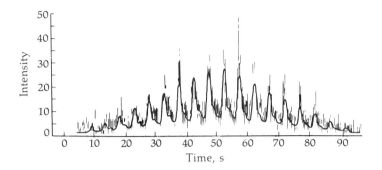

Figure 23.3 Pulsations in the x-ray flux of the Centaurus X-3 source.

eclipse of the x-ray star by the optical component ends, the original high level of x-ray flux is restored. A similar phenomenon has long been familiar in optical astronomy: we are speaking of the eclipsing variable stars, of which a typical representative is the famed star Algol.

Along with the $2^{d}_{.}087$ periodicity in the x-ray flux of the Centaurus X-3 source, another and far less superficial periodicity was discovered. The radiation of this source was found to have the character of a periodic pulsation with a period of only 4.84239 seconds (Figure 23.3). In the intervals between these very short pulses the x-ray flux drops by nearly a factor of 10. Accurate observations have revealed that the pulsation period itself varies smoothly according to a sinusoidal law with a period of $2^{d}_{.}08707$ (Figure 23.4). These tiny but regular variations in the pulsation period are readily explained by the Doppler effect resulting from the orbital motion of a source with a constant pulsation period. Such a behavior is proved just by the observed fact that the rate of change in the pulsation period becomes zero when the eclipse reaches its midpoint—that is, when the x-ray source is moving in its orbit perpendicular to the line of sight (Figure 23.5). From the amplitude of the variations in the pulsation period induced by the orbital motion of the x-ray star we find, using the standard Doppler effect equation, that the orbital velocity is 415 km/s.

It should be noted here that the pulsating component of the x-ray source often disappears for a few days. When this happens the x-ray flux from Centaurus X-3 diminishes by an order of magnitude and becomes approximately the same as during the eclipses, with the radiation of the source remaining more or less constant at the low level. Then the brief pulses are restored without any break in phase. These complicated effects are evidently associated with the actual mechanism by which Centaurus X-3 emits x rays. The transition between the two radiation states does not

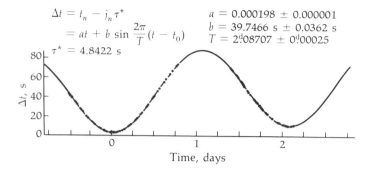

$$\Delta t = t_n - j_n \tau^*$$

$$= at + b \sin \frac{2\pi}{T}(t - t_0)$$

$$\tau^* = 4.8422 \text{ s}$$

$a = 0.000198 \pm 0.000001$
$b = 39.7466 \text{ s} \pm 0.0362 \text{ s}$
$T = 2^{\text{d}}08707 \pm 0^{\text{d}}00025$

Figure 23.4 Radial-velocity curve for the pulsating x-ray source Centaurus X-3.

take place abruptly, but lasts about an hour. During this short transition interval the spectrum becomes considerably harder. Such behavior suggests that the optical component of the binary system is surrounded by a fairly extended atmosphere which comes in front of the x-ray source and produces absorption before the source is occulted by the opaque disk of the star. The length of the transition interval varies, indicating that the gaseous envelope around the optical star is unstable.

Despite difficulties, the Centaurus X-3 source was identified with a 13^m spectroscopic binary star. This variable star possesses a number of peculiar properties. The identification was based on the good agreement in position (to 1'), but particularly on the periodic radial-velocity variations indicated by the spectral lines of this star: the period coincides exactly with the orbital period of Centaurus X-3. Without doubt these periodic variations in the radial velocity of the spectral lines arise from the orbital motion of the optical star.

By analyzing the x-ray and optical data one can establish the characteristics of the Centaurus X-3 binary system. It is a very close double star having a nearly circular orbit with a radius of 6×10^{11} cm, which is only 8.7 times the radius of the sun. The optical star is a rather luminous object whose mass is about 15 $\mathfrak{M}_\odot$ and whose radius is 5×10^{11} cm $= 7.2\ R_\odot$. The x-ray source is therefore only 1.5 $R_\odot$ away from the photosphere of the optical star. Although the x-ray component is diminutive, its mass is comparable to that of the sun. With such a relatively large mass ratio for the two members of the binary system, the optical star will fill its Roche lobe (Chapter 14) and a stream of gas should flow from a small region on its surface, forming a flattened disk around the compact x-ray component.

376

This model of the Centaurus X-3 source follows logically from the fact that the x-ray star belongs to a close binary system. Double stars have played a leading role in the development of astronomy. For example, only for binary systems can the mass of a star be determined with confidence. The phenomenon of novae is closely related to their membership in binary systems (Chapter 14). On the other hand, the evolution of binary-system stars is itself a distinctive process (once again see Chapter 14). The discovery of an x-ray star belonging to a close binary system not only furnishes most valuable information on the basic properties of such stars, but also affords an opportunity for understanding their nature. In particular, an analogy at once suggests itself between the $4^{s}\!.84$ pulsation period of the x rays coming from Centaurus X-3 and the radio waves from radio pulsars. In fact, objects such as the Centaurus X-3 source were promptly called x-ray pulsars. The analogy between x-ray and radio pulsars evidently has a deep significance—a point to which we shall return shortly.

Centaurus X-3 is not by any means the only x-ray source found to be a member of a close binary system. More than 10 such sources are known at the present time. Indeed, the only x-ray source yet encountered in the closest dwarf irregular galaxy, the Small Magellanic Cloud, belongs to a binary system and has an orbital period of $3^{d}\!.4$ (Figure 23.6). Remarkably enough, this one star alone contributes more than 80 percent of all the x-ray power of the Small Magellanic Cloud, which although a dwarf galaxy nevertheless contains several billion stars.

Figure 23.5   Comparison of the radial-velocity curve and the eclipse curve of the pulsating source Centaurus X-3.

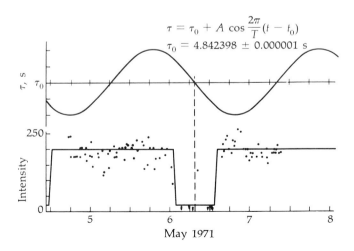

$$\tau = \tau_0 + A \cos \frac{2\pi}{T}(t - t_0)$$
$$\tau_0 = 4.842398 \pm 0.000001 \text{ s}$$

May 1971

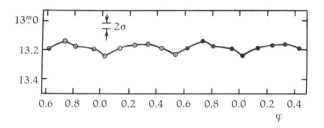

Figure 23.6 Optical light curve of the x-ray star in the Small Magellanic Cloud.

The shortest period on record has been observed in the extraordinary x-ray source Cygnus X-3, about which we shall have more to say. Its period is only $4^h.8$. But perhaps the most interesting binary x-ray source is Hercules X-1. This source was the next x-ray pulsar discovered after Centaurus X-3. Its pulsation period is $1^s.2378$, considerably shorter than that of Centaurus X-3, while its orbital period, as derived both from an analysis of x-ray eclipses (Figure 23.7) and from the periodic variations in the pulsation period due to the orbital motion of the pulsating source (Figure 23.8), is $1^d.70016$. In addition to these two characteristic periods (orbital motion and pulsation), the Hercules X-1 source has a third, much longer period. Its length is 35 days (Figure 23.9). For 11 or 12 days the x-ray source can be recorded very clearly, and its flux varies with the $1^d.70$ eclipse period. But then it turns off completely for 24 days. When the period of invisibility ends, the x-ray source turns on again very quickly, in just a few hours—an interesting feature. For the next three days the x-ray flux continues to grow, but quite gradually; then it begins to fall just as slowly and deliberately to zero, after which another 24-day invisibility phase commences.

Just before the onset of primary minimum (which represents eclipse of the little x-ray star by the optical star) the Hercules X-1 source exhibits a subsidiary minimum whose depth depends on the stage of the 11-day visibility cycle of the source (see Figure 23.9). Strong changes are then observed in the spectrum of the source: its radiation becomes considerably harder. Spectrum changes of a similar kind are also observed when the source turns on, but when it turns off the spectrum remains unaffected. All this behavior indicates that the system of the Hercules X-1 source contains gas which strongly absorbs soft x rays. Unlike Centaurus X-3, where the absorbing gas is concentrated in the atmosphere around the optical star, in the case of Hercules X-1 the absorption takes place in gas streams that participate in the orbital motion, and also perhaps in a gaseous disk surrounding the x-ray source. At the times when gas absorption does not alter

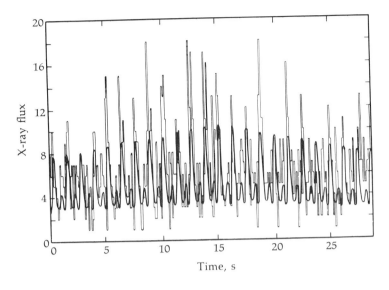

Figure 23.7   Uhuru x-ray light curve of the Hercules X-1 source.

it, the spectrum of the source is observed to be intrinsically quite hard. If it results from the thermal radiation of plasma, its temperature should exceed 50 million degrees at the lowest. Most likely, however, the x rays from Hercules X-1 (and from Centaurus X-3 as well) are not thermal in origin.

Soon after these surprising properties were recognized, the Hercules X-1 source was identified with the variable star HZ Herculis, whose brightness fluctuates within the range 13–15^m. These variations are accompanied

Figure 23.8   Variability in the pulsation period of Hercules X-1.

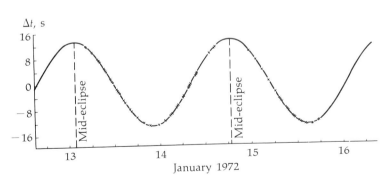

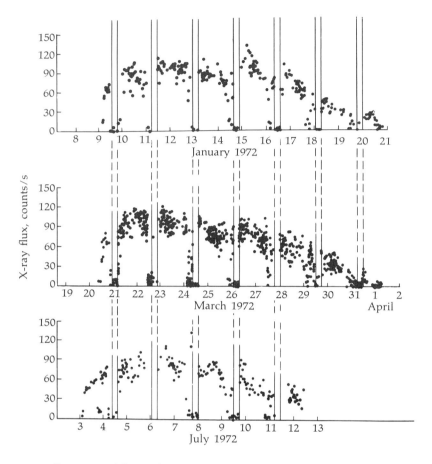

Figure 23.9 The 35-day cycle of visibility for the Hercules X-1 source, as recorded by *Uhuru*.

by concurrent changes in the spectrum. When the star is brightest it is also bluest. Subsequent observations, by Soviet astronomers in particular, have shown that the light variability of HZ Herculis is periodic in character, with a period exactly equal to the $1.\!^{d}70$ orbital period of the Hercules X-1 source. Minimum optical brightness corresponds to minimum x-ray flux. This circumstance suggests that on the surface of the optical star (which is of spectral type F) there is always a fairly large hot spot facing the x-ray source. The origin of the spot can be explained in a perfectly natural way: the powerful x rays heat the surface layers of the optical star that face the x-ray source.

380

That the orbit of the x-ray star around the optical component is nearly circular may be inferred from an analysis of the observational data. The variations in the $1\overset{s}{.}24$ pulsation period of the x-ray source that take place with a period of $1\overset{d}{.}70$ yield an orbital velocity of 169 km/s. The radius of the orbit is about 4×10^{11} cm or $5.7\ R_\odot$, and the radius of the optical star in the HZ Herculis system is about $2\ R_\odot$. The mass of that star is $2\ \mathfrak{M}_\odot$, and the mass of its x-ray companion is about $1\ \mathfrak{M}_\odot$. Knowing the radius of the optical star HZ Herculis and the temperature of its "dark" side, we can find the luminosity of the star and hence its absolute magnitude. A comparison of the absolute magnitude found in this way and the apparent magnitude gives the distance of HZ Herculis, which turns out to be around 2000 pc. Since Hercules X-1 is located at the rather high galactic latitude of 35°, we may draw the interesting conclusion that the distance of the x-ray source from the galactic plane is unusually great, more than 1000 pc! Interpretation of this circumstance should be inseparably associated with the question of the origin of the x-ray source Hercules X-1.

The observations indicate that the pulsed x rays from the Hercules X-1 pulsar (and from the Centaurus X-3 source as well) resemble the radio emission of ordinary pulsars in having a directive character. In such cases, just as with radio pulsars, the observed pulsation period represents the period of axial rotation of the radiating body. But probably the only object that can rotate in so short a period as $1\overset{s}{.}24$ is a neutron star. Thus the outward analogy between radio and x-ray pulsars may indeed reflect an identical nature: both types of pulsars may be neutron stars. However, x-ray pulsars are observed only in binary systems, whereas radio pulsars never belong to binary systems—except for the radio pulsar PSR 1913 + 16 (Chapter 21). And there is another important distinction between the two types of pulsars: the periods of radio pulsars steadily increase, at a rate depending only on the age of the pulsar, but in the Hercules X-1 source the pulsation period was found to have *decreased* after half a year of observation by about 10^{-5} of its value. This shortening of the period did not occur at a uniform rate at all. Similar behavior has been observed for the Centaurus X-3 source.

A special problem hinges on the absence of a 35-day cycle in the optical variability of HZ Herculis. For if the optical variations of this star result from its heating by a powerful flux of x rays from the secondary component, then why does this heating continue to operate for the 24 days of the 35-day period during which the x-ray source is turned off? Two alternative explanations may be offered, neither in any way excluding the other. In the first place, one may suppose that the beam pattern of the x-ray pulsar takes part in two motions. If the emission region does not coincide with the poles of the rotating neutron star (but is located, say, near the magnetic poles, as with radio pulsars), then as the star rotates about its axis the radiation beam will pass periodically across the observer. The geometry here is the same as for radio pulsars. Now imagine that the rotation axis itself describes a precessional motion (the free precession due to a slight asymmetry of the mass distribution within the neutron star) with a period of about 35 days. Then one could understand that for almost two-thirds of this period the radiation beam of the x-ray pulsar might not reach the earth at any phase of the axial rotation. At the same time, though, it would always be pointed toward some part of the surface of the optical star alongside, which is located so close that it fills a large solid angle.

This model has the shortcoming that it entails rather stringent geometric constraints. Also suspect is the fact that not one radio pulsar has been found to turn its pulses off periodically for a prolonged interval. Yet the phenomenon of free precession should not, one would think, depend on whether a neutron star is single or whether it belongs to a binary system. As an alternative, one might assume that near the compact x-ray source there is another source of soft x rays or ultraviolet radiation, more or less isotropic and not yet detected, which also heats the nearby optical star HZ Herculis. This second source could, for example, be the hot gaseous disk surrounding the rapidly spinning neutron star observed as the x-ray pulsar. Future observations of the Hercules X-1 source from above the atmosphere in the spectral region mentioned above should provide a decisive test of this hypothesis.

All the observational evidence indicates, then, that the x-ray sources belonging to binary systems represent extremely compact objects with a mass close to that of the sun. Very likely they are neutron stars spinning with great speed about their axes. And now we are ready to cope with the main question: what causes compact stars that are members of binary systems to emit such strong x rays? Nuclear energy sources cannot of course be involved here. There remain only two sources: the kinetic rotational energy of such a star and the potential gravitational energy released when masses of gas fall onto the surface of the neutron star. The latter mechanism is called accretion. It should be pointed out at once that if x-ray pulsars are indeed neutron stars, the first of these energy sources would not be effective. In fact, for the Centaurus X-3 source the equatorial velocity of the neutron star should be about 10 km/s. Hence the star would have a kinetic rotational energy of about 3×10^{44} erg. Since the power of the x rays of this source amounts to about 10^{37} erg/s, the supply of kinetic energy would be exhausted in only a year. And, generally speaking, if such an energy source were operative, x-ray pulsars ought to be decelerated and their rotation periods should steadily grow longer, contrary to the observations.

The infall of gas clouds and streams toward the surface of the neutron star represents a far more effective energy source. As such stars have a very small radius (about 10 km) while their mass is comparable to the sun's, matter may strike the surface of the star at a velocity as great as 100,000 km/s, or one-third the speed of light (the corresponding velocity for matter falling in toward the earth's surface is only 11 km/s, and toward the sun, 720 km/s). Falling at such a high velocity, one gram of matter would release $0.1\,c^2 \approx 10^{20}$ erg of energy. Thus in order to produce a power of 10^{37} erg/s, about 10^{17} g of gas would have to fall onto the surface of the neutron star every second.[1] This value is actually quite modest. At that rate no more than 0.001 the mass of the earth would fall onto the star each year. The source of this gas could only be the optical star located next to the neutron star. As we have seen above, the optical component of a binary system whose other component is a neutron star will fill its Roche lobe. Hence a stream of gas will flow continuously from the part of the surface of the optical star facing the neutron star.

Calculations show that this stream will feed the gaseous disk which is rotating rapidly around the neutron star.[2] Accelerated by the gravitational field of the star, gas will then fall from the disk onto the neutron star itself. The energy acquired by the gas as it falls onto the surface will be converted into radiation. However, the strong magnetic field of the neutron star complicates this picture of the motion of gas streams in a close binary system. The gas stream falling toward the neutron star will be halted a certain distance away, at the point at which the magnetic energy density becomes equal to the kinetic energy density of the gas stream. After that the gas will flow along the magnetic field lines until it reaches the surface of the neutron star. We would therefore expect the mass of ionized gas coming in from the optical star to strike the surface of the neutron star within two comparatively small areas around the magnetic poles. The size of these spots may be about 0.1 of the radius of the neutron star, or perhaps 1 km.

Within these small areas, processes of energy release will operate on a grand scale. Strong shock waves will be generated there; electrons will be accelerated very efficiently up to relativistic energies; and complex interactions will take place between the magnetic field and the plasma making up various condensations. The details of these processes are now under careful study by theoreticians, and much remains to be learned here. But the general picture of the production of strong x rays has already become clear. Relativistic and nonrelativistic electrons traveling in the intense magnetic field of the neutron star are emitting radiation. The energy source is the

---

[1] The hypothesis that x-ray stars are close binary systems whose active component is a neutron star in a state of accretion was formulated by the author of this book in 1967, three years before *Uhuru* was launched.

[2] The law of conservation of angular momentum implies that such a disk should form.

potential energy that infalling gas acquires in the extremely strong gravitational field of the neutron star. Finally, the source of the gas is the optical component of the close binary system, which fills its Roche lobe.

We do not yet know with assurance just what evolutionary processes lead to formation of a neutron star in a close binary system. The whole problem of stellar evolution in such systems was discussed in Chapter 14. Unquestionably the neutron star in a close binary system represents the final evolutionary product of the more massive component in that system. The formation of the neutron star should be preceded by an intensive transfer of mass from the evolving (and originally more massive) component to the other component. Presumably, after a substantial part of the mass of the evolving star has been transferred, a supernova outburst will occur, resulting in the formation of a neutron star. During the explosion up to 0.1 $\mathfrak{M}_\odot$ of gas could be ejected from the binary system at a velocity of thousands of kilometers per second. According to the law of conservation of momentum, the center of gravity of the binary system should receive an equal and oppositely directed momentum. Might this effect explain why the Hercules X-1 source is located so high above the galactic plane? It is interesting to note that the radial velocity of HZ Herculis, close to 60 km/s, is directed *toward* the galactic plane. The reason could be that the star has already receded to its maximum distance from the galactic plane and is now moving back down. In principle, such a system could execute several oscillations across the galactic plane with a characteristic period of roughly 100 million years.

An optically visible star filling its Roche lobe is not the only possible source of gas to be accreted by a neutron star in a binary system. A "stellar wind" might blow out from an optical component that was relatively distant from the neutron star and thus did not fill its Roche lobe. In this event the optical component would be a hot supergiant of spectral type O or B with a mass greater than 10 $\mathfrak{M}_\odot$. Stars of this kind can produce corpuscular radiation (that is, a stellar wind) of considerable power, say 10^{-7}–10^{-6} $\mathfrak{M}_\odot$/yr. Only a fraction of 1 percent of the corpuscular radiation leaving the supergiant in such a system would be intercepted by the neutron star, yet this amount would be quite adequate to generate x rays of the power observed. We arrive, then, at the idea that there ought to be two types of x-ray sources belonging to binary systems:

1. Sources in which the optical component is a hot, massive supergiant shedding a strong stellar wind. Centaurus X-3 is a typical example.

2. Sources in which the optical component, only slightly more massive than the sun, fills its Roche lobe. A typical representative would be Hercules X-1. These sources may be quite distant from the galactic plane, whil sources of the first type should be close to the plane.

That sources of both types might develop in close binary systems with massive components cannot be ruled out; but whereas in sources of the second type the components have similar masses, in sources of the first type the mass ratio exceeds 3. Calculations show that if the more massive component has $\mathfrak{M} > 10\ \mathfrak{M}_\odot$, then after the gas exchange a compact star of about 3 $\mathfrak{M}_\odot$ will remain, which can explode as a supernova and so turn into a neutron star. Otherwise the evolution process could yield only white dwarfs. If the mass ratio is fairly small, one finds that the system will progressively lose much of its material. When a supernova explodes in such a system the pair will usually break up; if it should remain intact, however, the system will acquire a substantial velocity, as is observed for the second type of source (see the end of Chapter 14).

One of the most distinctive features of x-ray sources is the very short pulsation period that in many cases accompanies the orbital period. We have already spoken in some detail of the $4^{s}.84$ pulsation period found in Centaurus X-3 and the $1^{s}.24$ period in Hercules X-1. But in 1975 the important discovery was made that some x-ray sources have comparatively long pulsation periods. For instance, the x-ray source 3U 0900 $-$ 40 has an orbital period of about 9^{d}; but it has also been found to pulsate with a period of fully 283^{s}. Several cases of long pulsation periods have been discovered in transient or nova-type x-ray sources (see below).

The transient source A1118 $-$ 61 has the longest known pulsation period, 405^{s}. New evidence indicates that it also has an 8^{d} orbital period. In fact, we can today assert with some confidence that all types of x-ray stars—whether constant, transient, or pulsating—are produced by the accretion of gas onto compact, fully evolved objects in binary systems.

These long pulsation periods are probably due to braking of the rotation of the neutron star by the magnetized plasma in which the binary system is embedded. The actual mechanism of this deceleration might be the generation of sound waves by the spinning neutron star, as well as ordinary viscosity. Thus a neutron star manifesting itself as an x-ray pulsar will adapt its rotation period to the physical characteristics of the binary system in which it is located (the period of orbital motion, the power of the stellar wind blowing from the optical component, and so on). The changes observed in the rotation periods of pulsars are probably caused in turn by variations in the power of the stellar wind that feeds the neutron star by accretion.

Transient x-ray sources are observed in the sky fairly often. Outwardly the phenomenon is very similar to a nova outburst. The sources are generally recorded for several weeks or months, after which they are extinguished. Some of them can reach an enormous brightness. Thus far the record is held by a source that appeared in the summer of 1975 not far from the constellation Orion. It flared up to an x-ray brightness an order of magnitude greater than that of Scorpius X-1, the brightest of the steady sources. The nature of the transient sources is not yet understood. Very likely at least some of them are binary systems in which the neutron star moves in an eccentric orbit, while the optical component sheds a stellar wind of widely varying power. In the case of the bright transient source A0535 + 26, for which a rotation period of 104^s was measured in 1975, a modulation of this period has now been discovered, indicating an orbital motion about the hot massive star with a period of either about 40 or 80 days.

Despite a number of attempts, for some years no proof was forthcoming that the brightest x-ray source of all, Scorpius X-1, is double. The problem was very difficult because irregular variations of large amplitude are superimposed on the expected, regular light variability of the optical star identified with this source. However, no periodicity in the x rays (similar to that observed for Centaurus X-3 and Hercules X-1) could be detected for Scorpius X-1. This circimstance was not, of course, an argument contravening the binary nature of the source, for it was entirely possible that the orbit plane might be inclined at a large angle to the line of sight.

Not until 1975 did a group of American astronomers analyzing optical observations manage to find an orbital period for Scorpius X-1; it turned out to be $0^d.78$. Each component has a mass below 2 $\mathfrak{M}_\odot$, while the orbital velocity of the system is very high, 145 km/s (compare the value for Centaurus X-3). Moreover, recently the bright source Cygnus X-2 has also been shown by an American group to be a low-mass binary system ($\mathfrak{M}_F \approx 2 \mathfrak{M}_\odot$ and $\mathfrak{M}_x \approx 1 \mathfrak{M}_\odot$, where $\mathfrak{M}_F$ is the mass of the optical component, which is of spectral type F). The orbital period of this system is $0^d.86$, and its distance is about 2000 pc. Thus Cygnus X-2, Scorpius X-1, and Hercules X-1 all appear to be objects of the same kind.

The radio emission of x-ray stars is a problem of special interest. Several of these objects, such as Scorpius X-1 and Cygnus X-1, have been recorded as sources of very weak, variable radio waves. In itself this circumstance does not pose any difficulty. Radio emission has been detected in recent years in several close binary systems, including Algol and β Lyrae. Powerful gas streams in such systems can produce considerable radio emission. However, in September 1972 a phenomenon quite out of the ordinary was observed. The very weak radio flux coming from the x-ray star Cygnus X-3 suddenly jumped by 2000 times! The flare lasted a few days and was repeated two weeks later. During these flares Cygnus X-3 was one of the brightest radio sources in the sky at centimeter wavelengths. In particular, it was possible on that occasion to use the 21-cm and 18-cm interstellar radio absorption lines impressed in the spectrum of the source to determine its distance, which turned out to be about 7000 pc. Strong bursts of radio waves can be attributed to the ejection of clouds of relativistic particles and plasma. Surprisingly, at the time of the radio flares in Cygnus X-3 the x rays did not show any change at all. Detailed studies of x-ray stars would appear to hold many other unexpected things in store for astronomers.

As a matter of fact a number of interesting new discoveries have been made very recently in x-ray astronomy, after the end of the *Uhuru* satellite mission. Outstanding results have been obtained by British researchers with the *Ariel* satellite, by the Dutch with the Astronomical Netherlands Satellite, by the Americans with the SAS-3, and with the international satellite *Copernicus*. Above all, substantial progress has been made with the problem of transient x-ray sources, which has awaited solution for some time. At the very beginning of x-ray astronomy it had been found that on occasion a prominent x-ray source will develop in the sky, radiate for a few months, and disappear. On the whole such events, as we have said, resemble the outbursts of novae, except that they take place in the x-ray range.

In late 1974 a transient x-ray source flared up in the constellation Centaurus in the immediate vicinity of the irregular long-period variable star RS Centauri. Remarkably, the x-ray source was found to be a pulsar with the extremely long period of 6.75 minutes. Another transient source, which appeared in the spring of 1975 in the constellation Taurus, also proved to be a long-period pulsar ($P = 104$ s), evidently matching an emission-line star of spectral type Be. These Be stars rotate so rapidly about their axes that they shed powerful gas streams from their equatorial zones. It seems all the more curious, then, that after having turned off for more than half a year the Taurus source again became observable in the late autumn of 1975. One cannot help but wonder whether a fully evolved compact object (a white dwarf or a neutron star) is moving in an eccentric orbit about a star with a strong stellar wind.

It is worth noting, incidentally, that intensive outflow of material is also observed in long-period giant variable stars. For example, in the prototype of this class of star, the celebrated variable Mira, a stellar wind is definitely blowing on the white dwarf companion in the system. Perhaps soft x rays may be discovered coming from Mira. In any event, one gains the impression that the transient x-ray sources are also binary systems, but with long periods and very elongated orbits instead of nearly circular ones. When the components of the system approach each other most closely, the rate at which the compact star accretes stellar wind will increase greatly, and an x-ray source will be formed.

Another phenomenon that might be responsible for the periodic behavior of transient x-ray sources is a pulsation of the extended atmosphere of a long-period variable star. This simple picture, however, would need support from further observations.

Long-period x-ray pulsars by no means occur only in transient sources. A wholly unexpected discovery was the 283ˢ pulsar period found in 3U 0900 − 40, the previously known x-ray source mentioned earlier. How are we to understand the origin of such a period in a close binary system that can hardly be more than 10 million years old? Newborn neutron stars ought to have far shorter rotation periods (Chapter 20). Apparently the severe deceleration might be produced if the neutron star has an abnormally intense magnetic field of perhaps 10^{15} gauss—far stronger even than the customary field of a neutron star—as the young Soviet theoretician Nikolai I. Shakura believes.

A most interesting but still enigmatic event was the detection of pulses of hard x rays with a cosmic origin. Although the first reports were published fairly recently (late in 1973), the discovery itself had been made in 1967. It has a most curious history.

Some years ago, as everyone knows, the Soviet Union and the United States concluded a treaty to suspend all nuclear explosions in the atmosphere and on the earth's surface. The great majority of countries (unfortunately not all) have joined in this agreement. In order to monitor such explosions the United States launched to a great height the *Vela* series of artificial satellites, equipped with special recording devices. One of the instruments was a detector of soft γ rays covering the 0.2–2 MeV range of photon energies. The sensitivity of this detector was almost independent of the direction of arrival of the γ rays; that is, the detector was practically isotropic. However, if the γ rays were pulsed in character (just what would be expected from a nuclear test!), the direction of arrival could be recorded, provided that the differences in the arrival times of the pulses at the several satellites of the monitoring system were known. The level of γ rays encountered by each satellite had to be continuously recorded, of course, and to an accuracy of better than 0.01 second.

Great must have been the consternation of those working on this patrol service when they realized that on occasion the instruments were recording brief, intense pulses of hard radiation associated neither with the earth nor with the sun. In those very rare cases when the coordinates could be determined (to about 5° accuracy), the sources of this puzzling radiation were found to have quite a high galactic latitude. This circumstance could mean either that the sources are located outside the Galaxy or that they are comparatively close to the sun (say no more than 100 pc away from us). Before examining these two possibilities, let us consider more fully the observed properties of the pulses, which still remain a riddle.

If these pulses could be detected at all with patrol equipment never intended for astronomical observations, the radiation flux must clearly be rather substantial. As a matter of fact, with the observed pulses lasting several tens of seconds, the flux density in the photon energy range mentioned above reached about 10^{-4} erg cm^{-2} s^{-1}, which is hundreds of times the total flux density of the strongest cosmic x-ray source, Scorpius X-1. That source, however, radiates mainly in the soft x-ray range, where the photons have energies of several electron kilovolts. But often at energies near 1 MeV the flux density received from the puzzling cosmic pulsed sources is even greater than that due to solar flares and exceeds the flux density of other cosmic sources by many orders of magnitude.

Naturally such a noteworthy phenomenon as pulses of cosmic γ rays has become an object of research with other satellites as well. And although only a few years have passed since the first news of the mysterious pulses, today the matter has already cleared up a little.

First of all it has been established that the spectrum of the cosmic pulses extends into a considerably softer region, at least down to 10 keV. It has also been learned that starting near a photon energy $E_1 \approx 200$ keV the spectrum falls off steeply (exponentially) toward higher E, while in the range $E < E_1$ the intensity obeys the milder power law $F \propto E^{-\alpha}$, where $\alpha \approx 0.5$. The spectrum varies both from source to source and with time for a given source. In the latter case the hardness of the spectrum evidently increases with the flux density.

The time structure of the pulses is very interesting. They consist of very strong, separate bursts lasting about 1 s and spaced about 10 s apart. Altogether the duration of the pulses is some tens of seconds. In an individual burst the radiation flux rises to its peak swiftly, always within 0.1 s. Accordingly, the emission region must be very small, less than 10,000 km across.

Bursts carrying an energy up to 10^{-4} erg/cm^2 are recorded once every few months. Presumably bursts of lower energy will occur considerably more often, since they should come from more distant and therefore more numerous sources.

The observed properties of the pulses of hard radiation are such that, in our view, they can hardly be regarded as extragalactic objects. In principle, powerful hard radiation can be expected from supernova outbursts. But no correlation has been found between pulses of hard radiation and supernovae exploding in other galaxies. For example, in 1972 a supernova appeared in the comparatively nearby galaxy NGC 5253 (Chapter 15), but no pulse of hard radiation was recorded at that time.

Some investigators believed that the sources of pulsed hard radiation are located in the Galaxy, a rather short distance away from us. It would be very natural to suppose that the pulses observed arise from the immense flares occurring on nearby UV Ceti dwarf stars (Chapter 1). If this is the case, the phenomenon should not differ in nature from solar flares, which are accompanied by optical, radio, and x-ray emission as well as γ rays.

Until recently only optical and radio emission had been observed during flares of UV Ceti stars. But why would not hard x rays also be emitted during such flares? To be sure, in assuming that the observed pulses of hard radiation are of this character, we would have to accept that the ratio of the x-ray power to the power in the optical and radio ranges is hundreds of times greater for solar flares than for these stellar flares. Of course this argument cannot serve as a refutation of the hypothesis of stellar flares as the origin of the observed pulsed x rays.

Natural though it may be, however, the flare star hypothesis certainly has not been proved. Support for it could come from an observation of simultaneous pulses and optical flares in some star whose position agrees closely with that of the hard-radiation source. Fairly accurate coordinates for sources of pulsed γ rays could be obtained by making simultaneous observations from spacecraft separated from each other by interplanetary distances. Plans for such observations have been worked out.

Other hypotheses attribute the γ-ray pulses to old neutron stars. Powerful processes can operate beneath the surface of these stars, accompanied by the release of a substantial amount of energy, particularly in the form of soft γ rays. If such activity is indeed taking place, then neutron stars would be experiencing not only starquakes but veritable volcanic eruptions.

With the Soviet satellite Kosmos 428, launched in 1971, a great many pulses lasting about 1 s have now been observed in the hard x-ray range, 40–300 keV. In December 1975, Soviet investigators first identified one of the pulsed sources as a globular star cluster, NGC 6624. Concurrently, Dutch scientists working in a softer energy range with the ANS, a satellite referred to above, detected two strong pulses from the same cluster.

In x-ray astronomy 1976 was a key year for x-ray pulses. Three satellites operating simultaneously (the ANS, *Ariel,* and SAS-3) continuously gathered a wealth of observational material. It has now been established that the pulses coming from NGC 6624 are almost periodic: a sequence of pulses separated by time intervals of $0\overset{d}{.}22$ has been recorded. But every month the interval grows markedly shorter. X-ray pulses were forthwith discovered in several other globular clusters, including NGC 1851, 6388, and 6541.

A very interesting source of x-ray pulses has been discovered near the galactic center. The x-ray pulses from this source exhibit the shortest "quasi-period" on record, about 17 s. Thousands of pulses have already been observed arriving from it. The longer the calm interval from one pulse to the next, the more powerful is the second pulse. It seems as though a store of energy builds up between pulses and is then abruptly set free. Gas evidently collects in the strong magnetic field of a neutron-star magnetosphere until the growing instability compels the gas to pour down to the surface, generating a brief pulse of x rays as it strikes the star. After the position of this source had been determined to an accuracy of about 1′, the noted American astronomer William Liller discovered at that site, in the red wavelength range, a globular cluster previously unrecognized because of strong interstellar absorption.

Thus identifications have been made between sources of pulsed x rays and globular clusters, or properly speaking the innermost parts of such clusters. What is surprising is the enormous peak power of these x-ray sources, up to 3×10^{40} erg/s according to the Soviet measurements, or a hundred times the total bolometric luminosity of a whole cluster, which numbers several hundred thousand stars! This power is tens of times that of the steady x-ray stars, whose spectrum is considerably softer. The author of this book has made the suggestion, supported with arguments, that the x-ray pulses being discussed here are identical to the γ-ray pulses described above. So far, however, not all investigators by any means share this point of view.

It is very important to understand that many pulses of hard radiation (perhaps even the majority) definitely cannot be attributed to globular clusters. There could be several times as many of these pulses as those identified with the clusters. But stars of types that commonly occur in globular clusters, such as short-period Cepheids, are perhaps 1000 times as numerous outside the clusters as in all the clusters combined. Accordingly, the phenomenon of pulsed hard radiation cannot represent a special property of some particular type of star population found in clusters. Instead, it should be associated with a cluster (or better, its nucleus) regarded as a whole. As for the pulsed sources outside of clusters, perhaps we are here confronted with objects of a new class, conceivably invisible cluster nuclei. At present, however, the situation is far from clear.

24

Black Holes and
Gravitational Waves

Sir Arthur Stanley Eddington, the illustrious English scientist, founder of the theory of internal stellar structure, was also a great expert in the general theory of relativity. At the time of the 1919 total solar eclipse he was the first to measure Einstein's predicted deflection of the light beam from a distant star in the gravitational field of the sun. All the more interesting is a remark by Eddington, filled with bitter pessimism, that the general theory of relativity is a pretty but fruitless blossom. In Eddington's day this remark was well justified. For if the special theory of relativity could literally conquer physics and then prevail even in technology, the fate of general relativity theory has been altogether different. It is almost as though the theory had been created prematurely by the genius of Einstein.

Sixty years have elapsed since the definitive version of the theory was published. This long span in the history of science may be divided into two parts: from 1916 to 1963 and after 1963. During the first period, general relativity theory occupied a very isolated position in physics and astronomy. This unusual situation was due to the tiny value of the corrections that general relativity contributes to the Newtonian theory of gravitation under normal laboratory or space conditions. These corrections are of the same order as the ratio of the Newtonian gravitational potential to the square of the velocity of light, or $G\mathfrak{M}/c^2R$. One can show that for almost all objects in the Galaxy the corrections are smaller than 10^{-5}; only for white dwarfs, with their comparatively high gravitational potential, do the corrections reach 3×10^{-4}. And the applications of general relativity effects to the standard Friedmann cosmology could not be checked by adequate observations of galaxies, because such measurements were restricted to red shifts $\Delta\lambda/\lambda = z \approx 10^{-2}$.

Essentially the whole majestic edifice of the general relativity theory rested on three effects it predicted—effects so minute that their measurement strained the capabilities of technology at that time. They are the deflection of a light ray in the solar gravity field, the gravitational red shift, and the very slow motion of the perihelion of Mercury. The disproportion between the grandeur of the theoretical concepts and the insignificance of the perceptible consequences was striking indeed.

The situation began to change sharply in 1963, when quasars were discovered with an enormous red shift, undoubtedly cosmological in origin. The variability in the optical and radio emission of these objects implies that they are compact, so in view of their large mass one would expect a large relativistic correction to their gravitational potential. In 1965 the primordial radiation of the universe was discovered, reflecting a physical state when the universe was tens of thousands of times younger than now. Relativistic cosmology thereby gained the foundation of actual astronomical observations. Then just two years later, in 1967, pulsars were discovered and soon proved to be neutron stars. For these objects the corrections to Newtonian gravitation theory because of general relativity can no longer be considered small. Finally, the discovery of x-ray stars in 1971 made it entirely realistic that black holes could be detected—objects that cannot be understood at all without the general theory of relativity.

All these outstanding discoveries of observational astronomy have at last made general relativity essential for the study and comprehension of the fundamental properties of the universe. On the other hand, the vigorous development of physical measurement techniques—a consequence of the scientific and technical revolution we are experiencing—has greatly improved the chances of verifying observationally the general relativity effects. While previously the experimental basis of relativity theory involved the measurement of three famous effects, as mentioned above, today at least 20 different experiments are feasible, of which about 15 have actually been performed.

Advances in radio astronomy, laser and space technology, and radar have been widely used in carrying out these very important experiments. For example, the difference to be expected in the rate of a clock on the ground and on a satellite moving in a synchronous orbit is $\Delta P/P \approx 5 \times 10^{-10}$, whereas hydrogen maser clocks remain stable to a precision of

$\Delta P/P \approx 10^{-13}$ for many months. Another example may be given. According to general relativity theory the distance between the earth and the moon should vary periodically with an amplitude of about 1 m, and modern laser techniques can now measure this distance to within 15 cm. And finally, it is worth noting that the deflection of a ray in the sun's gravity field is now measured most accurately by means of radio interferometry, with quasars as the radio sources. In the years to come the precision of these measurements will be raised to 10^{-3} of the quantity measured. New measurements of great accuracy will furnish refinements to the general theory of gravitation, which, like any vital branch of science, certainly cannot be thought of as having been perfected once and for all.

Let us now consider what general relativity effects ought to be expected during the terminal evolutionary phase of neutron stars. We are referring to what is possibly the hottest topic in modern astrophysics, the problem of black holes. As we have indicated several times, after a sufficiently massive star ($\mathfrak{M} > 2.5 \; \mathfrak{M}_\odot$) has exhausted its nuclear fuel supply it should contract catastrophically almost to a point, since no force could manage to counteract the gravity serving to contract the star. In principle, of course, such a star could at the close of its evolution (say during the supernova explosion process) throw off its extra mass, and then the catastrophically contracting star might stabilize as a neutron star. It is hard to imagine, however, that at this phase in its evolution the star would "know" exactly how much mass to eject in order to prevent itself from collapsing to a point. In any event, there is no evident reason why stars should not exist that are massive enough to contract without limit when their evolution ends.

During gravitational collapse the mechanical equilibrium of a star will be destroyed suddenly: the force of gravity will swiftly become a finite amount greater than the force due to the gas pressure differential. Hence the star will contract at practically the free-fall velocity. After a time interval $t_1 \approx (6\pi G\rho)^{-1/2}$ the star will have contracted so greatly, and the gravitational potential will have become so strong, that the need to include the corrections of general relativity theory becomes obvious. If, for instance, the mean density of the star at the onset of collapse is 10^6 g/cm^3, a typical density for the degenerate isothermal core in a fully evolved star, then $t_1 \approx 1$ s.

Immediately after the publication of Einstein's classical paper, the problem of what character the gravity field would have in the spherically symmetric case was accurately solved by the eminent German astrophysicist Karl Schwarzschild (the father of the now thriving Professor Martin Schwarzschild, who has contributed so much to the theory of stellar evolution). Using Karl Schwarzschild's solution, we can find the dependence of the radius of the collapsing star on time as it would appear if recorded by the clock of an external observer (say on the earth):

$$r = r_g + (r_1 - r_g)e^{-c(t-t_1)/2r_g}, \tag{24.1}$$

where $r_g = 2G\mathfrak{M}/c^2$ is the gravitational radius, and a sphere of radius r_g is called the Schwarzschild sphere. The sun has $r_g = 2.95$ km; the earth, $r_g = 0.89$ cm. The quantity r_1 is the radius of the star at time r_1, and in Equation 24.1 we assume that $r_1 - r_g \ll r_g$.

Application of Schwarzschild's solution to the problem of the collapse of a nonrotating star is perfectly legitimate, because we may regard the motion of every point on the surface of the collapsing star as free fall in a spherically symmetric gravity field. Equation 24.1 implies, then, that as r approaches r_g the rate of contraction perceived by an external observer will slow down asymptotically almost to zero. The external observer will never recognize a contraction of the star beneath the Schwarzschild sphere; according to his clock the contracting star would require an infinite time for that purpose. However, an imaginary observer located on the contracting star and falling along with it would experience no special effect as he crosses the Schwarzschild sphere. His clock would count the seconds until the star and he himself contract to a point. Here general relativity manifests itself in a most striking way. Roughly speaking, it has the effect of severely retarding the rate of all processes (according to the clock of an external observer) occurring in a very strong gravitational field.

To an external observer viewing a gravitational collapse, the luminosity of the star would seem to drop with catastrophic speed as the star's radius approaches the gravitational radius. This drop in luminosity represents the concurrent action of the gravitational red shift, the Doppler effect, and the aberration of light. Schwarzschild's theory yields the following expression for the time dependence of the luminosity of a collapsing star:

$$L = L_0 e^{-2c(t-t_1)/3\sqrt{3}r_g}. \tag{24.2}$$

In the limit as $t \to \infty$, the luminosity $L \to 0$, as does the frequency of the radiation. But to an observer located on the collapsing star, the luminosity (by his clock) may actually seem to grow. To the external observer, however, the collapsing star will practically cease to radiate, and it will stop contracting at a radius $r \approx r_g$ after a time (by *his* clock) of about r_g/c, or approximately 10^{-5} s. Not only the photon radiation but also the neutrino radiation of the collapsing star will exhibit this behavior. Vitalii L. Ginzburg has shown that to an external observer the magnetic field of a collapsing star will also seem to vanish as $r \to r_g$.

Thus in the very short time of about 10^{-5} s the external observer will perceive that the collapsing star has disappeared. Such an object has received the very graphic name of a *black hole*. No type of radiation—neither photon, neutrino, nor corpuscular—will be able to escape from the "hole."

395

In the world outside the only thing left of the star will be its gravitational field, as determined by its mass. If, for example, one of the components of a binary system were to collapse, that event would in no sense affect the motion of the other component.

Allowance for the rotation of the star complicates the picture of gravitational collapse but does not alter it qualitatively. We should emphasize at the outset that rotation can never, in the final analysis, avert the collapse. Collapse has to be the last step in the evolution of sufficiently massive objects after they have exhausted their nuclear energy supply. Fairly recently, in 1963, Roy P. Kerr gave an exact solution to the problem of general relativity theory for a spherically symmetric rotating body. This highly elegant solution opens up the possibility for some rather curious theoretical deductions. If the solution is applied to the problem of the collapse of a rotating star, the result is merely a certain departure of the properties of the gravitational field near the collapsed star from the Schwarzschild solution. The only characteristics of the collapsed star left for the external observer to perceive in this case will be its mass $\mathfrak{M}$ and its angular momentum K. To describe the typical effacement of individual collapsing stars as they asymptotically approach their gravitational radius, the noted American physicist John A. Wheeler uses the aphorism, "Black holes have no hair."

Theoreticians have lately given considerable attention to the abstract mathematical properties of black holes. In particular, they have investigated the possibility of collisions of black holes with ordinary stars and with one another. It turns out that such collisions can generate new black holes, and after the brief interval $r_g/c \approx 10^{-5}$ s these new objects will find themselves in a highly disturbed state characterized by intensive radiation of gravitational waves (see below), after which they will calm down again. Several basic mathematical theorems about black holes have been proved in very general form. Among them are:

1. If a black hole is produced in any manner it can never be destroyed.

2. A single black hole can never split into two black holes, although the converse process is possible.

Recently, however, the British theoretician Stephen W. Hawking has shown that, strictly speaking, Theorem 1 is incorrect: a black hole of very low mass will tend to evaporate as time passes. This curious phenomenon would seem to demolish all our views about black holes. Let us consider it more fully.

According to the concepts of modern physics, a vacuum does not at all represent an absolute void through which various bodies are moving. Instead, a vacuum is a huge reservoir filled with "virtual" particles and antiparticles of every kind. If such external agents as fields are absent, these virtual particles will not materialize, as though they simply did not exist. But sufficiently strong or variable fields (whether electromagnetic or gravitational) will induce a transformation of the virtual particles into material particles, which might well become observable.

Hawking has pointed out that a collapsing star will not be frozen perfectly stiff, so to speak. All processes will evidently have a characteristic time equal to the gravitational radius divided by the velocity of light:

$$\tau = r_g/c. \tag{24.3}$$

On the other hand, the amplitude of these processes will itself become very small as gravitational collapse continues; the star will literally congeal. But the essential feature is that the changing gravitational field will cause virtual particles (such as photons) to materialize only if their frequency is equal to the characteristic frequency of variation of the gravitational field:

$$\nu_g = 1/\tau = c/r_g. \tag{24.4}$$

Thus photons of frequency ν_g should continually be created in the gravitational field of the congealed star—the black hole. A distinctive resonance will develop.

According to Hawking's calculations, a black hole of mass $\mathfrak{M}$ will radiate like a black body at a temperature

$$T = \frac{hc}{8\pi^2 kr_g} = 1.2 \times 10^{26} \, \mathfrak{M}^{-1} \, °\text{K}, \tag{24.5}$$

where $h = 6.6 \times 10^{-27}$ erg s is the Planck constant and $k = 1.4 \times 10^{-16}$ erg/deg is the Boltzmann constant. The radiant power of the black hole will be approximately

$$L \approx \frac{hc^2}{r_g^2} \approx 10^{20} \left(\frac{10^{15} \, \text{g}}{\mathfrak{M}}\right)^2 \text{erg/s}. \tag{24.6}$$

Because of this radiation the black hole will lose mass, and it will steadily shrink. Equations 24.5 and 24.6 show that an ordinary black hole of stellar mass $\mathfrak{M} \approx \mathfrak{M}_\odot = 2 \times 10^{33}$ g would have a temperature $T \approx 10^{-7} \, °\text{K}$ and would emit radiation at the totally negligible rate $L \approx 10^{-16}$ erg/s. Such stellar black holes would radiate in the range of very long radio waves (kilometers in wavelength).

But affairs are different if the black hole has a much smaller mass. In principle, mini-holes might survive as vestiges of that remote epoch when the age of the universe was far less even than a nanosecond. As a miniature black hole melts away, or evaporates, the power and hardness of its radiation would grow until ultimately the tiny remnant of the hole explodes, producing a burst of hard γ rays. Curiously, the hole would radiate its last 10^9 g of mass in just 0.1 s, corresponding to an energy $\mathfrak{M}c^2 = 10^{30}$ erg, or millions of megaton hydrogen bombs.

The evaporation process will limit the lifetime of a black hole very roughly to

$$t \approx 10^{10}\left(\frac{\mathfrak{M}}{10^{15}\text{ g}}\right)^3 \text{ yr.} \tag{24.7}$$

Hence if black holes were formed some 10^{10} yr ago, when the universe was very minute and dense, then a hole could only have survived to the present epoch if its mass is greater than about 10^{15} g. These are the black holes that ought to be exploding today (provided they exist, of course). Since we observe no γ-ray bursts of this sort, we may infer that such relics are at least absent from the solar system, and that only a very minor part of the mass of the universe can now be in the form of small black holes. It is interesting to note that a black hole of mass 2×10^{15} g, or 2 billion tons, would have a gravitational radius r_g equal to the classical electron radius, 3×10^{-13} cm.

Of great theoretical interest is the character of the collapse as perceived by an imaginary observer located on the collapsing body. As we have mentioned, for such an observer nothing special will distinguish the time when the contracting star crosses the Schwarzschild sphere. Although theoreticians are not yet entirely clear about the fate of the contracting star, there appears to be no reason why it should not contract to a point. Some timid hope has been expressed that the situation might be different at a density of order 10^{93} g/cm³! At such densities quantum effects in a strong gravitational field should become important, although no one really knows yet what they are.

Of course, as emphasized above, such a situation can never be realized from the standpoint of an observer on the earth. But that does not mean that a discussion of the problem is devoid of any physical significance. For it certainly is not only stars that have a Schwarzschild sphere. Any mass—and in particular an arbitrarily large mass—has its own gravitational radius. Now it is well established that if the mean density of matter in the universe were to exceed about 10^{-29} g/cm³, the universe would be closed. But to say that the universe is closed is equivalent to saying that the whole universe is confined within its own gravitational radius.

At the prevailing level of observational astronomy one cannot exclude the possibility that, if not the whole universe, individual parts of it, quite large and massive, are contained by their own Schwarzschild spheres. For instance, some theoreticians believe that very massive black holes exist in the nuclei of galaxies. The mean density $\bar{\rho}$ of the matter inside a Schwarzschild sphere is proportional to $1/\mathfrak{M}^2$. Hence if the mass $\mathfrak{M}$ of a black hole is large enough, say 10^8–10^9 $\mathfrak{M}_\odot$, the mean density will be comparatively low, and in principle not only imaginary but perfectly real observers could be found there. It follows, then, that the question of whether a collapsing object will contract to a true point (that is, to an infinitely high density) or whether something will prevent that from happening is far from being only of abstract interest. We again wish to stress that science has not yet been able to give a definitive answer to this question.

Important as these problems are to astrophysicists (and not only astrophysicists), however, the main goal is to discover *real* black holes (live ones, even if they lack their own hair) in the universe. They might now be detected in at least three ways by astronomical observations:

1. searches for invisible black holes in binary (or multiple) stellar systems;

2. searches for black holes that are powerful sources of x rays in binary systems;

3. searches for the gravitational radiation that accompanies collapse.

With regard to searches for invisible but fairly massive members of binary systems, the problem happens to be just as difficult as it is uncertain. Although various authors have called attention in recent years to "suspicious" binary systems (including the remarkable β Lyrae system, as well as δ Geminorum, ε Aquarii, and some other objects), the results of their analyses are hardly unambiguous. Just because a massive component is invisible, it need not necessarily be a black hole. Stars, especially in binary systems (Chapter 14), exhibit a remarkable diversity of properties. Moreover, a suspected star might possibly be embedded in a dust cloud, making it invisible.

Efforts to detect black holes in close binary systems from x rays emitted by one component are considerably more promising. In Chapter 23 we discussed x-ray pulsars in some detail; they are neutron stars radiating in the x-ray range because of accretion. In just the same way we can imagine a close binary system with a black hole as one component. The optical member of such a system could fill its Roche lobe, and a powerful stream of gas would fall into the black hole. Since the gas stream would carry much angular momentum along with it, the gas would form a rapidly spinning disk around the black hole. The particles comprising the disk would orbit the black hole approximately according to Kepler's law. Viscosity would cause the particles of the disk to lose angular momentum continuously, and some of them would gradually settle into the black hole. One can show that during this settling process the gas would radiate a part of its gravitational potential energy into outer space.

As the gas sinks into the black hole the temperature of the inner zone of the disk will become very high. Such a disk could be a strong source of x rays. To a first approximation, the power and spectrum of the radiation would be the same as for neutron stars that are x-ray pulsars. Of course the x rays emitted when gas is accreted by a black hole could not produce strictly periodic pulses such as those observed for Hercules X-1 and Centaurus X-3. But not all x-ray pulsars representing neutron stars emit pulses at intervals of a second or so. Such pulses might, for example, be prevented by strong scattering or by an unfavorable orientation of the neutron star rotation axis with respect to an observer on earth. On the other hand, an x-ray source consisting of a hot, compact disk in rotation around a neutron star could, by its orbital motion around the optical component, be eclipsed periodically and so have the appearance of an x-ray pulsar.

In principle, then, the x-ray sources belonging to close binary systems might include black holes as well. A decisive test for distinguishing a black hole from a neutron star would involve determination of the mass of the x-ray source. But unfortunately this problem is far from simple. The time dependence of the radial velocity of the optical star due to its orbital motion about the center of gravity of the system provides us only with the mass function (Chapter 1), not the actual mass of the unseen x-ray source. If the x-ray source had a strictly periodic pulsating component, then in conjunction with an analysis of the radial velocity curve of the optical component one could determine the masses of both components. But in the case of an x-ray source representing a black hole, the x rays cannot have a pulsating component. In a situation of this kind one must resort to various indirect methods, which of course are not always fully trustworthy.

For several years now the possibility has been discussed that the bright x-ray source Cygnus X-1 may constitute a black hole. This source had been reliably identified with a bright type B star in which the wavelengths of the spectral lines vary with a period of 5.6 days. It was subsequently reported that the wavelength of the emission line of ionized helium in the spectrum of the star varies with the same period but in opposite phase. If these observations were confirmed, it would be natural to suppose that the emission line is formed not in the atmosphere of the optical star but in a gas stream near the x-ray source or in a disk surrounding it. This model would explain why the radial-velocity variations of the helium line are opposed in phase to the radial-velocity variations of the other lines (Figure 24.1). As can easily be seen, the measured ratio of radial-velocity amplitudes immediately yields the mass ratio. Since the mass of the optical B star is about 20 $\mathfrak{M}_\odot$ while the radial-velocity amplitudes seem to have a ratio of 1:2, we may draw the very important conclusion that the mass of the x-ray star is about 10 $\mathfrak{M}_\odot$. Neutron stars can only have a mass of about 2.5 $\mathfrak{M}_\odot$ at most, so it would follow that the Cygnus X-1 source is a black hole. At the present time most investigators

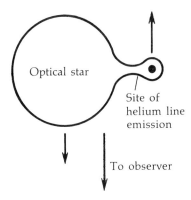

Figure 24.1   Why the radial velocity of the He II line at 4486 Å
varies in opposite phase with the absorption lines of the optical
component of the system.

believe that the compact x-ray component of the Cygnus X-1 system has a
mass in excess of 6 $\mathfrak{M}_\odot$, and accordingly should be a black hole.

As we have intimated, in addition to stellar black holes our Galaxy
might contain a fairly sizable number of black holes of considerably larger
mass. In particular, we have spoken in Chapter 23 of pulsed sources of hard
x rays that have been identified with globular clusters. Quite possibly these
pulses are produced when clumps of gas are accreted by very massive black
holes[1] ($\mathfrak{M} \approx$ 100–1000 $\mathfrak{M}_\odot$) located at the centers of the clusters. It is no
exaggeration to say that this absorbing problem is one of the most basic
confronting astrophysics today. It is related to the fundamental questions
of origin and evolution to which this book is devoted. But why the nucleus
of our Galaxy (where one would think the conditions for development of
a supermassive black hole would be especially favorable) is not a strong
source of pulsed x rays remains unclear. Perhaps, however, the difficulty
will be removed when the "feeding" of black holes with gas is better
understood.

The problem of supermassive black holes should be connected closely
with the whole question of activity in galaxy nuclei and quasars, a topic
that has received much attention in our science over the past decade.

It is time now to say a few words about the reception of gravitational
radiation as a prospect for detecting the collapse of stars. But first the reader
should have at least a general idea of what gravitational waves are.

---

[1]These values for the mass follow from the huge power of the pulsed x rays, which can reach
3×10^{40} erg/s.

Newton's law of universal attraction states that the gravitational field falls off with distance as r^{-2}. This law, however, presupposes that the body responsible for the attraction is a point or a spherical configuration. Now imagine that the attraction is caused by masses moving within a region whose size is small compared to the distance to the observer. In this event we can divide the attractive force at the observing point into two parts. The first and principal part is equal to $G\mathfrak{M}/r^2$, where $\mathfrak{M}$ is the combined mass of the bodies and r is the distance from the observer to the center of gravity of the system of masses responsible for the attraction. The second part of the attractive force, constituting a small increment, depends on the relative arrangement of the masses. One can show that this correction is of order $G\mathfrak{M}a^2/r^4$.

The very simple diagram of Figure 24.2 illustrates the situation. In this case the additional force will in fact be given by

$$\frac{G\mathfrak{M}_1}{(r+a)^2} + \frac{G\mathfrak{M}_2}{(r-a)^2} - \frac{G(\mathfrak{M}_1 + \mathfrak{M}_2)}{r^2} \approx \frac{6G\mathfrak{M}a^2}{r^4}, \tag{24.8}$$

where we have taken $\mathfrak{M}_1 = \mathfrak{M}_2 = \mathfrak{M}$ and a small compared to r. The quantity proportional to $\mathfrak{M}a^2$ is called the quadrupole moment. The quadrupole moment is different from zero not only for a system of bodies but also for any asymmetric body, such as a triaxial ellipsoid. It can vary with time, as will happen for a binary star system or a rotating asymmetric body. In these cases the time variation will be strictly periodic. Hence Newtonian gravitation theory predicts that the acceleration of a test particle at the observer attributable to the quadrupole moment will vary periodically with the same phase and without any lag. In fact, Newton's whole theory rests on the concept of *instantaneous* action at a distance.

We now observe that in a gravitational field recording instruments can measure only relative accelerations, that is, the difference in the accelerations at two points. The relative acceleration due to a spherically symmetric or point body depends on distance as $1/r^3$, a familiar expression for tidal forces. The quadrupole component of the gravitation from a system of bodies or an asymmetric body will induce a relative acceleration of order $G\mathfrak{M}a^2l/r^5$, where l is the distance between a pair of test particles. Notice that this relative acceleration decreases very rapidly with distance.

In this respect the relativistic theory of gravitation differs radically from Newtonian theory. According to general relativity theory, if $r > ct$ (where t is the characteristic time for a change in the quadrupole moment, such as the orbital period in a binary star system or the period of axial rotation for an asymmetric body) the relative acceleration due to the quadrupole moment will vary not as r^{-5} but as r^{-1}. And if the time variation of the quadrupole moment is periodic, the phase of the relative acceleration will

402

Figure 24.2   A gravitational quadrupole.

be shifted by the amount r/cP. This behavior means that the time-variable quadrupole moment of a gravitating body (or system of bodies) will produce at large distances its own distinctive gravitational field, with the character of a wave that is traveling at the speed of light. One can show that gravitational waves will be transverse and polarized.

The fundamental difference between the Einstein and Newtonian theories of gravitation is demonstrated by Keplerian motion in a binary star system. According to Newton's classical theory, such a system will conserve its energy for an indefinitely long time (if the stars are regarded as being of point size). But according to Einstein's theory of gravitation, the system should continuously lose energy by radiating gravitational waves. This effect will be especially strong for close binary systems (see Chapter 22, where we discussed the possibility of explaining pulsars by systems of binary neutron stars). At a great enough distance from the binary system, the relative acceleration due to the gravitational wave will exceed by many orders of magnitude the ordinary, static tidal acceleration induced by the system, which falls off as r^{-3}.

What sort of cosmic objects would be sources of gravitational radiation? Above all these would be close binary (or multiple) systems. The power of the gravitational radiation from a binary system, averaged over the orbital period, is given by the expression

$$L_g = \frac{32}{5} \frac{G^4}{c^5} \frac{\mathfrak{M}_1{}^2\mathfrak{M}_2{}^2(\mathfrak{M}_1 + \mathfrak{M}_2) f(e)}{a^5}, \qquad (24.9)$$

and the frequency of the gravitational radiation will be equal to twice the frequency of the orbital motion (or $4\pi/T$, where T is the period of the system). Here $\mathfrak{M}_1$ and $\mathfrak{M}_2$ are the masses of the components, a is the major semiaxis of the orbit, and $f(e)$ is a certain function of the orbital eccentricity e, increasing from 1 to very large values as e increases.

From Equation 24.9 we find that the power of the gravitational radiation generated as Jupiter moves in its orbit is $L_g = 5 \times 10^{10}$ erg/s, an altogether negligible amount (just 5 kilowatts). But in close binary systems the power of gravitational radiation is incomparably greater. For some systems, such as UV Leonis, it reaches 2×10^{32} erg/s, which is 5 percent of the luminosity of the sun. The flux density of the gravitational radiation we receive from such a stellar system should be almost 10^{-10} erg cm^{-2} s^{-1}, approximately the flux density of luminous energy from a 14^m star. The total flux density of the gravitational radiation reaching the earth from all the stars in the Galaxy should be about 10^{-9} erg cm^{-2} s^{-1}, and the frequency of this radiation, as determined by the average orbital period, would be roughly 10^{-4} Hz.

Another source of gravitational radiation is the axial rotation of stars with an asymmetric mass distribution (such as triaxial ellipsoids). Equation 22.3 gives the power of the gravitational radiation of such a star. The frequency of the gravitational waves radiated will be twice the frequency of axial rotation—a general property of quadrupole radiation.

The very strong dependence of L_g on the angular rotational velocity Ω is noteworthy. It implies that of all rotating stars, pulsars would be expected to emit the strongest gravitational radiation, especially the pulsar of very short period located in the Crab Nebula. In Chapter 20 we pointed out the disparity between the age of this pulsar, as derived from an analysis of its deceleration (1240 yr), and the actual age of the Crab Nebula (920 yr). The discrepancy can be resolved if we suppose that along with its magnetic dipole radiation and pulsar wind this pulsar is emitting gravitational waves, and that they contribute about 25 percent of the total radiated power, or around 8×10^{37} erg/s. According to Equation 22.3, this amount of gravitational radiation would require a "triaxiality parameter" of only $\varepsilon_e = 10^{-7}$, a completely negligible departure of the shape of the neutron star from a spheroid.

If the disparity between the computed and true ages of the pulsar NP 0532 indeed results from gravitational radiation, then the flux density of its gravitational waves at the earth would be of order 2×10^{-7} erg cm^{-2} s^{-1}, many hundreds of times that from the most "favorable" binary systems. In future attempts to detect gravitational radiation from this pulsar, our extremely accurate knowledge of the frequency of the radiation and the time variation of the frequency will be of great value. The author of this book is deeply convinced that if gravitational radiation is ever to be discovered arriving from any cosmic object, that object will prove to be the pulsar in the Crab Nebula. The Crab, we know, has already enriched astronomical science time and again with discoveries of the first rank. I trust that in the future this remarkable nebula will keep up its good tradition.

From what other objects might we hope eventually to detect gravitational radiation? A strong pulse of this radiation lasting less than a second would be expected at the instant when a star undergoes gravitational collapse and explodes as a supernova. Of course in our Galaxy such phenomena occur extremely rarely, about once a century. But the amount of energy in a gravitational pulse can be so great (fully 10^{50} erg!) that even if a supernova were to explode not in our Galaxy but in some other stellar system tens of millions of parsecs distant from us, a considerable flux would be received from it. The spectrum of this gravitational radiation would be quite wide.

Finally it could be that some special, as yet unexplained processes taking place in the nuclei of galaxies (including the nucleus of our own Galaxy) might generate bursts of gravitational radiation. Then from all directions in the sky very short pulses of gravitational radiation would be arriving from objects in remote parts of the universe. Naturally it would not be possible to predict the arrival times of the pulses. A special sky patrol service will be needed, with several gravitational detectors. These detectors will continuously register various types of noise, particularly microseisms, the continual oscillations of the earth's crust. But only pulses recorded at different stations at coincident times should be taken into account. Undoubtedly when gravitational radiation is ultimately detected from such objects as exploding stars and galaxy nuclei, astronomy will benefit from a fundamentally new research tool whose possibilities are presently very hard to foresee.

To conclude this chapter it is worthwhile to outline briefly the experimental techniques that would be used to measure gravitational radiation. We must emphasize at the outset the exceptional difficulty of the problem: the amount of relative acceleration of a test body that must be measured is unimaginably small. For instance, if two test bodies are separated from each other by a distance equal to the earth's diameter, 12,750 km, then the relative acceleration in the field of a gravitational wave emitted by a binary system located 10 pc away, having components of solar mass and an orbital period of 8 hours, would be about 10^{-24} cm/s^2. No measurement technique presently available can permit one even to dream of recording such a tiny value. Nonetheless, receivers of gravitational radiation (designed, of course, to detect an incomparably greater power) have been developed, and the first experiments have already been carried out.

Many types of gravitational wave detectors are conceivable. All the schemes proposed share a maximum utilization of every capability of modern measurement technology, based to a large extent on electronics. But still no genuine measurements of gravitational radiation have been made that may be accepted as entirely trustworthy.

The experiments by the American physicist Joseph Weber, who has devoted many years to the problem of detecting gravitational waves, have been of great influence. The basic idea of these experiments is the following. A solid aluminum cylinder about 2 m long, weighing almost a ton, is used as the receiver of gravitational radiation. This cylinder is suspended by a wire in a vacuum chamber so as to isolate it mechanically from the outside world. The slightest vibrations of the cylinder are recorded by piezoelectric sensors attached to it and connected in an electrical circuit sensitive to the natural vibration frequency of the cylinder, 1660 Hz. Any natural mechanical vibrations of the cylinder will decay in about 10 s. During this interval the cylinder will be able to make Q vibrations. The value $Q \approx 10^5$ represents an intrinsic property of the cylinder. Weber observed that from time to time, twice a month on the average, spontaneous vibrations developed in the cylinder without any external cause. But the significant feature was that two such cylinder receivers were operating concurrently, separated from each other by about 1000 km, and vibrations were seemingly observed with both cylinders simultaneously, to within about 0.5 second!

The accuracy with which these vibrations are measured is so high that the energy recorded in each vibration is 6×10^{-14} erg, only three times the thermal noise background. Oscillations of such small energy have an amplitude of just 10^{-13} cm, a value smaller than the radius of an electron! Of course so small a displacement in the surface of the cylinder cannot be measured directly. It is the mechanical stress induced in the cylinder by these vibrations that is recorded by the pressure-sensitive detectors.

How much of the energy carried by a gravitational wave could be recorded by the Weber cylinders? Calculations show that the proportion is exceedingly small. The effective cross section of these cylinders for gravitational radiation is only about 3×10^{-19} cm^2. But the surprising result follows that the energy of the gravitational radiation should be fantastically high, of the order of 10^5 erg cm^{-2} s^{-1}! This value is a million billion times the flux density expected from the nearest binary star systems emitting strong gravitational radiation!

Alas, despite many attempts, no one has succeeded in repeating Weber's experimental findings. The great majority of investigators believe that Weber's apparatus could not actually have been recording gravitational waves. We have nevertheless described these experiments rather fully, as they are a technique of great interest.

Uncertain though the situation may be, one thing is clear: gravitational astronomy has already taken its first step. We can only hope that it will not stop at this point.

Further Reading

I. GENERAL INTRODUCTIONS TO ASTRONOMY

G. O. Abell. *Exploration of the Universe,* 3rd ed. Holt, Rinehart (1975).

F. N. Bash. *Astronomy.* Harper and Row (1977).

J. C. Brandt and S. P. Maran. *New Horizons in Astronomy.* Freeman (1972).

L. W. Fredrick and R. H. Baker. *Astronomy,* 10th ed. Van Nostrand Reinhold (1976).

C. P. Gaposchkin and K. Haramundanis. *Introduction to Astronomy,* 2nd ed. Prentice-Hall (1970).

F. Hoyle. *Astronomy and Cosmology.* Freeman (1975).

I. R. King. *The Universe Unfolding.* Freeman (1976).

D. H. Menzel, F. L. Whipple, and G. de Vaucouleurs. *Survey of the Universe.* Prentice-Hall (1970).

L. Motz and A. Duveen. *Essentials of Astronomy,* 2nd ed. Columbia University Press (1977).

J. M. Pasachoff. *Contemporary Astronomy.* Philadelphia: Saunders (1977).

E. v. P. Smith and K. C. Jacobs. *Introductory Astronomy and Astrophysics.* Philadelphia: Saunders (1973).

S. P. Wyatt. *Principles of Astronomy,* 3rd ed. Boston: Allyn and Bacon (1977).

II. TOPICAL SURVEYS

D. A. Allen. *Infrared: the New Astronomy.* Wiley (1975).

C. P. Gaposchkin. *Stars in the Making.* Harvard University Press (1952).

W. A. Fowler. *Nuclear Astrophysics.* Philadelphia: American Philosophical Society (1965).

T. Page and L. W. Page, eds. *Starlight.* Macmillan (1967).

————. *The Evolution of Stars.* Macmillan (1968).

R. J. Tayler. *The Stars: Their Structure and Development.* Springer (1970).

A. J. Meadows. *Stellar Evolution.* 2nd ed. Pergamon (1978).

W. Baade. *Evolution of Stars and Galaxies.* Harvard University Press (1963).

L. H. Aller. *Atoms, Stars, and Nebulae,* 2nd ed. Harvard University Press (1971).

J. G. Taylor. *Black Holes.* Random House (1974).

H. L. Shipman. *Black Holes, Quasars, and the Universe.* Boston: Houghton Mifflin (1976).

W. J. Kaufmann. *The Cosmic Frontiers of General Relativity* [black holes]. Boston: Little, Brown (1977).

III. SELECTED ARTICLES

The arrangement of topics parallels the text. Periodicals are abbreviated as follows:

> *AS* *American Scientist*
> *PT* *Physics Today*
> *QJ* *Quarterly Journal of the Royal Astronomical Society* (London)
> *SA* *Scientific American*
> *ST* *Sky and Telescope*

Stellar populations.[1,2,3] G. R. and E. M. Burbidge. *SA* (Nov. 1958) [203].
Understanding the main sequence stars. S.-S. Huang. *ST* (March 1963).
The Hertzsprung–Russell diagram today. M. Hack. *ST* (May–June 1966).
Late-type stars. M. Hack. *ST* (Feb.–March 1967).
Interpreting early-type stellar spectra. D. Mihalas. *ST* (Aug. 1973).
Interpretation of the Be stars. S.-S. Huang. *ST* (June 1975).
The magnetic and related stars. M. Hack. *ST* (July–Aug. 1968).
The oldest star clusters. O. Struve. *ST* (Nov. 1962).
Globular-cluster stars.[2] I. Iben. *SA* (July 1970).
Subdwarf stars. E. M. and G. R. Burbidge. *SA* (June 1961).
A new scale of stellar distances.[3] O. C. Wilson. *SA* (Jan. 1961) [254].

Interstellar grains.[1] J. M. Greenberg. *SA* (Oct. 1967).
Bok globules [star formation].[3] R. L. Dickman. *SA* (June 1977). [366].
Radio signals from hydroxyl radicals. A. H. Barrett. *SA* (Dec. 1968).
Interstellar masers. D. F. Dickinson *et al.* *ST* (Jan. 1970).
Molecules in the interstellar medium. L. E. Snyder and D. Buhl. *ST* (Nov.–Dec. 1970).
Microwave celestial water-vapor sources. K. J. Johnston *et al.* *ST* (Aug. 1972).
Interstellar molecules. P. M. Solomon. *PT* (March 1973).
Interstellar molecules.[2] B. E. Turner. *SA* (March 1973).

---

[1]Reprinted in *Frontiers in Astronomy,* Freeman (1970).

[2]Reprinted in *New Frontiers in Astronomy,* Freeman (1975).

[3]Offprint [number bracketed] available from W. H. Freeman and Company.

Molecules and evolution in the Galaxy. D. Buhl. *ST* (March 1973).

Molecules in space. A. H. Cook. *QJ* **16,** 21–38 (1975).

The formation of interstellar molecules. E. Herbst and W. Klemperer. *PT* (June 1976).

Interstellar clouds and molecular hydrogen. M. Jura. *AS* (July 1977).

The structure of the interstellar medium.[3] C. Heiles. *SA* (Jan. 1978) [394].

Ultraviolet spectroscopy with Copernicus. T. P. Snow. *ST* (Nov. 1977).

T Tauri stars and associated nebulosities. O. Struve. *ST* (Oct. 1961).

Infrared stars. H. L. Johnson. *ST* (Aug. 1966).

The infrared sky. G. Neugebauer and R. B. Leighton. *SA* (Aug. 1968).

The brightest infrared stars.[2] G. Neugebauer and E. E. Becklin. *SA* (April 1973).

The age of the Orion Nebula. P. O. Vandervoort. *SA* (Feb. 1965).

Charles Darwin and the problem of stellar evolution. O. Struve. *ST* (March 1959).

Ages of the stars. O. Struve. *ST* (Sept. 1960).

Early solar evolution. D. Ezer and A. G. W. Cameron. *ST* (Dec. 1962).

The youngest stars.[1,2] G. H. Herbig. *SA* (Aug. 1967).

The birth of stars.[2] B. J. Bok. *SA* (Aug. 1972).

The early evolution of stars. S. E. and K. M. Strom. *ST* (May–June 1973).

The origin and evolution of the solar system. A. G. W. Cameron. *SA* (Sept. 1975).

Some problems of stellar rotation. O. Struve. *ST* (Dec. 1960–Jan. 1961).

The rotation of stars. H. A. Abt. *SA* (Feb. 1963).

Stellar rotation and atmospheric motions. M. Hack. *ST* (Aug.–Oct. 1970).

Birth of the elements. A. G. W. Cameron. *ST* (May 1963).

Light elements in stellar atmospheres. M. Hack. *ST* (June–July 1965).

The origin of the elements. D. D. Clayton. *PT* (May 1969).

The abundance of helium in the cosmos. M. Hack. *ST* (Sept.–Oct. 1972).

Nuclear and differentiated abundance patterns. C. R. Cowley. *ST* (Oct.–Nov. 1976).

Energy production in stars. H. A. Bethe. *PT* (Sept. 1968).

Energy in the universe.[2] F. J. Dyson. *SA* (Sept. 1971).

High temperature nuclear astrophysics. W. A. Fowler. *QJ* **15,** 82–106 (1974).

Neutrino astronomy.[3] P. Morrison. *SA* (Aug. 1962) [283].

The detection of solar neutrinos. H. Reeves. *ST* (May 1964).

Neutrinos from the sun.[1,2] J. N. Bahcall. *SA* (July 1969).

Ordinary stars, white dwarfs, and neutron stars. L. C. Green. *ST* (Jan. 1971).

Evolution of red giant stars. A. V. Sweigart. *PT* (Jan. 1976).

The structure of emission nebulae. J. S. Miller. *SA* (Oct. 1974).

The planetary nebulae, I–XIV. L. H. Aller. *ST* (May 1969–July 1970).

Recent findings about planetary nebulae. Y. Terzian. *ST* (Dec. 1977).

Flare stars. O. Struve. *ST* (Sept. 1959).

Runaway stars. O. Struve. *ST* (March 1961).

Radio-emitting flare stars. A. C. B. Lovell. *SA* (Aug. 1964).

Are most stars members of double-star systems? L. Winkler. *PT* (Sept. 1967).
The origin of binary stars. S.-S. Huang. *ST* (Dec. 1967).
Stellar evolution in double stars. B. P. Flannery. *AS* (Nov. 1977).
Cepheid pulsations. A. N. Cox and J. P. Cox. *ST* (May 1967).
Pulsating stars. J. R. Percy. *SA* (June 1975).
The anomalous Cepheid masses. A. N. Cox. *ST* (Feb. 1978).
The dwarf novae. G. S. Mumford. *ST* (Feb.–March 1962; Oct. 1963).
Exploding stars. R. P. Kraft. *SA* (April 1962).
After the supernova, what? J. C. Wheeler. *AS* (Jan. 1973).
Historical supernovae. F. R. Stephenson and D. H. Clark. *SA* (June 1976).
Supernovae in other galaxies. R. P. Kirshner. *SA* (Dec. 1976).
Supernovae today. L. C. Green. *ST* (July 1977).

Cassiopeia A. O. Struve. *ST* (Oct. 1960).
The filamentary nebula S147. O. Struve. *ST* (Oct. 1962).
Supernova remnants.[2] P. Gorenstein and W. H. Tucker. *SA* (July 1971).
The Gum Nebula—a new kind of astronomical object. S. P. Maran *et al.* *PT* (Sept. 1971).
The Gum Nebula. S. P. Maran. *SA* (Dec. 1971).
X rays from supernova remnants. P. A. Charles and J. L. Culhane. *SA* (Dec. 1975).
Cassiopeia A—an unseen supernova. K. Kamper and S. van den Bergh. *ST* (April 1976).
The astrophysics of cosmic rays. V. L. Ginzburg. *SA* (Feb. 1969).
Origins of cosmic rays. R. Cowsik and P. B. Price. *PT* (Oct. 1971).
Cosmic rays—astronomy with energetic particles. P. Meyer *et al.* *PT* (Oct. 1974).
The Crab Nebula.[1] J. H. Oort. *SA* (March 1957).

Dying stars.[1,3] J. L. Greenstein. *SA* (Jan. 1959) [216].
Gravitational collapse. K. S. Thorne. *SA* (Nov. 1967).
Pulsars today. L. C. Green. *ST* (Nov.–Dec. 1970).
The nature of pulsars.[2] J. P. Ostriker. *SA* (Jan. 1971).
Solid stars. M. A. Ruderman. *SA* (Feb. 1971).
Starquakes: have they been observed? L. C. Green. *ST* (Feb. 1971).
Rotation in high-energy astrophysics. F. Pacini and M. J. Rees. *SA* (Feb. 1973).

The identification of the x-ray source in Scorpius. H. Gursky. *ST* (Nov. 1966).
X-ray stars. R. Giacconi. *SA* (Dec. 1967).
The x-ray sky. H. W. Schnopper and J. P. Delvaille. *SA* (July 1972).
Progress in x-ray astronomy. R. Giacconi. *PT* (May 1973).
X-ray sources and their optical counterparts. C. Jones *et al.* *ST* (Nov. 1974–Jan. 1975).
X-ray emitting double stars.[2] H. Gursky and E. P. J. van den Heuvel. *SA* (March 1975).
The galactic x-ray sources. A. P. Willmore. *QJ* **17**, 400–421 (1976).
Some recent advances in x-ray astronomy. A. P. Lightman. *ST* (Oct. 1976).
X-ray stars in globular clusters.[3] G. W. Clark. *SA* (Oct. 1977) [385].
Cosmic gamma-ray bursts. I. B. Strong and R. W. Klebesadel. *SA* (Oct. 1976).

Introducing the black hole. R. Ruffini and J. A. Wheeler. *PT* (Jan. 1971).

Black holes. R. Penrose. *SA* (May 1972).

Black holes: new horizons in gravitational theory. P. C. Peters. *AS* (Sept. 1974).

The search for black holes.[2] K. S. Thorne. *SA* (Dec. 1974).

Black holes and their astrophysical implications. D. L. Block. *ST* (July–Aug. 1975).

The quantum mechanics of black holes.[3] S. W. Hawking. *SA* (Jan. 1977) [349].

Particle creation near black holes. R. Wald. *AS* (Sept. 1977).

The detection of gravitational waves. J. Weber. *SA* (May 1971).

Gravitational waves—a progress report. J. L. Logan. *PT* (March 1973).

Sources of gravity waves. T. J. Sejnowski. *PT* (Jan. 1974).

Gravitational wave astronomy. R. W. P. Drever. *QJ* **18**, 9–27 (1977).

IV. SPECIALIZED TEXTS

S. A. Kaplan. *Interstellar Gas Dynamics.* Pergamon (1966).

N. C. Wickramasinghe. *Interstellar Grains.* Wiley (1967).

B. M. Middlehurst and L. H. Aller, eds. *Nebulae and Interstellar Matter (Stars and Stellar Systems,* Vol. 7). Univ. Chicago Press (1968).

L. Spitzer. *Diffuse Matter in Space.* Wiley (1968).

S. A. Kaplan and S. B. Pikel'ner. *The Interstellar Medium.* Harvard University Press (1970).

B. T. Lynds, ed. *Dark Nebulae, Globules, and Protostars.* Univ. Arizona Press (1971).

A. H. Cook. *Celestial Masers.* Cambridge University Press (1977).

V. C. Reddish. *Stellar Formation.* Pergamon (1977).

T. de Jong and A. Maeder, eds. *Star Formation* (Intl. Astron. Union Sympos. No. 75). Reidel (1977).

T. L. Swihart. *Astrophysics and Stellar Astronomy.* Wiley (1968).

H.-Y. Chiu *et al.,* eds. *Stellar Astronomy,* 2 vols. Gordon and Breach (1969).

M. Harwit. *Astrophysical Concepts.* Wiley (1973).

W. K. Rose. *Astrophysics.* Holt, Rinehart (1973).

J. L. Greenstein, ed. *Stellar Atmospheres (Stars and Stellar Systems,* Vol. 6). Univ. Chicago Press (1960).

A. B. Underhill. *The Early Type Stars.* Reidel (1966).

E. Novotny. *Introduction to Stellar Atmospheres and Interiors.* Oxford University Press (1973).

D. F. Gray. *The Observation and Analysis of Stellar Photospheres.* Wiley (1976).

J. H. Jeans. *Astronomy and Cosmogony.* Cambridge University Press (1929); Dover (1961).

A. S. Eddington. *The Internal Constitution of the Stars.* Cambridge University Press (1930); Dover (1959).

S. Chandrasekhar. *An Introduction to the Study of Stellar Structure.* Univ. Chicago Press (1939); Dover (1957).

O. Struve. *Stellar Evolution.* Princeton University Press (1950).

M. Schwarzschild. *Structure and Evolution of the Stars.* Princeton University Press (1958).

L. H. Aller and D. B. McLaughlin, eds. *Stellar Structure* (*Stars and Stellar Systems,* Vol. 8). Univ. Chicago Press (1965).

H.-Y. Chiu. *Stellar Physics,* Vol. 1. Waltham: Blaisdell (1968).

D. D. Clayton. *Principles of Stellar Evolution and Nucleosynthesis.* McGraw-Hill (1968).

J. P. Cox and R. T. Giuli. *Stellar Structure,* 2 vols. Gordon and Breach (1968).

H. Reeves. *Stellar Evolution and Nucleosynthesis.* Gordon and Breach (1968).

T. C. Weekes. *High-Energy Astrophysics.* London: Chapman and Hall (1969).

L. Motz. *Astrophysics and Stellar Structure.* Waltham: Ginn (1970).

V. C. Reddish. *The Physics of Stellar Interiors.* Edinburgh University Press (1974).

D. E. Osterbrock. *Astrophysics of Gaseous Nebulae.* Freeman (1974).

Z. Kopal. *Close Binary Systems.* Wiley (1959).

L. Binnendijk. *Properties of Double Stars.* Univ. Pennsylvania Press (1960).

A. H. Batten. *Binary and Multiple Systems of Stars.* Pergamon (1973).

C. P. Gaposchkin. *The Galactic Novae.* Amsterdam: North-Holland (1957).

J. S. Glasby. *Variable Stars.* Harvard University Press (1969).

———. *The Dwarf Novae.* American Elsevier (1970).

———. *The Nebular Variables.* Pergamon (1974).

B. V. Kukarkin, ed. *Pulsating Stars.* Wiley (1975).

D. H. Clark and F. R. Stephenson. *The Historical Supernovae.* Pergamon (1977).

I. S. Shklovskii. *Supernovae and Related Problems.* 1st ed., Wiley (1968); 2nd ed., Princeton University Press (1979).

———. *Cosmic Radio Waves.* Harvard University Press (1960).

J. L. Steinberg and J. Lequeux. *Radio Astronomy.* McGraw-Hill (1963).

J. D. Kraus. *Radio Astronomy.* McGraw-Hill (1966).

A. G. Pacholczyk. *Radio Astrophysics.* Freeman (1970).

G. L. Verschuur and K. I. Kellermann, eds. *Galactic and Extragalactic Radio Astronomy.* Springer (1974).

R. N. Manchester and J. H. Taylor. *Pulsars.* Freeman (1977).

F. G. Smith. *Pulsars.* Cambridge University Press (1977).

R. Giacconi and H. Gursky. *X-Ray Astronomy.* Reidel (1974).

M. Rees, R. Ruffini, and J. A. Wheeler. *Black Holes, Gravitational Waves, and Cosmology.* Gordon and Breach (1974).

H. Ögelman and R. Rothschild, eds. *Recognition of Compact Astrophysical Objects.* NASA Special Publ. No. 421 (1977).

V. RESEARCH REVIEWS

Publications are abbreviated as follows:

AR   *Annual Review of Astronomy and Astrophysics,* Palo Alto, California [reprint number bracketed]

412

ARNS   *Annual Review of Nuclear Science*
FAp   *Frontiers of Astrophysics,* Harvard University Press (1976)
RPP   *Reports on Progress in Physics,* London
SSR   *Space Science Reviews,* D. Reidel Publishing Company
VA   *Vistas in Astronomy,* Pergamon Press

The nearby stars. P. van de Kamp. AR **9**, 103–126 (1971) [2017].
Spectral classification. W. W. Morgan and P. C. Keenan. AR **11**, 29–50 (1973) [2042].

OH molecules in the interstellar medium. B. J. Robinson and R. X. McGee. AR **5**, 183–212 (1967).
Structure and dynamics of the interstellar medium. S. B. Pikel'ner. AR **6**, 165–194 (1968).
Radiofrequency recombination lines. A. K. Dupree and L. Goldberg. AR **8**, 231–264 (1970) [2009].
Physical conditions and chemical constitution of dark clouds. C. Heiles. AR **9**, 293–322 (1971) [2024].
Heating and ionization of H I regions. A. Dalgarno and R. A. McCray. AR **10**, 375–426 (1972) [2039].
Coherent molecular radiation. M. M. Litvak. AR **12**, 97–112 (1974) [2060].
Interstellar OH and H_2O masers, and H_2CO masers and dasers. D. ter Haar and M. A. Pelling. RPP **37**, 481–561 (1974).
Radio radiation from interstellar molecules. B. Zuckerman and P. Palmer. AR **12**, 279–313 (1974) [2066].
Ultraviolet studies of the interstellar gas. L. Spitzer and E. B. Jenkins. AR **13**, 133–164 (1975) [2078].
Molecule formation in the interstellar gas. A. Dalgarno and J. H. Black. RPP **39**, 573–612 (1976).
Interstellar radio spectrum lines. W. B. Somerville. RPP **40**, 483–565 (1977).

Interstellar grains. J. M. Greenberg. AR **1**, 267–290 (1963).
Interstellar dust. B. T. Lynds and N. C. Wickramasinghe. AR **6**, 215–248 (1968).
Interstellar grains. P. A. Aannestad and E. M. Purcell. AR **11**, 309–362 (1973) [2053].
Formation and destruction of dust grains. E. E. Salpeter. AR **15**, 267–293 (1977) [2115].

Orion 2: first scientific results [ultraviolet stellar spectroscopy]. G. A. Gurzadyan. SSR **18**, 95–139 (1975).
Infrared sources of radiation. G. Neugebauer and E. E. Becklin. AR **9**, 67–102 (1971) [2016].

Evolution of protostars. C. Hayashi. AR **4**, 171–192 (1966).
Theories of star formation. D. McNally. RPP **34**, 71–108 (1971).
Processes in collapsing interstellar clouds. R. B. Larson. AR **11**, 219–238 (1973) [2050].
Young stellar objects and dark interstellar clouds. S. E. Strom *et al.* AR **13**, 187–216 (1975) [2080].
Star formation and the early phases of stellar evolution. S. E. Strom. FAp, 95–117 (1976).

Nuclear astrophysics. G. R. Burbidge. ARNS **12**, 507–576 (1962).

413

The origin of the elements. R. J. Tayler, *RPP* **29**, 489–538 (1966).
Theories of nucleosynthesis. J. W. Truran. *SSR* **15**, 23–49 (1973).
Stellar opacity. T. R. Carson. *AR* **14**, 95–117 (1976) [2092].

Neutrinos in astrophysics and cosmology. H.-Y. Chiu. *ARNS* **16**, 591–618 (1966).
Solar neutrinos. J. N. Bahcall and R. L. Sears. *AR* **10**, 25–44 (1972) [2028].
Neutrinos in the universe. T. de Graaf. *VA* **15**, 161–183 (1973).
Neutrino processes in stellar interiors. Z. Barkat. *AR* **13**, 45–68 (1975) [2075].
Neutrinos from the sun. B. Kuchowicz. *RPP* **39**, 291–343 (1976).

White dwarfs. W. J. Luyten. *Advances Astron. Astrophys.* **2**, 199–218 (1963).
White dwarfs. V. Weidemann. *AR* **6**, 351–372 (1968).
Recent developments in the theory of degenerate dwarfs. J. P. Ostriker. *AR* **9**, 353–366 (1971) [2026].

Some observational aspects of stellar evolution. O. J. Eggen. *AR* **3**, 235–274 (1965).
Stellar evolution within and off the main sequence. I. Iben. *AR* **5**, 571–626 (1967).
Stellar evolution. R. J. Tayler. *RPP* **31**, 167–223 (1968).
Stellar evolution toward the main sequence. P. Bodenheimer. *RPP* **35**, 1–54 (1972).
Evolution of rotating stars. K. J. Fricke and R. Kippenhahn. *AR* **10**, 45–72 (1972) [2029].
Post main sequence evolution of single stars. I. Iben. *AR* **12**, 215–256 (1974) [2064].
Endpoints of stellar evolution. A. G. W. Cameron. *FAp,* 118–146 (1976).

Emission nebulae at radio wavelengths. Y. Terzian. *VA* **16**, 279–307 (1974).
Planetary nebulae. M. J. Seaton. *RPP* **23**, 313–354 (1960).
Planetary nebulae. D. E. Osterbrock. *AR* **2**, 95–120 (1964).
Central stars of planetary nebulae. E. E. Salpeter. *AR* **9**, 127–146 (1971) [2018].
Planetary nebulae. J. S. Miller. *AR* **12**, 331–358 (1974) [2068].
The planetary nebulae as radio sources. A. R. Thompson. *VA* **16**, 309–328 (1974).

Evolutionary processes in close binary systems. B. Paczyński. *AR* **9**, 183–208 (1971) [2020].
Consequences of mass transfer in close binary systems. H.-C. Thomas. *AR* **15**, 127–151 (1977) [2110].
Mass loss from stars. R. Weymann. *AR* **1**, 97–144 (1963).
Pulsating stars. J. P. Cox. *RPP* **37**, 563–698 (1974).
The structure of cataclysmic variables. E. L. Robinson. *AR* **14**, 119–142 (1976) [2093].

Supernovae and supernova remnants. R. Minkowski. *AR* **2**, 247–266 (1964).
Supernova remnants. L. Woltjer. *AR* **10**, 129–158 (1972) [2032].
Explosive nucleosynthesis in stars. W. D. Arnett. *AR* **11**, 73–94 (1973) [2044].
The spectra of supernovae. J. B. Oke and L. Searle. *AR* **12**, 315–329 (1974) [2067].
The interaction of supernovae with the interstellar medium. R. A. Chevalier. *AR* **15**, 175–196 (1977) [2112].

Cosmic radio waves and their interpretation. J. L. Pawsey and E. R. Hill. *RPP* **24**, 69–115 (1961).

Cosmic magnetobremsstrahlung (synchrotron radiation). V. L. Ginzburg and S. I. Syrovat-skii. *AR* **3**, 297–350 (1965).

The polarization of cosmic radio waves. F. F. Gardner and J. B. Whiteoak. *AR* **4**, 245–292 (1966).

Pulsars. A. Hewish. *AR* **8**, 265–296 (1970) [2010].

Pulsars: structure and dynamics. M. Ruderman. *AR* **10**, 427–476 (1972) [2040].

Pulsars. F. G. Smith. *RPP* **35**, 399–461 (1972).

On the pulsar emission mechanisms. V. L. Ginzburg and V. V. Zheleznyakov. *AR* **13**, 511–535 (1975) [2087].

Recent observations of pulsars. J. H. Taylor and R. N. Manchester. *AR* **15**, 19–44 (1977) [2106].

Superdense stars. J. A. Wheeler. *AR* **4**, 393–432 (1966).

Neutron stars. A. G. W. Cameron. *AR* **8**, 179–208 (1970) [2007].

Neutron stars. Gr. Baym and C. Pethick. *ARNS* **25**, 27–77 (1975) [5558].

Neutron stars, black holes, and supernovae. H. Gursky. *FAp,* 147–202 (1976).

X rays from the stars. H. Friedman. *ARNS* **17**, 317–346 (1967).

Extrasolar x-ray sources. P. Morrison. *AR* **5**, 325–350 (1967).

Optical observations of extrasolar x-ray sources. W. A. Hiltner and D. E. Mook. *AR* **8**, 139–160 (1970) [2005].

Compact x-ray sources. G. R. Blumenthal and W. H. Tucker. *AR* **12**, 23–46 (1974) [2057].

X-ray astronomy. J. L. Culhane. *VA* **19**, 1–67 (1975).

Celestial binary x-ray sources. K. M. V. Apparao and S. M. Chitre. *SSR* **19**, 281–404 (1976).

Soft x-ray sources. P. Gorenstein and W. H. Tucker. *AR* **14**, 373–416 (1976) [2102].

X-ray astronomy (special issue). *SSR* **20**, 687–888 (Sept. 1977).

Gamma-ray astronomy. C. E. Fichtel. *SSR* **20**, 191–234 (1977).

Astrophysical processes near black holes. D. M. Eardley and W. H. Press. *AR* **13**, 381–422 (1975) [2085].

Gravitational-wave astronomy. W. H. Press and K. S. Thorne. *AR* **10**, 335–374 (1972) [2038].

R. B. R.

415

Name Index

417

419

Subject Index

A stars, 100, 190, 228

A0535 + 26. 386

A1118 − 61. 385

Aberration of light, 395

Absolute magnitude *M*, 17, 20. *See also* Luminosity

Absorption
atmospheric, 4
coefficient κ, 79–80, 129–131, 245
interstellar, 31–36, 39–41, 59, 66, 69, 100, 181, 219, 222, 243–244, 279, 391
lines, 18–19, 32–35, 109
negative, 80, 82, 84
of x rays, 378

Acceleration
of charged particles, 144, 257–259, 355–356, 364, 383
and gravity, 402, 405

Accretion
by black holes, 399–401
by neutron stars, 339, 382–385, 387

Activated medium, 80, 82

Adiabatic processes, 132–133, 232, 234, 255, 257

Age
clusters, 189–193
galaxies, 299
Galaxy, 52, 60, 175, 190
maser sources, 90–91, 107
meteorites, 288
planetary nebulae, 196, 198
pre-supernovae, 225–226
protostars, 110–111
pulsars, 322–325, 332, 337, 339
stars, 28, 60, 104, 116, 136, 183, 185, 188–193, 299
sun, 127, 137, 151, 175, 190, 288
supernova remnants, 234, 236, 242, 244, 255, 257–260, 322–324

Aldebaran, 217

Algol, 208, 213, 375, 386

Alpha particles, 59, 139–142, 145, 149, 282, 293–294

Altair, 22

429

R. B. R.